Magnetism and Metallurgy

Volume 2

CONTRIBUTORS

S. CHIKAZUMI
E. W. COLLINGS
V. V. DAMIANO
G. DE VRIES
C. DOMENICALI

L. J. DYKSTRA
C. D. GRAHAM, JR.
J. S. KOUVEL
G. W. RATHENAU
D. S. RODBELL

HERMANN TRÄUBLE

Magnetism and Metallurgy

EDITED BY

AMI E. BERKOWITZ
General Electric Research
and Development Center
Schenectady, New York

ECKART KNELLER
Institut Werkstoffe der Elektrotechnik
Ruhr Universität
Bochum, Germany

VOLUME 2

1969

ACADEMIC PRESS   New York and London

COPYRIGHT © 1969, BY ACADEMIC PRESS, INC.

ALL RIGHTS RESERVED
NO PART OF THIS BOOK MAY BE REPRODUCED IN ANY FORM, BY PHOTOSTAT, MICROFILM, RETRIEVAL SYSTEM, OR ANY OTHER MEANS, WITHOUT WRITTEN PERMISSION FROM THE PUBLISHERS. RIGHTS FOR PRIVILEGED USE BY THE UNITED STATES GOVERNMENT FOR SPECIAL PURPOSES ARE RESERVED.

ACADEMIC PRESS, INC.
111 Fifth Avenue, New York, New York 10003

United Kingdom Edition published by
ACADEMIC PRESS, INC. (LONDON) LTD.
Berkeley Square House, London W1X 6BA

LIBRARY OF CONGRESS CATALOG CARD NUMBER: 69-13488

PRINTED IN THE UNITED STATES OF AMERICA

List of Contributors

Numbers in parentheses indicate the pages on which the authors' contributions begin.

S. CHIKAZUMI (577), The Institute for Solid State Physics, University of Tokyo, Tokyo, Japan

E. W. COLLINGS* (689), The Franklin Institute Research Laboratory, Philadelphia, Pennsylvania

V. V. DAMIANO (689), The Franklin Institute Research Laboratory, Philadelphia, Pennsylvania

G. DE VRIES (749), Natuurkundig Laboratorium der Universiteit van Amsterdam, Amsterdam, the Netherlands

C. DOMENICALI (689), Temple University, Philadelphia, Pennsylvania

L. J. DYKSTRA (513)

C. D. GRAHAM, JR. (577, 723), General Electric Research and Development Center, Schenectady, New York

J. S. KOUVEL[†] (523), General Electric Research and Development Center, Schenectady, New York

G. W. RATHENAU (749), Philips Research Laboratories, N. V. Philips' Gloeilampenfabrieken, Eindhoven, the Netherlands and Natuurkundig Laboratorium der Universiteit van Amsterdam, Amsterdam, the Netherlands

D. S. RODBELL (815), General Electric Research and Development Center, Schenectady, New York

HERMANN TRÄUBLE[‡] (621), Max Planck Institute for Metals Research, Institute of Physics, Stuttgart, Germany

* Present address: Battelle Memorial Institute, Columbus, Ohio.
† Present address: Department of Physics, University of Illinois, Chicago, Illinois.
‡ Present address: Max-Planck Institut für Physikalische Chemie, Göttingen, Germany.

Preface

These volumes are intended to serve as both tutorial and reference texts for the study of the correlations between the magnetic properties of metals and alloys and their metallurgical structure. The high level of activity in this field during the past fifty years is primarily due to the wide and still expanding range of applications of magnetic materials. Another stimulus for these investigations is the recognition that the magnetic effects accompanying structural changes can be used as tools for the analysis of physical structures. In recent years, our quantitative understanding of these phenomena has advanced to the point where it seemed appropriate to review the field in a manner that would be useful to students as well as to those active in research.

The choice of topics considered in the individual chapters resulted from an attempt to present a self-contained as well as comprehensive treatment of the field. In the first section, the intrinsic and phenomenological concepts of magnetism are developed. In the second section, modern experimental methods are reviewed. The third section contains the chapters correlating specific structural features with the accompanying magnetic behavior. In these latter chapters, the authors have generally developed a formal treatment of their subjects and have then discussed the most pertinent investigations reported in the literature. The coverage of the literature is selective rather than exhaustive. For obvious practical reasons, there has been no attempt to impose a uniformity of style or notation.

The editors are very grateful to the individual authors for the considerable expenditure of time and effort that their contributions represent. We also appreciate the highly competent translations made by Mr. Edward K. Parker, Manager of the IBM Laboratory Library at Kingston, New York. The preparation of these volumes was initiated under the auspices of the United States Air Force, Wright-Patterson Air Force Base. The Franklin Institute, Philadelphia, was the contracting laboratory. Furthermore, we would like to express our appreciation to the IBM Laboratories at Yorktown Heights, New York and Burlington, Vermont for their generous assistance during the editing of these volumes. Finally, the helpful cooperation of Academic Press is gratefully acknowledged.

A. BERKOWITZ
E. KNELLER

August 1969

Contents

List of Contributors v

Preface vii

Contents of Volume 1 xiii

PART C **Relation between Magnetic and Structural Properties**

(*continued*)

Chapter X. **Nonferromagnetic Precipitate in a Ferromagnetic Matrix**
L. J. Dykstra

Text... 513
References 521

Chapter XI. **Effects of Atomic Order–Disorder on Magnetic Properties**
J. S. Kouvel

Introduction 523
General Theoretical Considerations 528
Discussion of Specific Alloy Systems 544
References 573

Chapter XII. Directional Order
S. Chikazumi and C. D. Graham, Jr.

Introduction and Phenomenology	577
Directional Ordering	580
Roll Magnetic Anisotropy	602
References	617

Chapter XIII. The Influence of Crystal Defects on Magnetization Processes in Ferromagnetic Single Crystals
Hermann Träuble

Introduction	622
Dislocations in Single Crystals	625
Interactions between Dislocations and Magnetization	628
Domain Structure	646
Theory of the Magnetization Curve	650
Experimental Results	667
References	685

Chapter XIV. Recovery and Recrystallization
V. V. Damiano, C. Domenicali, and R. W. Collings

Survey of Magnetic and Structural Concepts	689
Ferromagnetic Metals and Alloys	697
Nonferromagnetic Metals and Alloys	712
Significance and Uses of Magnetic Analysis for Metallurgy	718
References	720

Chapter XV. Textured Magnetic Materials
C. D. Graham, Jr.

Introduction	723
Textures in Metals	724
Permanent Magnet Materials	730
Nickel–Iron Alloys	732
Silicon–Iron Alloys	734
Magnetic Control of Texture	744
References	745

Chapter XVI. *Diffusion*
G. W. Rathenau and G. de Vries

Survey of Diffusion Phenomena	749
Magnetic Detection of Diffusion Using Directional Order	751
Directional Order in Other Systems, Especially Substitutional Alloys	785
Concluding Remarks	810
References	810

Chapter XVII. *Interpretation of Magnetic Resonance Measurements in Metals*
D. S. Rodbell

Ferromagnetic Resonance	816
Antiferromagnetic Resonance	831
Nuclear Resonance	831
References	837

Author Index, Volume 2	*1*
Subject Index, Volumes 1 and 2	*11*

Contents of Volume 1

Part A. Principles of Magnetism

Magnetic Moments in Solids
S. H. Charap

Principles of Ferromagnetic Behavior
W. J. Carr, Jr.

Magnetic Resonance
P. E. Seiden

Part B. Experimental Methods

Direct Current Magnetic Measurements
T. R. McGuire and P. J. Flanders

Alternating Current Magnetic Measurements
H. J. Oguey

Part C. Relation between Magnetic and Structural Properties

Magnetic Moments and Transition Temperatures
E. Vogt

Constitution of Multiphase Alloys
A. E. Berkowitz

Fine Particle Theory
Eckart Kneller

Alnico Permanent Magnet Alloys
  K. J. de Vos

AUTHOR INDEX — SUBJECT INDEX

CHAPTER X

Nonferromagnetic Precipitate in a Ferromagnetic Matrix

L. J. DYKSTRA

1. Precipitation of a finely dispersed, nonferromagnetic phase in a ferromagnetic matrix is usually accompanied by an increase in hysteresis and a decrease in permeability, known as magnetic aging. Most of the work on this subject refers to alloys in the annealed condition and to continuous precipitation, i.e., the major part of the reaction takes place by growth of a large number of stable nuclei distributed in a rather uniform fashion throughout the interior of the grains.

The basic cause of the magnetic aging effect is believed to be well understood. As explained in principle in Chapter XIII, the precipitated particles interact with the Bloch walls present in the matrix for a number of reasons [1–4]. Because of this interaction they form obstacles which hinder the free movement of these walls and macroscopically this is reflected in the observed increase in magnetic hardness. Magnetic aging is a rather complex phenomenon, the main reason being that the changes in the bulk magnetic properties not only depend on the amount and intrinsic magnetization of precipitated phase but also on numerous other parameters which define the form of the precipitate on a micro scale such as particle size, size distribution, degree of randomness of the particle distribution in the matrix, particle shape, particle orientations for nonspherical shapes, and on the degree of coherency with the matrix. In principle a theoretical treatment of magnetic aging seems possible using as a starting point the concepts as outlined in Chapter XIII. However, none of the various parameters necessary for the calculation of H_c and χ_0 or μ_0 is generally exactly known. Systematic experimental work on the influence of these parameters has been limited thus far to particle size and amount of second phase [5–7].

Magnetic aging studies are commonly based on measurements of the coercive force H_c and several theoretical attempts [1–5, 8] have been made to relate quantitatively the increase in coercive force connected with precipitation to the amount of precipitated phase and the particle size and shape. In these analyses the particle distribution in the matrix is highly idealized, and this is one of the reasons why close quantitative agreement between these theoretical approaches and experiment can hardly be expected [9]. Of particular interest in the analysis of magnetic aging data is the effect of particle size. According to theory [1, 5], the change in coercive force, ΔH_c, corresponding to a given, small volume fraction of precipitate is relatively minor when the precipitated phase is present in an extremely finely dispersed form such that the average particle diameter is small compared to the wall thickness, δ. However, ΔH_c increases rapidly with increasing particle diameter until a maximum value $\Delta H_{c,\max}$ is reached for an optimum particle size of the order of δ. A gradual decline in coercive force may be expected [5] when the particle size exceeds the optimum value because of the formation of secondary domains or so-called spikes around the particles.

Systematic aging data on the initial permeability μ_0 are scarce. Also, the theory of μ_0 concerning the effect of nonferromagnetic precipitates [10, 11] is in a less developed stage than that of H_c.

2. Detailed studies of magnetic aging connected with precipitation of a nonferromagnetic phase have been made only on a number of iron rich alloys of the binary systems Fe–C [5, 12], Fe–N [6, 13, 14], and Fe–Cu [7] having compositions in the α-solid solution range. The solubility of these elements in the α phase is generally small [7, 15] and decreases rapidly with decreasing temperature. Supersaturated solutions are readily obtained by rapid quenching from a high-temperature region where the alloy is homogeneous. Subsequent aging at a low or intermediate temperature leads to precipitation and magnetic hardening. In view of the low solubilities, the volume fractions of precipitate involved are always small and do not exceed a few percent. The studies in these systems are based on combined measurements of precipitation rate and coercive force. Precipitation rates in the Fe–C and Fe–N alloys were determined by means of the internal friction technique based on the Snoek effect of stress-induced interstitial diffusion of solute atoms in bcc lattices [16]. In the Fe–Cu alloys, precipitation rates were established on the basis of electrical resistivity measurements. Both techniques measure the amount of solute present and are little or not at all influenced by the presence of the precipitated phase.

3. Figure 1 shows a typical set of precipitation and aging isotherms of an Fe–C alloy [12] plotted on a logarithmic time scale. The precipitated phase is an iron carbide of composition Fe_3C. Results obtained by different workers [5, 12] vary slightly in the observed reaction rates. This presumably is

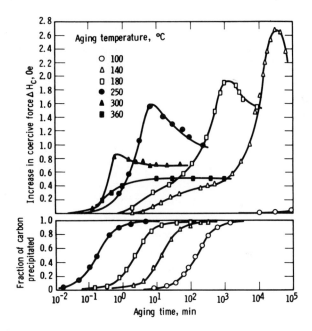

FIG. 1. Amount of carbon precipitated as measured by internal friction and increase in coercive force as a function of aging time at various aging temperatures for an Fe–C alloy containing 0.019 wt % C. (After Nacken and Heller [12].)

related to differences in the number of effective nuclei present in the quenched material. The magnetic aging isotherms in Fig. 2 refer to an Fe–Cu alloy [7]. Here the precipitate is nearly pure Cu. Precipitation rates are not shown but the resistivity measurements indicated that at temperatures around 700°C precipitation of the second phase was practically complete after a few minutes aging time, before any perceptible change in H_c occurred. The aging studies in this system include measurements of the initial permeability μ_0. The response of μ_0 to small amounts of the second phase appears to be sharp; in fact, the solubility curve of Cu in α-Fe was determined ac-

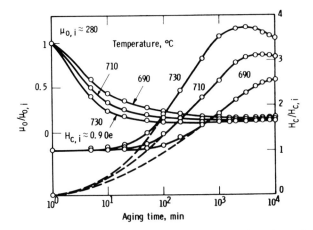

FIG. 2. Influence of aging time and temperature on the ratios of coercive force H_c to its initial value $H_{c,i}$ (0.9 Oe) and of initial permeability μ_0 to its initial value $\mu_{0,i}$ (280) for an Fe–Cu alloy containing 2.14 wt % Cu at various aging temperatures. (After Quereshi [7].)

curately from aging measurements of μ_0. At temperatures around 700°C to which the data in Fig. 2 refer, the solubility is about 0.4 wt %.

4. In a first approximation the aging curves in Figs. 1 and 2 can be explained in terms of variations in the amount and the average particle size of the precipitated phase [5, 7, 12]. Of interest is the observed increase in coercive force, $\Delta H_c'$, reached at the end of precipitation. For the Fe–C alloy (Fig. 1) $\Delta H_c'$ is a strong function of aging temperature. At 100°C or below hardly any change in coercive force is observed; however, $\Delta H_c'$ rises sharply and passes through a maximum as the aging temperature is increased. The amount of precipitate is essentially constant over the range of aging temperatures employed and the observed trend in $\Delta H_c'$ is generally interpreted as an effect of particle size [5, 12]. A gradual increase in particle size with rising reaction temperature, reflecting a decrease in the number of effective nuclei, is normal metallurgical experience and qualitatively the observed variation of $\Delta H_c'$ is therefore in accordance with the ΔH_c–particle size relationship predicted by theory. In the Fe–Cu system, on the other hand, no increase in H_c was observed at the end of precipitation in the concentration and temperature range employed [7]. Apparently, the density of effective nuclei was extremely high, such that the Cu-rich particles were still far below optimum size at the completion of the actual precipitation stage. This was confirmed by electron microscopic data on particle dimensions measured during aging.

5. The changes in H_c observed in both alloy systems (Figs. 1 and 2) after precipitation has been completed are generally attributed to coalescence of the particles. Coalescence primarily occurs in the postprecipitation stages and involves a long-range diffusion process in which the larger, thermodynamically more stable particles grow at the expense of the gradually dissolving, smaller particles. In the Fe–C alloy (Fig. 1) coalescence causes a considerable additional increase in H_c at aging temperatures below 250°C at which particle dimensions at the end of precipitation are still subcritical as a result of the large number of effective nuclei. This increase is followed by an ultimate decline as a result of *overaging* when the iron carbide particles exceed critical dimensions. At 100°C coalescence is extremely sluggish because of the low diffusion rate. A well-developed coalescence stage, clearly separated from the actual precipitation stage, is shown by the 140°C isotherm. On the other hand, at high aging temperatures the number of effective nuclei has been reduced to such an extent that the particles are already oversize at the end of precipitation and under these conditions coalescence, if it occurs, will only lead to further overaging. In the Fe–Cu alloys (Fig. 2) magnetic aging is exclusively connected with coalescence. Electron microscopic observations [7] on a specimen containing 2.14 wt % Cu aged at 710°C confirmed that as a result of coalescence, the particle density dropped roughly from 10^{16} to 10^{13} cm^{-3} between aging times of 4 and 10^4 min, corresponding to an approximate increase in average particle diameter from 10^2 to 10^3 Å.

6. The effects of aging on μ_0 and H_c do not necessarily run parallel as is evident from Fig. 2 (and Fig. 4) in which μ_0 shows a much "earlier" response than H_c but is relatively constant in the region where H_c shows its largest variation. Whether this difference in behavior of H_c and μ_0, which is not readily explained by present theory, is a general phenomenon is unknown since concurrent aging measurements of H_c and μ_0 in other alloy systems have not been reported.

The results of magnetic aging studies in Fe–N alloys [6, 13, 14] are rather similar to those obtained in the Fe–C system and for this reason are not reproduced here. Aging in the Fe–N alloy is slightly more complicated due to the fact that precipitation normally occurs in two successive stages corresponding to two nitrides which differ in chemical composition and structure [16]. The highest coercive force values are associated with a metastable nitride [17] of composition $Fe_{16}N_2$ which precipitates out first. The final precipitate of stable Fe_4N is usually very coarse and its appearance is therefore accompanied by a sharp drop in H_c. Magnetic aging in dual solutions of C and N in α-Fe has also been investigated [6, 18]. The results indicate that the rate of the precipitation of either element is influenced by the presence of the other.

7. Figures 3 and 4 show the dependence of H_c and μ_0 on particle size. The data refer to a constant volume fraction of precipitate of 0.003 for the Fe–C alloy [5] and 0.013 for the Fe–Cu alloy [7]. In the Fe–Cu alloy particle diameters were obtained from an analysis of electron microscopic data, assuming a uniform particle size. For Fe–C alloys systematic electron

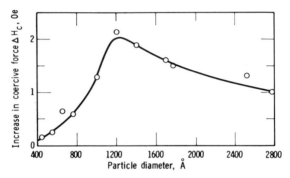

FIG. 3. Increase in coercive force as a function of particle diameter for a volume fraction of iron carbide precipitate equal to 0.003. (After Dijkstra and Wert [5].)

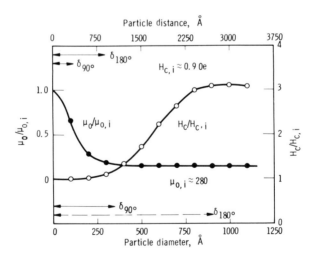

FIG. 4. Ratios of coercive force H_c to its initial value $H_{c,i}$ (0.9 Oe) and of initial permeability μ_0 to its initial value $\mu_{0,i}$ (280) as a function of average particle diameter and average particle separation for a precipitate in an Fe–Cu alloy containing 2.14 wt % Cu, aged at 710°C. (After Quereshi [7].)

microscopic data on particle size variations during aging are not available. In this case particle diameters were estimated indirectly on the basis of a theoretical analysis of the rate of precipitation [19] which was rather successful in explaining the shape of the precipitation isotherms in these alloys. Assuming that (1) the shapes of the precipitated particles are spherical and (2) the number of particles is constant during the major part of the actual precipitation process and all particles grow at equal rate, this analysis leads to the following expression for the number of particles per cubic centimeter of N which participate in the reaction:

$$N = \tfrac{3}{4}\pi(2D\tau)^{-3/2}\{(c_0 - c_1)/(c - c_1)\}^{1/2} \tag{7.1}$$

where D is the diffusion constant of the solute atoms, τ is the characteristic time at which the fraction of precipitation completed is $1 - 1/e$, c and c_1 are the initial and equilibrium concentrations, respectively, of the solute atoms in the matrix, and c_0 is the concentration of solute atoms in the precipitate. Values for N found in the Fe–C alloys are in the range 10^{12}–10^{13} cm^{-3} and, as anticipated, decrease with rising aging temperature [19]. The particle diameter d then follows from the relation $d = (6\alpha/\pi N)^{1/3}$, where α is the volume fraction of precipitate. Following this procedure, the set of diameters used in Fig. 3 was derived from a series of isotherms similar to those in Fig. 1. The general shape of the curves of ΔH_c vs. d and the location of their maxima at 1000–1200 Å, close to the theoretically estimated 180° wall thickness of 1000 Å, are in accordance with theory.

There are indications that the assumptions (1) and (2) on which Eq. (7.1) is based are rather crude approximations for the precipitate of Fe$_3$C. Original electron microscopic results [20] which seemed to indicate a spherical shape were not confirmed in later work [21, 22] in which there was clear evidence of a platelike precipitate along the (100) planes of the parent lattice. Furthermore, the data in Fig. 1 indicate, and the results obtained elsewhere [5] corroborate that $\Delta H_{c,\max}$ reached at the end of precipitation at 250°C is considerably smaller than $\Delta H_{c,\max}$ obtained after prolonged aging at 140°C when considerable coalescence has occurred. This is hard to reconcile with assumption (2). In fact, there is strong experimental evidence [12] that a certain degree of continuous nucleation occurs during the precipitation reaction which would result in a considerable spread in particle size. The high degree of coalescence observed in the Fe–Cu alloys suggests that also here the particle size is nonuniform. For these reasons, the particle diameters plotted in Figs. 3 and 4 merely are certain "average" values and the graphs represent the true dependence of H_c (and μ_0) on particle size only to a first approximation.

8. A fact usually ignored is that Fe_3C is strongly ferromagnetic with a saturation magnetization at room temperature of 1200 cgs units and a Curie point of 200°C. The magnetic properties of $Fe_{16}N_2$ are unknown but it is reasonable to assume that in view of its structure and high Fe content, the saturation moment of this compound is close to that of pure Fe. Nevertheless, both precipitates seem to behave like nonferromagnetic inclusions [5] (see Fig. 4). This would seem to require some further justification.

9. Figure 5 shows on a double logarithmic scale a plot of $\Delta H_{c,\max}$ against the volume fraction α of various nonferromagnetic precipitates in α-Fe. The values of $\Delta H_{c,\max}$ are the highest values observed for a given alloy. They do not necessarily represent absolute maxima, as is evident from Fig. 2, e.g., in which the maxima in the isotherms (H_c) increase monotonically

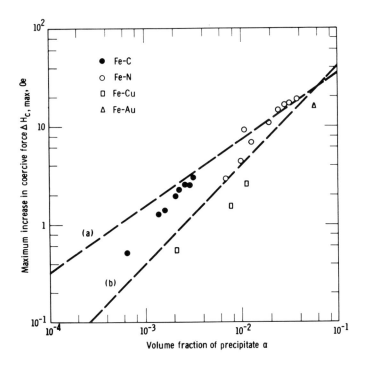

FIG. 5. Maximum observed increase in coercive force $\Delta H_{c,\max}$ as a function of the volume fraction α of various precipitates in α-iron [5, 7, 12–14, 23]. The volume fractions of $Fe_{16}N_2$ are based on an x-ray density [24] of 7.5 g cm^{-3}; the volume fraction of precipitate in the Fe–Au alloy was calculated from solubility and x-ray data [25]. The dashed curves represent theoretical estimates according to Kersten [1] [curve (a): $\Delta H_{c,\max} \sim \alpha^{2/3}$] and to Néel [2] [curve (b): $\Delta H_c \sim \alpha$].

with aging temperature between 690 and 730°C. For comparison, the dashed lines in Fig. 5 represent the theoretical contribution to H_c estimated for two special cases, (1) according to the inclusion model of Kersten [1] for spherical particles arranged on a cubic lattice ($H_{c,\max} \sim \alpha^{2/3}$) and (2) according to Néel's theory [2] based on the effect of internal magnetic poles and valid for a particle distribution which is ideally random ($H_c \sim \alpha$). In both theories the contribution to ΔH_c due to coherency strains between precipitate and matrix has been neglected. However, this term is presumably minor as mechanical overaging is observed to occur long before H_c starts to increase [7, 16, 23].

Quantitative agreement between experiment and either theory can hardly be expected for several reasons. First, in plotting the experimental data in Fig. 5, the influence of variations in parameters other than α, such as size distribution, spatial distribution, and particle shape, has been completely ignored. Further, that ΔH_c rather than H_c has been plotted is somewhat arbitrary as it is not at all clear whether the contributions to H_c from various sources are simply additive or not. Finally, as already mentioned, Fe_3C and $Fe_{16}N_2$ are not truly nonferromagnetic. Nevertheless, the results in Fig. 5 seem to indicate that the physical basis is essentially correct on which the magnetic aging effect of "nonferromagnetic precipitates" is explained.

REFERENCES

1. M. Kersten, "Grundlagen einer Theory der ferromagnetischen Hysterese und der Koerzitivkraft." Hirzel, Leipzig, 1943.
2. L. Néel, *Ann. Univ. Grenoble* **22**, 299 (1945/46).
3. H. J. Williams and W. Shockley, *Phys. Rev.* **75**, 178 (1949).
4. E. I. Kondorskii, *Dokl. Akad. Nauk SSSR* **68**, 37 (1949).
5. L. J. Dijkstra and C. Wert, *Phys. Rev.* **79**, 979 (1950).
6. W. Köster and L. Bangert, *Arch. Eisenhuettenw.* **25**, 231 (1954).
7. A. H. Quereshi, *Z. Metallk.* **52**, 799 (1961).
8. R. Brenner, *Z. Angew. Phys.* **7**, 391 (1955).
9. R. Brenner, *Z. Angew. Phys.* **7**, 499 (1955).
10. M. Kersten, *Physik. Z.* **44**, 63 (1943).
11. M. Kersten, *Z. Angew. Phys.* **7**, 397 (1955).
12. M. Nacken and W. Heller, *Arch. Eisenhuettenw.* **31**, 153 (1960).
13. J. Kerr and C. Wert, *J. Appl. Phys.* **26**, 1147 (1955).
14. M. Nacken and J. Rahmann, *Arch. Eisenhuettenw.* **33**, 131 (1962).
15. M. Hansen, "Constitution of Binary Alloys." McGraw-Hill, New York, 1958.
16. C. Wert, *in* "Thermodynamics in Physical Metallurgy." Am. Soc. Metals, Cleveland, Ohio, 1950.
17. K. H. Jack and D. Maxwell, *J. Iron Steel Inst.* (*London*) **170**, 254 (1952).
18. C. Wert, *Acta Met.* **2**, 361 (1954).

19. C. Wert and C. Zener, *J. Appl. Phys.* **21**, 5 (1950).
20. J. Radavich and C. Wert, *J. Appl. Phys.* **22**, 367 (1951).
21. A. L. Tsou, J. Nutting, and J. W. Menter, *J. Iron Steel Inst.* (*London*) **172**, 163 (1952).
22. W. Pitch, *Acta Met.* **5**, 175 (1957).
23. W. Köster and E. Braun, *Z. Metallk.* **41**, 238 (1950).
24. K. H. Jack, *Proc. Roy. Soc.* (*London*) **A208**, 216 (1951).
25. E. Raub and W. Walter, *Z. Metallk.* **41**, 234 (1950).

CHAPTER XI

Effects of Atomic Order-Disorder on Magnetic Properties

J. S. KOUVEL*
*General Electric Research and Development Center
Schenectady, New York*

INTRODUCTION	523
GENERAL THEORETICAL CONSIDERATIONS	528
Band Theory: Collective Electron Model	528
Phenomenological Model: Atomic Moments Variable with Local Environment	531
Exchange Anisotropy and Exchange Polarization Effects..	543
DISCUSSION OF SPECIFIC ALLOY SYSTEMS	544
Iron–Cobalt	544
Nickel–Iron	549
Nickel–Manganese	555
Iron–Aluminum	561
Iron–Palladium and Iron–Platinum	569
References	573

Introduction

1. It is classic metallurgical experience that alloys can change profoundly in many of their physical properties when subjected to various thermal and mechanical treatments. However, it was only with the advent of x-ray diffraction techniques and their application to metal systems that many of

* Present address: Department of Physics, University of Illinois, Chicago, Illinois.

the observed changes in properties could be correlated with the atomic rearrangements normally associated with order–disorder processes. Among the different types of physical behavior, none has been found to be more sensitive to atomic ordering than the magnetic properties of certain alloys at or near particular stoichiometric compositions. Indeed, the main purpose of this chapter is to examine the close interplay between atomic order–disorder and magnetic behavior and show its important consequences regarding our understanding of magnetism and magnetic materials.

According to general scientific usage, magnetic materials are defined as those that are ferromagnetic, antiferromagnetic, or ferrimagnetic below a *magnetic ordering (Curie or Néel) temperature*, which is conceptually distinct from (but in some cases difficult to separate experimentally from) the critical temperature for atomic ordering. Apart from the Curie (or Néel) temperature, the most significant intrinsic property of a ferromagnet or ferrimagnet is its *saturation magnetization*, which determines the maximum flux carrying capacity and, more importantly from the scientific viewpoint, gives a measure of the vector sum of the individual atomic moments. In an antiferromagnet, in which the vector sum of the moments is zero (in zero field) by definition, the corresponding intrinsic property of note is the *magnetic susceptibility*, which characteristically rises to a maximum at or near the Néel temperature. Any list of primary intrinsic properties of a magnetic material should also include the *Curie constant* which describes the rate of decrease of the paramagnetic susceptibility above the Curie (or Néel) point and gives a measure of the average atomic moment that is independent of the type of magnetic order below this point.

At the root of all these properties are the so-called *exchange interactions* which arise from a quantum mechanical description of the forces between electrons. Although these interactions exist in all systems, their magnetic manifestations are particularly pronounced in the different transition group elements (the iron, palladium, and platinum groups, and the lanthanide and actinide series) and in their alloys and compounds. In these materials, by virtue of the high density of states of their unfilled d or f electron bands, the exchange interactions have the unique opportunity to produce a magnetic moment on each transition group atom, couple it with the moments of other transition group atoms in the lattice, and thus generate an ordered magnetic state. In most cases of metals, exchange interactions are considered to be isotropic in that the energy associated with them depends solely on the relative orientation of the interacting moments. Thus, the exchange energy associated with a positive (ferromagnetic) or negative (antiferromagnetic) coupling has its minimum value when the moments are aligned parallel or antiparallel to each other, respectively, with little or no regard for their alignment relative to the crystal axes.

In most crystalline solids, there is a single dominant interaction between the moments of neighboring atoms, and this gives rise to a simple overall arrangement of parallel or antiparallel moments characterizing the ordered magnetic state. Departures from a simple magnetic configuration can be expected when there is more than one strong interaction, particularly if the interactions make conflicting demands on the orientations of the different atomic moments. Indeed, this is thought to be the basic reason for the complex magnetic states of many important alloys. A similar reason, moreover, is given for the fact that the magnetic states of these alloys depend very sensitively on the degree of atomic order, which obviously would affect any delicate balance between competing interactions. In simpler cases, where the orientations of the atomic moments involve no conflict between the various exchange interactions (for example, when all the interactions are positive), the magnetic state of the material is qualitatively unaffected by order–disorder. Even then, however, there may still be significant quantitative changes in the Curie temperature and other basic properties simply because the rearrangement of the different atoms changes the statistical number of interactions (and, hence, the strength of the net coupling) between their moments.

Although the exchange forces between the individual atomic moments of most magnetic materials are essentially isotropic, anisotropy does arise from an indirect coupling of the moments with the electric fields of their crystalline environment. This phenomenon, known as *magnetocrystalline (or crystal) anisotropy*, manifests itself in a preference of the moments to lie in certain crystallographic directions. Consequently, in a ferromagnetic crystal, this preference is exhibited by the bulk magnetization, and the energy required to turn the magnetization from an "easy" direction to a relatively "hard" direction by means of an externally applied magnetic field is called the crystal anisotropy energy. In metallic systems, the crystal fields are reduced by the screening action of the conduction electrons; the crystal anisotropy forces are correspondingly diminished and are usually much weaker than the exchange forces. Nevertheless, the crystal anisotropy often decides the ease with which a ferromagnetic material can be magnetized, and it becomes important to know how it is influenced by various physical factors. For an alloy, a principal factor is the degree of atomic order, which can have a profound effect on the local symmetry at the lattice site of each moment-bearing atom and, consequently, on the overall crystal anisotropy.

From the strain dependence of the crystal anisotropy energy originate all the magnetomechanical properties of a material, the most important of which is *magnetostriction*. Operationally, the magnetostriction of a ferromagnetic body is defined as the change in its physical dimensions resulting from a change in the magnitude or direction of an applied magnetic field. Intrinsically, however, it is the response of the magnetization to the change in field that determines the magnetostriction. Since magnetostriction and crystal anisotropy are intimately but indirectly related, it is safe to predict that if one of them is affected by atomic order, the other will also be affected, although not necessarily in the same way.

Since all the previously mentioned properties originate from various interactions on an atomic scale, they depend on the features that characterize a material on the same submicroscopic scale, such as its crystal structure, chemical composition, and atomic order. However, these intrinsic magnetic properties cannot be properly evaluated experimentally unless the material is reasonably homogeneous on a grosser scale. For instance, any macroscopic fluctuations in the degree of atomic order can cause a spread in the value of the Curie temperature as it pertains to different parts of a ferromagnetic material, making the results of a bulk measurement rather difficult to interpret. Indeed, as later discussion will show, this particular example refers to a real problem encountered with many ferromagnetic alloy systems in which atomic order develops inhomogeneously and whose Curie points (and other intrinsic magnetic properties) depend very sensitively on the atomic order.

When it comes to deciding the technological possibilities of a given magnetic substance, the intrinsic properties described above may do no more than set the stage. The ultimately decisive role is often played by the so-called *structure-sensitive properties*, such as remanence, coercivity, and other detailed aspects of the magnetization processes in a ferromagnet. The magnitude of these properties depends very critically on any gross inhomogeneities (of the type mentioned earlier) but is generally determined by the finer "structure" that is present to some extent in all real materials but which has little effect on the intrinsic (or structure-insensitive) properties. The term *structure* is defined here as all the microscopic imperfections in crystal structure, chemical composition, and atomic order, and since these are amenable to thermal and mechanical treatment, the structure-sensitive magnetic properties can often be empirically set to meet prescribed technological needs.

Over recent years, however, much of the empiricism with regard to structure-sensitive magnetic properties has gradually given way to more detailed knowledge concerning the exact structural and chemical constitution of various practical magnetic materials, combined with a growing understanding of the basic magnetization processes in inhomogeneous systems. It is now known, for example, that many of the best permanent magnet alloys have several coexisting phases which differ from each other structurally and chemically (in some cases, as the result of order–disorder) and which therefore also differ in their intrinsic magnetic properties. Since the phases may coexist as microscopically small but well-defined regions and since only one of them may be appreciably ferromagnetic at room temperature, it can be supposed that each of the small ferromagnetic regions is a uniformly magnetized, single domain. Hence, the magnetization cannot reverse by domain boundary motion, which in a homogeneous bulk ferromagnet normally only requires a very small coercive field. Instead, it must reverse by rotation against the restraining forces of anisotropy, which may be magnetocrystalline in origin or may derive from the nonspherical shape of the ferromagnetic regions. The latter possibility, known as *shape anisotropy*, is particularly important because this type of anisotropy can often be developed by suitable heat treatment in a magnetic field. In such a solid state reaction, especially if it simply involves order–disorder, all the ferromagnetic regions tend to elongate along the field axis in order to lower their magnetostatic energy. The end result is a highly coercive permanent magnet material.

As this introductory discussion has tried to show, there exists a broad spectrum of magnetic behavior ranging from the most intrinsic properties, which ultimately derive from electronic interactions on an atomic scale and are basically important to the science of magnetism, to the most structure-sensitive technological properties dependent on gross macroscopic features of magnetic materials. Moreover, there is a corresponding range in the type of influence atomic ordering can exert on magnetic behavior. At one end of this range, the order–disorder process is microscopically homogeneous (long-range order developing uniformly out of short-range order) and its most telling effect is on the intrinsic properties. At the other extreme, the process is completely inhomogeneous (macroscopic ordered regions growing out of a matrix of essentially disordered material) and relates most directly to the structure-sensitive properties.

In the following section of this chapter, the variation of different intrinsic magnetic properties with atomic order–disorder is examined within a formal theoretical context, which it is hoped will provide a plausible rationale for some of the effects observed experimentally. The measured results for these and other effects in various alloy systems, as reported by previous researchers, are compiled and discussed in the final section of this chapter.

General Theoretical Considerations

Band Theory: Collective Electron Model

2. The intrinsic magnetic properties of a material being mainly electronic in origin, what is needed for a truly fundamental understanding of these properties is a detailed knowledge of the energy distribution of all the electronic states in the system which constitute its band structure. Unfortunately, this type of information is hard to deduce from first principles except in the case of insulators, where the "bands" are sharply defined energy levels. Even in the case of the pure transition group metals, the presence of conduction electrons and the overlap of their broad energy bands with the narrower (though far from sharp), unfilled d or f bands introduce formidable complications into any basic calculation. Various approximate methods have been resorted to but have generally disagreed in their detailed predictions about electronic band structure and its manifestations in magnetic behavior. Indeed, as classic a problem as the ferromagnetism of iron has yet to be resolved in satisfactory detail. This being the situation with all the transition group elements, it can be appreciated that the even more complex problem posed by their alloys—whether ordered or disordered—is almost intractable at this fundamental level.

It should be noted, however, that much theoretical progress has recently been made in the relatively simpler problem of explaining the magnetic (and associated electrical) properties of *very dilute* alloys of a transition group element in a nonmagnetic host metal, such as copper or gold [1]. Generally, these theories cannot be extended with any rigor to alloys of higher solute concentration, which tend more readily towards atomic order and thus concern us here more directly. It nevertheless remains that the experimental information obtained on the very dilute alloys can be highly pertinent and helpful to our empirical understanding of the more concentrated alloys.

At a somewhat less fundamental level but still based on the band theory of metals, the *collective electron model* developed by Stoner [2] in the 1930's established a useful relationship between many of the intrinsic magnetic (and thermal) properties of the transition group metals and certain specific features of the electronic band structure. Assuming a simple band structure with various adjustable parameters, Stoner, Wohlfarth, and others [3] derived a plausible explanation for the measured properties of all the pure metals of the iron, palladium, and platinum groups. Most noteworthy was the rationale provided for the fact that the low-temperature saturation

XI. ATOMIC ORDER–DISORDER AND MAGNETIC PROPERTIES

magnetizations of the ferromagnetic metals—iron, nickel, and cobalt—correspond on a per-atom basis to nonintegral numbers (about 2.2, 0.6, and 1.7, respectively) of electron moments or Bohr magnetons (μ_β). This was shown to arise immediately from the energy overlap of the $3d$ and $4s$ (conduction) electron bands, both of which are occupied up to a certain energy level (the Fermi level); thus, the total integral number of electrons per atom available for both these bands can split up into a nonintegral number in each of them. This situation exists in all the transition group metals, but in those that are ferromagnetic the exchange forces are strong enough to cause an energy shift of the $3d$ states with a magnetic polarization in one direction relative to those with an oppositely directed polarization. The consequent imbalance in the number of electrons occupying states of different polarization gives rise to a spontaneous magnetization which we identify with ferromagnetism.

Although the collective electron model also gives reasonable predictions for other properties of the pure ferromagnetic metals, such as the variation of their saturation magnetizations and paramagnetic susceptibilities with temperature, its greatest potential has been thought to lie in its elucidation of ferromagnetic *alloys*. One of its earliest applications, in fact, was to the nickel–copper system. Within the simple scheme of this model, it was predicted that the extra electron provided by each copper atom added to nickel would occupy one of the empty $3d$ states of a given polarization, all the $3d$ states of opposite polarization being considered already occupied (and most of the empty $4s$ states having much higher energies). Hence, the net $3d$ electron polarization (per atom) responsible for the spontaneous ferromagnetic moment would start at $0.6\mu_\beta$ for pure nickel, decrease linearly at the rate of $0.01\mu_\beta$ per at. % copper (i.e., $1\mu_\beta$ per 100 at. % per copper), and finally reach zero at about 60 at. % copper. This prediction, valid specifically at 0°K, is illustrated in Fig. 1a and is very close to what is observed experimentally [4]. Similarly, as shown in this figure, the addition of zinc, with two extra electrons per atom, to nickel should cause the ferromagnetic moment to decrease twice as rapidly, and this also is in accord with experiment [5]. Conversely, since cobalt has one less electron per atom than nickel, it is consistent with this model that the ferromagnetic moment of Ni–Co is found [6] to increase with Co concentration at about the rate that the moment of Ni–Cu decreases with Cu concentration. Hence, when the observed ferromagnetic moment per atom (at 0°K) is plotted as a function of the electron-to-atom ratio, its variation for these three nickel alloy systems describes essentially a single straight line with a slope of $-1\mu_\beta$ per electron, as shown in Fig. 1b.

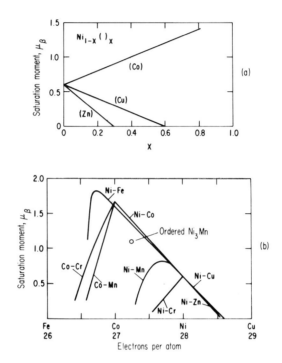

FIG. 1. Saturation moment in Bohr magnetons per atom: (a) plotted versus atomic concentration of Cu, Zn, and Co in Ni; (b) plotted versus electron concentration in various disordered Ni-base and Co-base alloys and in ordered Ni_3Mn.

The universal curve thus described would appear to give firm support to the collective electron model in its simplest form known as the *rigid band model*, in which it is assumed that the band structure (i.e., the density of allowable states at different energies) is rigidly fixed and that a change in the number of electrons per atom simply causes an adjustment of the Fermi energy level. Unfortunately, for this simple model the universality of the curve is soon violated when one takes into account other alloys of nickel and cobalt, even just those having close-packed crystal structures. This is demonstrated in Fig. 1b by the curves representing the magnetization data (extrapolated to 0°K) for the alloys with iron, manganese, and chromium. For Ni–Fe, even though up to about 50 at % Fe the change in moment is roughly at the $-1\mu_\beta$ per electron rate, at higher Fe concentrations the rate of change reverses sign and becomes positive [7, 8]. A similar reversal in the variation of the ferromagnetic moment is shown by the Ni–Mn alloys at fairly low Mn concentrations [9]. In Ni–Cr [5], Co–Mn [10], and Co–Cr [11], the change in moment with electron concentration is positive at all compositions, which departs from the rigid band model completely.

Moreover, even if the rigid band model gave an accurate description of all the ferromagnetic solid solution alloys, which it clearly does not, it would be of no practical help to our understanding of the magnetic effects of order–disorder since it is inherently unable to distinguish between ordered and disordered states of an alloy. This model cannot explain, for instance, why the ferromagnetic moment of ordered Ni_3Mn, represented by the isolated point in Fig. 1b, is so much larger than the moment of the same alloy disordered [9].

Phenomenological Model: Atomic Moments Variable with Local Environment

3. Saturation Magnetization and Order–Disorder. In view of the inadequacy of the rigid band model when applied to magnetic alloys and in the absence of a more rigorous band theory (except for the very dilute alloys mentioned earlier), we are forced to seek an alternative approach on a more phenomenological level. Fortunately, there is a very promising approach pointed out by the results of some fairly recent neutron diffraction experiments. In these experiments, first carried out by Shull and Wilkinson [12], measurements were made of the diffuse scattering of neutrons from various ferromagnetic, solid solution alloys. Since neutrons interact not only with the atomic nuclei but also with the magnetic moments of the atoms, the spatial atomic disorder of a binary ferromagnetic alloy gives rise to two components of diffuse neutron scattering. One of them is due to the difference between the nuclear scattering amplitudes for the two kinds of atoms and the other to the difference between the magnetic scattering amplitudes, which are proportional to the atomic moments. Specifically, for a disordered ferromagnetic alloy of composition $A_{1-x}B_x$, the cross section for the magnetic diffuse scattering is expressible as

$$d\sigma_M \propto x(1-x)(\mu_A - \mu_B)^2, \qquad (3.1)$$

when appropriate "form factor" adjustments are made; μ_A and μ_B are the magnetic moments of the two atomic species. From the change in $d\sigma_M$ with magnetic field one can separate it from all other diffuse scattering sources, none of which are field sensitive. The experimental value for $d\sigma_M$ is then combined with that of the saturation moment, which represents the compositional average

$$\bar{\mu} = (1-x)\mu_A + x\mu_B, \qquad (3.2)$$

and values for μ_A and μ_B are thus determined individually. However, because $d\sigma_M$ depends only on the absolute magnitude of $\mu_A - \mu_B$, Eqs. (3.1) and (3.2) will yield two sets of solutions for μ_A and μ_B. This ambiguity is inherent in the use of unpolarized neutrons and can be quite serious, but in most cases one set of solutions is much more plausible than the other.

To illustrate the results of this diffuse neutron scattering technique, which will ultimately form the basis of our phenomenological description, we choose two of the cases considered in the original work of Shull and Wil-

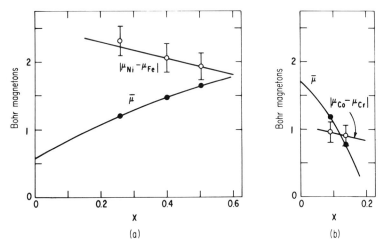

FIG. 2. Saturation moment ($\bar{\mu}$) and atomic moment difference versus atomic composition of disordered Ni–Fe and Co–Cr alloys: (a) $Ni_{1-x}Fe_x$, (b) $Co_{1-x}Cr_x$.

kinson [12], namely, the alloys of Ni–Fe and Co–Cr. This particular choice is made because, as discussed earlier, the variations of the ferromagnetic moments of these two systems with composition (ignoring Fe-rich Ni–Fe) are very different. The moments $\bar{\mu}$ of the alloys studied at room temperature by Shull and Wilkinson are shown in Fig. 2, together with the values for $|\mu_A - \mu_B|$ deduced via Eq. (3.1) from their neutron diffraction data. Using Eq. (3.2) and combining these results, they obtained for the constituent atoms of each alloy the two sets of moment values plotted in Fig. 3 (where a negative value means the moment of that atom is aligned antiparallel to that of the other). The points joined by solid curves represent the solutions with the lesser variation with composition. Indeed, in the case of Ni–Fe, this favored set of solutions for μ_{Ni} and μ_{Fe} is nearly composition independent. Together with the fact that $\mu_{Fe} - \mu_{Ni}$ is always about $2\mu_\beta$, this provides an immediate quantitative explanation for the saturation moments ($\bar{\mu}$) of the fcc Ni–Fe alloys. In the case of Co–Cr, the more drastic changes in $\bar{\mu}$ are explained empirically by the rapid decrease of μ_{Co} with increasing Cr concentration (even for the favored solution, as shown in Fig. 3).

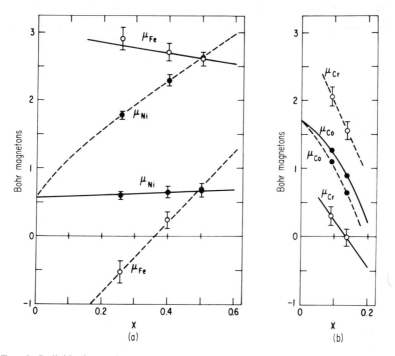

Fig. 3. Individual atomic moments versus atomic composition of disordered Ni–Fe and Co–Cr alloys. The solid and dashed lines drawn through alternate solutions are derived from data represented in Fig. 2: (a) $Ni_{1-x}Fe_x$, (b) $Co_{1-x}Cr_x$.

Thus, we see that neutron diffraction experiments of this kind can offer a simple empirical interpretation for the changes in saturation moment with alloy composition. Moreover, the difference in magnetic moment between the constituent atomic species of an alloy, which these experiments reveal most clearly, is completely outside the scope of the rigid band model and thus demonstrates the severe limitations of this model, even with regard to "well-behaved" alloys like Ni–Fe. It does *not* follow, however, that we must adopt the extreme viewpoint that all atoms of a given kind in an alloy of specified composition have the same magnetic moment, regardless of any differences in their local environments. In fact, various recent studies of dilute ferromagnetic alloys have shown that this viewpoint is seldom if ever strictly justified. The most definitive of these studies are the low-angle neutron scattering experiments of Low and Collins [13–15], whose results give the size as well as the strength of the magnetic moment disturbance associated with a solute atom. Their experiments on the Ni-base alloys, for instance, show that the disturbance around a Cr solute atom extends out to the Ni nearest neighbor atoms (whose moments are less than half those of the more distant Ni atoms), but an Fe solute atom produces a disturbance limited essentially to itself (i.e., the moments of neighboring Ni atoms are virtually unaffected). Similar conclusions have been drawn about these and other dilute alloy systems from nuclear magnetic resonance [16] and Mössbauer effect [17] studies, in both of which the measured hyperfine field at the atomic nucleus is assumed to be directly related to the local electronic moment. In the general picture that emerges, the solute atoms (even if they are nonmagnetic, like Al or Si) always have some, however little, effect on the moments of the neighboring host atoms; alternatively, from the point of view of a host atom, the magnitude of its moment depends on how close it is to a solute alloy.

Theoretically, it should be possible to trace this variability in magnetic moment to certain fairly localized exchange interaction effects on the band structure. However, it is simpler and more immediately profitable here to incorporate this variability phenomenologically into an atomistic model that encompasses the more concentrated ferromagnetic alloys and will eventually allow us to predict some of the effects of order–disorder. We do this by generalizing the above conclusions on the dilute alloys to a ferromagnetic binary alloy of atomic composition $A_{1-x}B_x$ (considered first at $0°K$) and assuming that the A and B atoms have magnetic moments, μ_A and μ_B, both of whose magnitudes are dependent on their local environments. For simplicity, we will suppose that the local environment of an atom can be meaningfully specified solely by the number of A and B atoms occupying its nearest neighbor sites. In so doing, we will be ignoring not only the influence of more distant neighbors but also the fact that all the A or B nearest neighbor atoms are not equivalent since *their* local environments may be different.

Initially, let us consider a completely disordered alloy in which all lattice sites are equivalent. The probability that any atom has n_A A-type neighbors and n_B B-type neighbors may be written alternatively as $P(n_A, 1 - x)$ or $P(n_B, x)$, where

$$P(n, x) = N!(1 - x)^{N-n} x^n / (N - n)! n! \qquad (3.3)$$

and $N = n_A + n_B$, the total number of nearest neighbors. Hence, if the

A atoms are identified by the number of B neighbors, the average of their individual moments $\mu_A(n_B)$ for a given alloy composition is

$$\bar{\mu}_A(x) = \sum_{n_B=0}^{N} P(n_B, x)\mu_A(n_B), \quad (3.4a)$$

and analogously,

$$\bar{\mu}_B(x) = \sum_{n_A=0}^{N} P(n_A, 1-x)\mu_B(n_A). \quad (3.4b)$$

Clearly, $\bar{\mu}_A(x)$ and $\bar{\mu}_B(x)$ combine as follows:

$$\bar{\mu}(x) = (1-x)\bar{\mu}_A(x) + x\bar{\mu}_B(x), \quad (3.5)$$

to give $\bar{\mu}(x)$, the average moment per atom, which is directly related to the saturation magnetization of the alloy as a whole; this expression simply generalizes Eq. (3.2) for the variability in μ_A and μ_B. Thus, in principle, if we know the functions $\mu_A(n_B)$ and $\mu_B(n_A)$, we can calculate $\bar{\mu}_A(x)$ and $\bar{\mu}_B(x)$ and, in turn, $\bar{\mu}(x)$ for the disordered alloy of any composition. However, if our starting information is $\bar{\mu}_A(x)$ and $\bar{\mu}_B(x)$, which is the operational case discussed later, it is not obvious from Eqs. (3.4a) and (3.4b) that we can easily work back to a unique solution for $\mu_A(n_B)$ and $\mu_B(n_A)$.

It will now be shown that certain analytical properties of the distribution function given in Eq. (3.3) make this problem readily soluble in either direction. Specifically, we have found that

$$\sum_{n=0}^{N} n(n-1) \cdots (n-l+1)P(n, x) = N(N-1) \cdots (N-l+1)x^l \quad (3.6a)$$

and

$$\sum_{n=0}^{N} n(n-1) \cdots (n-l+1)P(n, 1-x) = N(N-1) \cdots (N-l+1)(1-x) \quad (3.6b)$$

for any integer l between 1 and N, inclusive. These relationships, together with the normalization property

$$\sum_{n=0}^{N} P(n, x \text{ or } 1-x) = 1, \quad (3.7)$$

can be used in conjunction with Eqs. (3.4a) and (3.4b) to operate directly on the power series expressions,

$$\mu_A(n_B) = a_0 + a_1 n_B + a_2 n_B^2 + \cdots + a_N n_B^N \quad (3.8a)$$

and
$$\mu_B(n_A) = b_0 + b_1 n_A + b_2 n_A^2 + \cdots + b_N n_A^N, \tag{3.8b}$$

each of whose $N + 1$ coefficients are uniquely determined from the $N + 1$ values of μ_A or μ_B. In this way, we obtain after some manipulation

$$\begin{aligned}\bar{\mu}_A(x) = a_0 &+ (a_1 + a_2 + a_3 + a_4 + a_5 + \cdots)Nx \\ &+ (a_2 + 3a_3 + 7a_4 + 15a_5 + \cdots)N(N-1)x^2 \\ &+ (a_3 + 6a_4 + 25a_5 + \cdots)N(N-1)(N-2)x^3 \\ &+ (a_4 + 10a_5 + \cdots)N(N-1)(N-2)(N-3)x^4 \\ &+ (a_5 + \cdots)N(N-1)(N-2)(N-3)(N-4)x^5 + \cdots \end{aligned} \tag{3.9a}$$

and

$$\begin{aligned}\bar{\mu}_B(x) = b_0 &+ (b_1 + b_2 + b_3 + b_4 + b_5 + \ldots)N(1-x) \\ &+ (b_2 + 3b_3 + 7b_4 + 15b_5 + \cdots)N(N-1)(1-x)^2 \\ &+ (b_3 + 6b_4 + 25b_5 + \cdots)N(N-1)(N-2)(1-x)^3 \\ &+ (b_4 + 10b_5 + \cdots)N(N-1)(N-2)(N-3)(1-x)^4 \\ &+ (b_5 + \cdots)N(N-1)(N-2)(N-3)(N-4)(1-x)^5 + \cdots, \end{aligned} \tag{3.9b}$$

extending out to terms in x^N and $(1-x)^N$, respectively. Conversely, if we start with

$$\bar{\mu}_A(x) = a_0' + a_1' x + a_2' x^2 + a_3' x^3 + \cdots \tag{3.10a}$$

and

$$\bar{\mu}_B(x) = b_0' + b_1'(1-x) + b_2'(1-x)^2 + b_3'(1-x)^3 + \cdots, \tag{3.10b}$$

we find from Eqs. (3.4), again using Eqs. (3.6) and (3.7), that

$$\mu_A(n_B) = a_0' + a_1' \frac{n_B}{N} + a_2' \frac{n_B(n_B-1)}{N(N-1)} + a_3' \frac{n_B(n_B-1)(n_B-2)}{N(N-1)(N-2)} + \cdots \tag{3.11a}$$

and

$$\mu_B(n_A) = b_0' + b_1' \frac{n_A}{N} + b_2' \frac{n_A(n_A-1)}{N(N-1)} + b_3' \frac{n_A(n_A-1)(n_A-2)}{N(N-1)(N-2)} + \cdots, \tag{3.11b}$$

where again only terms of Nth order or less are allowed. Thus, by means of Eqs. (3.8)–(3.11), $\mu_A(n_B)$ and $\mu_B(n_A)$ can be converted into or deduced from $\bar{\mu}_A(x)$ and $\bar{\mu}_B(x)$ with comparable ease.

XI. ATOMIC ORDER–DISORDER AND MAGNETIC PROPERTIES

Let us now briefly apply these considerations to a simple, hypothetical case of order–disorder. We will compare a ferromagnetic alloy of atomic composition $A_{0.75}B_{0.25}$ having a completely disordered fcc structure with the same alloy when it has perfect Cu_3Au-type order, as illustrated in

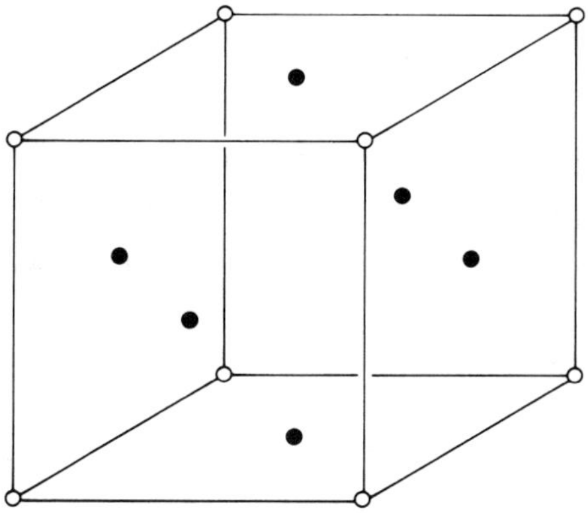

FIG. 4. Cubic, Cu_3Au-type ordered structure A_3B: (●) A; (○) B.

Fig. 4. For the disordered state, it is immediately found from Eq. (3.10) for $x = \frac{1}{4}$ that

$$\bar{\mu}_A(\tfrac{1}{4}) = a_0' + \tfrac{1}{4}a_1' + (1/16)a_2' + \cdots,$$
$$\bar{\mu}_B(\tfrac{1}{4}) = b_0' + \tfrac{3}{4}b_1' + (9/16)b_2' + \cdots,$$

$$\begin{aligned}\bar{\mu}(\text{disorder}) &= \tfrac{3}{4}\bar{\mu}_A(\tfrac{1}{4}) + \tfrac{1}{4}\bar{\mu}_B(\tfrac{1}{4}) \\ &= \tfrac{3}{4}a_0' + (3/16)a_1' + (3/64)a_2' + \cdots + \tfrac{1}{4}b_0' + (3/16)b_1' \\ &\quad + (9/64)b_2' + \cdots .\end{aligned} \quad (3.12)$$

This result can be regarded as arising from the individual atomic moments varying according to Eq. (3.11), where n_A and n_B follow the distribution functions shown in Fig. 5, as derived from Eq. (3.3). In the ordered state (Fig. 4) the distribution functions are unique: all the A atoms have four B (and eight A) neighbors and all the B atoms have 12 A (and no B) neighbors. Hence, we obtain from Eq. (3.11),

$$\mu_A(4) = a_0' + \tfrac{1}{3}a_1' + (1/11)a_2' + \cdots,$$
$$\mu_B(12) = b_0' + b_1' + b_2' + \cdots,$$

$$\begin{aligned}\bar{\mu}(\text{order}) &= \tfrac{3}{4}\mu_A(4) + \tfrac{1}{4}\mu_B(12) \\ &= \tfrac{3}{4}a_0' + \tfrac{1}{4}a_1' + (3/44)a_2' + \cdots + \tfrac{1}{4}b_0' + \tfrac{1}{4}b_1' + \tfrac{1}{4}b_2' + \cdots .\end{aligned} \quad (3.13)$$

Assuming that the moment parameters (a_0', a_1', etc.) are not themselves affected by the ordering process (due to any changes in lattice spacing, for instance), one deduces from Eqs. (3.12) and (3.13) that the ordering in general will cause the saturation moment of the alloy to change. Moreover, the magnitude and sign of this change will depend on the values of all the moment parameters that reflect the variation of the individual atomic moments with local environment (i.e., all the parameters except a_0' and b_0').

A similar type of approach to the relationship between saturation moment and atomic order has previously been taken by Goldman and Smoluchowski [18]. The basis of statistical analysis in this earlier work was the average magnetic moment of a spherical cluster of atoms consisting of a central atom and all its nearest neighbors (and, in most cases, all its second-nearest

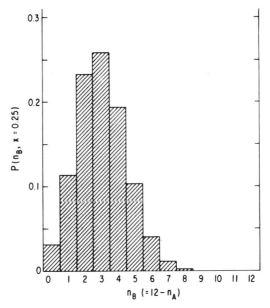

FIG. 5. Nearest-neighbor atomic distribution function for a disordered fcc system of composition $A_{0.75}B_{0.25}$.

neighbors). Furthermore, this average local moment was assumed to depend in an arbitrary quantitative fashion on the atomic composition of the cluster. Predictions made of the saturation moments of ordered and disordered FeCo and Ni_3Fe were in reasonable agreement with experiment. However, this agreement was tenuous since the quantitative information fed into the analysis had little theoretical or experimental justification. In

XI. ATOMIC ORDER–DISORDER AND MAGNETIC PROPERTIES   539

a more recent analysis of this problem by Sato [19], the statistics used are much too schematic for meaningful calculation.

The present treatment represents a significant improvement over these earlier analyses in that its general relationships, Eqs. (3.8)–(3.11), can incorporate the quantitative results of different types of experiments and can thus lead to an empirically consistent view of the effects of ordering on saturation moment. The potential of these relationships is most fully realized when they can be applied to available data from diffuse neutron scattering experiments on disordered ferromagnetic systems. First, however, it is necessary to assume that the scattering cross section in these experiments, for an alloy of composition $A_{1-x}B_x$, is given by

$$d\sigma_M \propto x(1-x)\{\bar{\mu}_A(x) - \bar{\mu}_B(x)\}^2, \qquad (3.14)$$

which extends Eq. (3.1) to the case where μ_A and μ_B, the individual atomic moments, are functions of their environments and should therefore be replaced by their average values, $\bar{\mu}_A(x)$ and $\bar{\mu}_B(x)$, as defined in Eq. (3.4). This assumption is almost certain to be a reasonable one. Indeed, Marshall [19a] has recently derived a fairly rigorous expression for the diffuse magnetic scattering cross section, which does not depart significantly from Eq. (3.14) unless the dependence of the atomic moments on local environment is very strong. Hence, Eq. (3.14) can be regarded as an important auxiliary to Eqs. (3.8)–(3.11).

To demonstrate the application of these relationships to a simple real case, we return to the results for the disordered Ni–Fe alloys shown in Figs. 2 and 3, where μ_{Ni} and μ_{Fe} should now be interpreted as the average quantities, $\bar{\mu}_{Ni}(x)$ and $\bar{\mu}_{Fe}(x)$, pertinent to the composition $Ni_{1-x}Fe_x$. In the Ni-rich (fcc) region, we find that to a good approximation,

$$\bar{\mu}_{Ni}(x) \approx 0.57 + 0.20x, \qquad \bar{\mu}_{Fe}(x) \approx 2.20 + 0.80(1-x) \qquad (3.15)$$

in Bohr magnetons (μ_β), for the solutions represented by the solid curves in Fig. 3. Using the conversion from Eq. (3.10) to Eq. (3.11), with $N = 12$, we obtain

$$\mu_{Ni}(n_{Fe}) \approx 0.57 + 0.20 n_{Fe}/12, \qquad \mu_{Fe}(n_{Ni}) \approx 2.20 + 0.80 n_{Ni}/12 \qquad (3.16)$$

in Bohr magnetons, for the individual atomic moments as a function of their nearest neighbor occupation. Since the alloy of composition Ni_3Fe ($x = \frac{1}{4}$), when suitably annealed, orders in a Cu_3Au-type structure (see Fig. 4), the hypothetical case discussed earlier becomes immediately pertinent. Hence, we apply Eqs. (3.12) and (3.13), with the values for the a' and b' coefficients given by a comparison of Eqs. (3.15) and (3.16) with Eqs. (3.10) and (3.11), and calculate for the disordered state of Ni_3Fe,

$$\bar{\mu}_{Ni}(\tfrac{1}{4}) \approx 0.62\mu_\beta, \quad \bar{\mu}_{Fe}(\tfrac{1}{4}) \approx 2.80\mu_\beta, \quad \bar{\mu}(\text{disorder } Ni_3Fe) \approx 1.16\mu_\beta, \quad (3.17)$$

and for the ordered state,

$$\mu_{Ni}(4) \approx 0.637\mu_\beta, \quad \mu_{Fe}(12) \approx 3.00\mu_\beta, \quad \bar{\mu}(\text{order } Ni_3Fe) \approx 1.22\mu_\beta. \quad (3.18)$$

Thus, a simplified but consistent use of the experimental information in Figs. 2 and 3 on various disordered Ni–Fe alloys yields the prediction that the saturation moment of Ni_3Fe increases by about 5% with ordering, which is in very reasonable agreement with direct magnetization measurement [20]. Qualitatively, this result stems from the fact that Eq. (3.16) has both the Ni and Fe moments increasing slowly with the number of unlike neighbors, which is on the average raised by the ordering.

Further discussion of Ni–Fe will be given in the following section. However, two general comments should be made here about the neutron diffraction experiments on whose results the above calculations were based. First, had these experiments been performed with polarized rather than unpolarized neutrons, there would have been only one solution for the average atomic moments, corresponding to one of the two solutions represented in Fig. 3. This advantage in the use of polarized neutrons has been exploited in some recent experiments on Fe–Co as well as Ni–Fe, which are discussed in the next section. Second, the Shull and Wilkinson experiments were carried out at room temperature, which for the Ni-rich Ni–Fe alloys is well below the Curie temperature; hence, the results are reasonably representative of very low temperatures for which the above saturation moment analysis is specifically valid. However, in the event that the temperature of measurement is relatively close to the Curie point, the validity of this type of analysis will break down for the following reasons: (1) in a given alloy (ordered or disordered), the individual atomic moments will be thermally reduced from their low-temperature values by different proportional amounts; (2) the Curie point and, hence, the magnitude of this thermal effect will vary with composition and degree of order.

4. Curie Temperature and Order–Disorder. The effect of atomic ordering on the Curie point of a ferromagnetic alloy is an important subject in itself, and we will now consider it in a simple extension of our previous arguments.

It will be adequate for this purpose to use the molecular field model, suitably modified so that the individual atomic moments are still allowed to vary in magnitude with their local environment. Furthermore, only nearest neighbor interactions will be taken into account. For a disordered alloy of composition $A_{1-x}B_x$, the contribution of each nearest neighbor B-type atom to the molecular field at an A-type atom will be represented by $\frac{1}{2}J_{AB}\bar{\mu}_B(x)$, where J_{AB} is the exchange coefficient and $\bar{\mu}_B(x)$ is the average moment for all B-type atoms, as defined in Eq. (3.4); an analogous approximation will be made for the molecular field contributions from AA, BA, and BB interactions. These assumptions lead to the following expressions for the individual atomic moments at temperature T:

$$\mu_A(n_B) = \mu_A^{(0)}(n_B)B_j(y_A),$$
$$\mu_B(n_A) = \mu_B^{(0)}(n_A)B_j(y_B), \qquad (4.1)$$
$$y_A \equiv [(N - n_B)J_{AA}\bar{\mu}_A(x) + n_B J_{AB}\bar{\mu}_B(x)]\mu_A^{(0)}(n_B)/2kT,$$
$$y_B \equiv [n_A J_{BA}\bar{\mu}_A(x) + (N - n_A)J_{BB}\bar{\mu}_B(x)]\mu_B^{(0)}(n_A)/2kT,$$

where $\mu_A^{(0)}(n_B)$ and $\mu_B^{(0)}(n_A)$ are the moments at 0°K, which may still be thought to be describable by Eqs. (3.8) or (3.11), and B_j is the Brillouin function defined as

$$B_j(y) = \frac{2j+1}{2j}\coth\left(\frac{2j+1}{2j}y\right) - \frac{1}{2j}\coth\left(\frac{1}{2j}y\right), \qquad (4.2)$$

the index j being the spin quantum number whose value is an integer or half-integer. Substituting the above expressions into Eq. (3.4), we obtain

$$\bar{\mu}_A(x) = \sum_{n_B=0}^{N} P(n_B, x)\mu_A^{(0)}(n_B)B_j(y_A)$$
$$\bar{\mu}_B(x) = \sum_{n_A=0}^{N} P(n_A, 1-x)\mu_B^{(0)}(n_A)B_j(y_B), \qquad (4.3)$$

which are highly implicit expressions for the average moments, $\bar{\mu}_A(x)$ and $\bar{\mu}_B(x)$, at any given temperature up to the Curie point. In the general case, they are extremely difficult to solve, but if one of the components of the binary alloy is nonmagnetic (e.g., if $\mu_B^{(0)} = 0$), they reduce to a single transcendental relationship that is readily soluble.

For the effect of ordering on the Curie temperature T_C, an explicit answer can be arrived at, even in the complicated case when both components of a binary alloy are magnetic. At temperatures approaching T_C, all the atomic moments are decreasing to zero and, consequently, so are y_A and y_B, the arguments of the Brillouin functions in Eq. (4.3). Since from Eq. (4.2),

$$B_j(y) \to (j+1)y/3j \quad \text{as} \quad y \to 0, \tag{4.4}$$

it follows from Eq. (4.3) for $T \approx T_C$ that

$$\begin{aligned}\bar{\mu}_A(x) &= \{T'_{AA}\bar{\mu}_A(x) + T'_{AB}\bar{\mu}_B(x)\}/T_C \\ \bar{\mu}_B(x) &= \{T'_{BA}\bar{\mu}_A(x) + T'_{BB}\bar{\mu}_B(x)\}/T_C,\end{aligned} \tag{4.5}$$

where

$$\begin{aligned} T'_{AA} &\equiv \frac{j+1}{6jk} J_{AA} \sum_{n_B=0}^{N} (N - n_B)P(n_B, x)\{\mu_A^{(0)}(n_B)\}^2 \\ T'_{AB} &\equiv \frac{j+1}{6jk} J_{AB} \sum_{n_B=0}^{N} n_B P(n_B, x)\{\mu_A^{(0)}(n_B)\}^2 \\ T'_{BA} &\equiv \frac{j+1}{6jk} J_{BA} \sum_{n_A=0}^{N} n_A P(n_A, 1-x)\{\mu_B^{(0)}(n_A)\}^2 \\ T'_{BB} &\equiv \frac{j+1}{6jk} J_{BB} \sum_{n_A=0}^{N} (N - n_A)P(n_A, 1-x)\{\mu_B^{(0)}(n_A)\}^2. \end{aligned} \tag{4.6}$$

Equation (4.5) leads to a quadratic equation in T_C, the Curie temperature of the disordered alloy, whose physically meaningful solution is

$$T_C(\text{disorder}) = \tfrac{1}{2}(T'_{AA} + T'_{BB}) + \{\tfrac{1}{4}(T'_{AA} - T'_{BB})^2 + T'_{AB}T'_{BA}\}^{1/2}, \tag{4.7}$$

in terms of the interaction temperatures defined in Eq. (4.6), each of which is a function of x. For the ordered state, we consider the common (though not general) situation in which n_A and n_B are each the same for all B-type and A-type atoms, respectively. Hence, $\bar{\mu}_A(x)$ and $\bar{\mu}_B(x)$ in Eq. (4.1) should be replaced by $\mu_A(n_B)$ and $\mu_B(n_A)$, respectively, and if we again apply Eq. (4.4) to go to the limit of $T \to T_C$, we obtain

$$\begin{aligned}\mu_A(n_B) &= \{T''_{AA}\mu_A(n_B) + T''_{AB}\mu_B(n_A)\}/T_C \\ \mu_B(n_A) &= \{T''_{BA}\mu_A(n_B) + T''_{BB}\mu_B(n_A)\}/T_C,\end{aligned} \tag{4.8}$$

and, in turn,

$$T_C(\text{order}) = \tfrac{1}{2}(T''_{AA} + T''_{BB}) + \{\tfrac{1}{4}(T''_{AA} - T''_{BB})^2 + T''_{AB}T''_{BA}\}^{1/2}, \tag{4.9}$$

with

$$T''_{AA} \equiv \frac{j+1}{6jk} J_{AA}(N - n_B)\{\mu_A^{(0)}(n_B)\}^2$$

$$T''_{AB} \equiv \frac{j+1}{6jk} J_{AB} n_B \{\mu_A^{(0)}(n_B)\}^2$$

$$T''_{BA} \equiv \frac{j+1}{6jk} J_{BA} n_A \{\mu_B^{(0)}(n_A)\}^2 \quad (4.10)$$

$$T''_{BB} \equiv \frac{j+1}{6jk} J_{BB}(N - n_A)\{\mu_B^{(0)}(n_A)\}^2$$

We observe that Eqs. (4.7) and (4.9) for the Curie temperatures of the disordered and ordered systems, although formally the same, involve interaction temperatures which, as defined in Eqs. (4.6) and (4.10), may be very different. This difference arises from the changes with order–disorder in the relative numbers of A–A, A–B, and B–B nearest neighbor atom pairs, either directly from the corresponding changes in the relative numbers of the various exchange couplings or, more indirectly, from the variations in atomic moment and their effect on the strength of each coupling. As shown in the following section, the atomic moment variations in Ni–Fe can be neglected and reasonable predictions can still be made about the Curie point, whereas in Fe–Al these variations are more important and should be taken into account.

Exchange Anisotropy and Exchange Polarization Effects

5. Brief mention should also be made here of two special features of general interest that emerge from detailed considerations in the next section. The first is the very striking contrast between the magnetic properties of ordered and disordered Ni_3Mn. When disordered, this alloy exhibits unusual changes of magnetization with field and temperature (including asymmetrical hysteresis loops, when cooled in a field) which are indicative of a mixed magnetic state consisting of submicroscopic regions of ferromagnetically and antiferromagnetically aligned atomic moments. The antiferromagnetic interactions that are partly responsible for this so-called *exchange anisotropy* state are believed to be those between nearest neighbor Mn–Mn pairs. Such atom pairs are absent in perfectly ordered Ni_3Mn which, consistently enough, shows a fairly simple ferromagnetic behavior. This situation, it should be noted, is outside the scope of the analysis given above, where all the atomic moments are assumed to be aligned parallel to each other in both the ordered and disordered states.

The second item of interest is with reference to the alloys of the iron group metals with palladium group or platinum group metals. Recent work has made it clear that a special phenomenon, which we shall call *exchange polarization*, may be uniquely pertinent to the basic magnetic properties of this family of alloys. This phenomenon was originally recognized in the dilute ferromagnetic alloys of iron or cobalt in palladium, whose saturation magnetizations are far too large to be reasonably ascribable to the iron or cobalt moments alone. Indeed, it was demonstrated by diffuse neutron scattering experiments (of the type described earlier) that a substantial moment contribution in Fe–Pd comes from the Pd atoms surrounding each of the Fe atoms; consequently, the Pd moments can be thought to be polarized into existence by the effective exchange field from the neighboring Fe moments. Hence, in the ordered alloys, where each of the Pd group or Pt group atoms will generally have one or more Fe group neighbors, they can each be expected to have an induced moment if their neighboring Fe group moments subject them to a net exchange field. This obviously will always be the case when the ordered alloy is ferromagnetic (e.g., CoPt, $FePd_3$) and all the Fe group moments are parallel; in all such cases that have been investigated by neutron diffraction a sizable moment on the Pd group or Pt group atoms has been observed. However, if the ordered alloy is antiferromagnetic (e.g., $FePt_3$) such that the moments on the Fe group neighbor atoms add vectorially to zero, the net exchange field experienced by each Pd group or Pt group atom will also be zero, and there should therefore be no induced moment; this also has been observed experimentally. Clearly, exchange polarization can play a crucial role in determining the magnetic effects of ordering in this interesting class of materials.

Discussion of Specific Alloy Systems

Iron–Cobalt

6. That atomic ordering takes place in iron–cobalt alloys near the composition FeCo was first proposed on the basis of their electrical resistivity behavior [21], but is also reflected in a small change in saturation magnetization [22]. X-ray [23] and neutron [24] diffraction studies have revealed the formation of an ordered structure of the cubic CsCl type (see Fig. 6) in annealed alloys ranging in composition from about 40 to 60 at. % Co. Alloys quenched from above the ordering temperature (730°C for stoichiometric FeCo) are retained in a substantially disordered bcc structure [23–25],

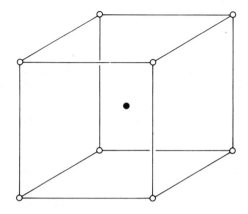

FIG. 6. Cubic, CsCl-type ordered structure AB: (●) A; (○) B.

upon which the ordered CsCl-type structure is based. The ordering process is unusual in several respects: it not only results in a slightly larger lattice parameter [23] (whereas most other systems contract upon ordering), but it appears to have a subsidiary transformation temperature near 550°C (marked by anomalies in magnetic coercivity [26], specific heat [25], and electrical resistivity [27]), at which point the long-range order parameter shows an unusual temperature dependence [28].

The alloy FeCo is ferromagnetic up to about 980°C, where it transforms martensitically from bcc to fcc and its saturation magnetization drops abruptly to zero. A reasonable extrapolation of the saturation magnetization

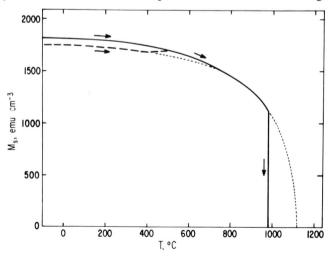

FIG. 7. Saturation magnetization versus temperature for ordered and disordered FeCo (solid and dashed curves, respectively) upon slow heating.

(M_s) from lower temperatures gives approximately 1120°C as the "virtual" Curie point of the disordered bcc phase [21], as indicated in Fig. 7. This figure also shows that M_s for the annealed state is slightly higher than that for the quenched (from above 730°C) state and that upon slow heating, the latter rises and merges with the former near 550°C as atomic ordering takes place; the slight kink in the curve at about 730°C marks the atomic disordering temperature [29]. From this behavior it is impossible to estimate the effect of order–disorder on the Curie temperature accurately, but it is clearly very small.

The results of some recent neutron diffraction measurements [30] on Fe–Co alloys allow us to examine the influence of atomic ordering on the saturation moment of these alloys within the analytical framework described in the previous section. In these experiments, polarized neutrons were used in order to avoid any ambiguity in the assignment of moments to the component atoms. The striking conclusion to be drawn from the results is that while the average iron moment ($\bar{\mu}_{Fe}$) rises from $2.2\mu_\beta$ in pure iron to just over $3\mu_\beta$ in alloys of 50 at. % or more Co, the average cobalt moment ($\bar{\mu}_{Co}$) stays essentially constant at $1.8\mu_\beta$ over the entire bcc composition range of 0–75 at. % Co. The constancy of $\bar{\mu}_{Co}$ implies that the magnitude of the individual cobalt moments, $\mu_{Co}(n_{Co})$ is independent of the atomic environment, which we designate by the number of Co nearest neighbors, n_{Co}. Hence, the change in the total saturation moment ($\bar{\mu}$) with atomic ordering, as well as its change with alloy composition, can be regarded as being entirely due to the variability of the individual iron moments, $\mu_{Fe}(n_{Co})$, with respect to their environment (again specified by n_{Co}). This being the case, we can take available data for $\bar{\mu}(x)$ for disordered alloys of composition $Fe_{1-x}Co_x$ and convert each value to a corresponding value for $\bar{\mu}_{Fe}(x)$ by means of Eq. (3.5) in which A = Fe, B = Co, and $\bar{\mu}_{Co}(x)$ is considered constant at $1.8\mu_\beta$. We have done this with published magnetization data [31] and the results are shown in Fig. 8, where the values for $\bar{\mu}_{Fe}(x)$ agree quite closely with those deduced from the neutron diffraction data [30]. From the following conditions suggested by these $\bar{\mu}_{Fe}(x)$ points

$\bar{\mu}_{Fe} = 2.22\mu_\beta$ and $d\bar{\mu}_{Fe}/dx = 1.82\mu_\beta$ at $x = 0$,

$\bar{\mu}_{Fe} = 3.02\mu_\beta$ at $x = 0.5$,

$\bar{\mu}_{Fe} = 3.15\mu_\beta$ and $d\bar{\mu}_{Fe}/dx = 0$ at $x = 1$,

coefficients were determined for the power series expression,

$$\bar{\mu}_{Fe}(x) = 2.22 + 1.82x + 0.87x^2 - 3.48x^3 + 1.72x^4, \tag{6.1}$$

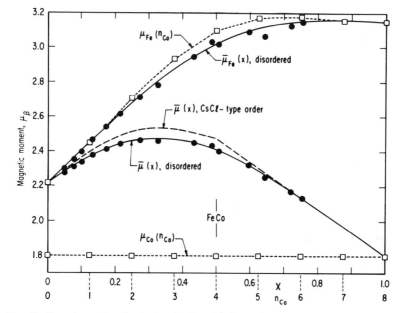

FIG. 8. Experimental and calculated values (circles and curves, respectively) for atomic moments in disordered and ordered Fe–Co alloys. The average moments are plotted against alloy composition; individual moments are plotted against the number of Co nearest neighbor atoms.

in Bohr magnetons. The curve derived from this expression, as drawn in Fig. 8, clearly matches the $\bar{\mu}_{\text{Fe}}(x)$ points quite well. Equivalently, a close fit to the $\bar{\mu}(x)$ data was obtained with the curve (also shown in Fig. 8) calculated from this expression combined with Eq. (3.5), $\bar{\mu}_{\text{Co}}$ being again fixed at $1.8\mu_\beta$. Equation (6.1) was then identified formally with Eq. (3.10a) and converted to an expression for $\mu_{\text{Fe}}(n_{\text{Co}})$ that is analogous to Eq. (3.11a), with $N = 8$. The $\mu_{\text{Fe}}(n_{\text{Co}})$ values were computed and are plotted in Fig. 8, where they are seen to rise almost linearly from $2.22\mu_\beta$ to about $3.1\mu_\beta$ as n_{Co} increases from 0 to 4 and remain nearly constant for larger n_{Co}. This variation of $\mu_{\text{Fe}}(n_{\text{Co}})$, which is interesting in itself, was the basis for computing the saturation moment for the fully ordered state. This was done not just for stoichiometric FeCo but for all $\text{Fe}_{1-x}\text{Co}_x$ compositions for which the Fe and Co sites in the ordered structure shown in Fig. 6 are occupied as follows:

$$[\text{Fe}]_{\text{Fe sites}}[\text{Fe}_{1-2x}\text{Co}_{2x}]_{\text{Co sites}}, \quad x \leq \tfrac{1}{2}$$

and

$$[\text{Fe}_{2-2x}\text{Co}_{2x-1}]_{\text{Fe sites}}[\text{Co}]_{\text{Co sites}}, \quad x \geq \tfrac{1}{2}.$$

Since all nearest neighbors of Fe sites are Co sites and vice versa, the saturation moment per atom may be expressed as

$$\bar{\mu}(x) = \tfrac{1}{2}\bar{\mu}_{\text{Fe}}(2x) + (\tfrac{1}{2} - x)\mu_{\text{Fe}}(n_{\text{Co}} = 0) + x\mu_{\text{Co}}, \qquad x \leq \tfrac{1}{2} \qquad (6.2a)$$

and

$$\bar{\mu}(x) = (1 - x)\mu_{\text{Fe}}(n_{\text{Co}} = 8) + x\mu_{\text{Co}}, \qquad x \geq \tfrac{1}{2}, \qquad (6.2b)$$

where $\bar{\mu}_{\text{Fe}}(2x)$ is to be evaluated from Eq. (6.1), $\mu_{\text{Fe}}(n_{\text{Co}} = 0) = 2.22\mu_\beta$, $\mu_{\text{Fe}}(n_{\text{Co}} = 8) = 3.15\mu_\beta$, and $\mu_{\text{Co}} = 1.80\mu_\beta$. The saturation moments thus obtained for Fe–Co alloys having the maximum CsCl-type order permitted by their composition are represented by the long-dashed curve in Fig. 8, which lies consistently above the curve for the disordered alloys. The largest difference between the curves occurs from about 30 to 50 at. % Co and is about $0.07\mu_\beta$ (i.e., ~3%), in good agreement with experiment [22]. Since the shapes of these curves are very similar, the calculated difference between them is valid even if the supposedly disordered alloys had some retained order. Moreover, the fact that the magnetizations of the "disordered" alloys were measured at room temperature does not weaken these calculations; their Curie points are so high that any variation in them would have negligible effect at room temperature.

The room temperature moment of about $2.5\mu_\beta$ per atom for ordered FeCo corresponds to a saturation magnetization of just under 2000 emu cm^{-3}, which is about the largest value known for any material at this temperature.

Torque measurements have been made of the magnetocrystalline anisotropy coefficient K_1 for disordered (quenched) and ordered Fe–Co alloys at room temperature [32]. With increasing Co concentration, K_1 for the disordered state reverses in sign from positive to negative near 40 at. % Co, reaching a value of approximately -2×10^5 ergs cm^{-3} at 50 at. % Co. The effect of atomic ordering is to raise K_1 so that it remains positive up to the stoichiometric FeCo composition, where it then goes through zero. It is suggested [32] that the low anisotropy of ordered (annealed) FeCo may account, at least in part, for its high magnetic permeability, for which it is known commercially as *Permendur* [33].

The saturation magnetostriction (λ_s) of a "well-ordered" polycrystalline FeCo sample was found by strain gage measurements at room temperature to be 9.2×10^{-5}, which represented a 40% increase from the measured value for a "slightly ordered" (quenched) alloy [18]. More recently, similar measurements [32] on ordered FeCo single crystals gave for the magnetostriction coefficients, $\lambda_{100} \approx 14 \times 10^{-5}$ and $\lambda_{111} \approx 4 \times 10^{-5}$, so that $\lambda_s = (2\lambda_{100} + 3\lambda_{111})/5 \approx 8 \times 10^{-5}$, in fair agreement with the earlier result; however, λ_{100} and λ_{111} for disordered crystals were found to be the same (within ~10%) as those for the ordered crystals, which differs markedly from the previous measurement.

Nickel–Iron

7. Until fairly recently, the only firm evidence for atomic ordering in the Ni–Fe system was that for the cubic, Cu_3Au-type order (see Fig. 4) occurring in fcc alloys at and about the stoichiometric composition Ni_3Fe at temperatures near 500°C [12, 34, 35]. Indeed, most of the following discussion will be given to the many property changes found to accompany this well-known ordering process. In 1962, however, it was reported that fcc alloys of composition NiFe, when subjected to neutron [36] or electron [37] irradiation and, simultaneously, to a magnetic field, developed a tetragonal, CuAu-type order, the largest effect (with neutrons) taking place at about 300°C. Various other techniques [38] have been subsequently used to accelerate the sluggish reactions associated with both these types of order (i.e., Ni_3Fe as well as NiFe); the phase equilibrium diagram based on this work is shown in Fig. 9 and represents a considerable modifica-

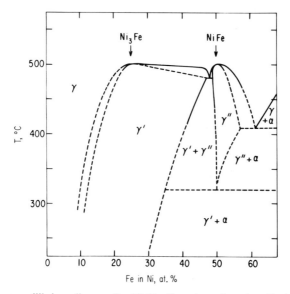

FIG. 9. Phase equilibrium diagram for Ni–Fe alloys (γ = fcc, γ' = Cu_3Au-type order, γ'' = CuAu-type order, α = bcc).

tion of the previously accepted diagram [39]. However, the latter gives the practical near-equilibrium conditions pertinent to the following discussion.

Atomic ordering (Cu_3Au type) has been found to raise the saturation moment of the ferromagnetic alloys having 65–80 at. % Ni [20, 35, 40]. The largest increase, about 5%, occurs in stoichiometric Ni_3Fe, whose saturation moment reaches $\sim 1.2\mu_\beta$ per atom (i.e., a magnetization of ~ 1000 emu cm^{-3}). This increase in moment was shown in the previous section to be consistent with the composition dependence of the average moments on the Ni and Fe atoms in the disordered alloys, as deduced from neutron diffraction data [12]. Moreover, the values calculated for the Ni and Fe moments in ordered Ni_3Fe, which are given in Eq. (3.18), agree very closely with the neutron diffraction results [12]: $\mu_{Ni} = 0.62\ (\pm 0.05)\mu_\beta$ and $\mu_{Fe} = 2.97\ (\pm 0.15)\mu_\beta$ aligned ferromagnetically with respect to each other. An alternative ferrimagnetic solution (with μ_{Fe} antiparallel to μ_{Ni}) of the same diffraction data can now be disallowed on the basis of recent measurements with polarized neutrons [30]. Indeed, these polarized neutron experiments have also removed the ambiguity concerning the atomic moments in the disordered alloys by favoring the values given in Eq. (3.15) and represented by solid curves in Fig. 3.

From the slow decrease in the magnetization of ordered Ni_3Fe measured during rapid heating [35, 40, 41], it was concluded that its Curie point lies well above the order–disorder temperature of about 775°K (~ 500°C) and even exceeds the Curie point of the disordered alloy (generally placed near 875°K) by as much as 125° [40]. This conclusion was based on the assumption of a fixed shape for the normalized magnetization–temperature curves, which would overestimate the Curie temperature of the ordered phase if its curve is more concave towards the temperature axis than that of the disordered phase [42]. However, any disordering of the specimen during these measurements would have a compensating effect; therefore, the original conclusion that the Curie point of Ni_3Fe increases with ordering by about 100° may be fairly accurate.

The change in the Curie temperature of Ni_3Fe with atomic order can be examined within the context of the theoretical discussion in the previous section. We have already shown in Eq. (3.16) that the magnitudes of the individual atomic moments in Ni–Fe are only mildly sensitive to their local environment, and if their variation is neglected entirely, the Curie point calculations are enormously simplified. In particular, if $\mu_A^{(0)}(n_B)$ and $\mu_B^{(0)}(n_A)$ in Eq. (4.6) are replaced by their mean values $\bar{\mu}_A^{(0)}$ and $\bar{\mu}_B^{(0)}$, it is found [by virtue of Eqs. (3.6) and (3.7)] that the expressions for the exchange temperatures for the disordered state reduce to

$$T'_{AA} = T_{AA}(1-x), \quad T'_{AB} = T_{AB}x, \quad T'_{BA} = T_{BA}(1-x), \quad T'_{BB} = T_{BB}x, \quad (7.1)$$

pertinent to the composition $A_{1-x}B_x$, where

$$T_{AA} = \frac{j+1}{6jk} NJ_{AA}\{\bar{\mu}_A^{(0)}\}^2, \quad T_{AB} = \frac{j+1}{6jk} NJ_{AB}\{\bar{\mu}_A^{(0)}\}^2,$$
$$T_{BA} = \frac{j+1}{6jk} NJ_{BA}\{\bar{\mu}_B^{(0)}\}^2, \quad T_{BB} = \frac{j+1}{6jk} NJ_{BB}\{\bar{\mu}_B^{(0)}\}^2. \quad (7.2)$$

Similarly, for the fully ordered A_3B state, in which every A atom has four B nearest neighbors and every B atom has 12 (i.e., all) A nearest neighbors (see Fig. 4), we obtain from Eq. (4.10)

$$T''_{AA} = \tfrac{2}{3}T_{AA}, \quad T''_{AB} = \tfrac{1}{3}T_{AB}, \quad T''_{BA} = T_{BA}, \quad T''_{BB} = 0, \quad (7.3)$$

where T_{AA}, T_{AB}, and T_{BA} are again defined in Eq. (7.2).

Identifying Ni with A and Fe with B, we will now try to match the experimental Curie point (T_C) vs. composition (x) curve for disordered $Ni_{1-x}Fe_x$ with a theoretical curve based on Eq. (4.7) in which T'_{AA}, etc., are given in Eq. (7.1). As shown in Fig. 10, a typical experimental curve [39] has T_C increasing from $\sim 630°K$ at $x = 0$ (pure Ni) to a peak value near $885°K$ at $x \approx 0.33$, then descending rapidly to values below $500°K$ at $x \approx 0.7$, where it normally meets the fcc $\to$ bcc phase transformation temperature.

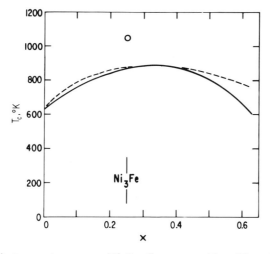

FIG. 10. Curie temperature versus Ni–Fe alloy composition. The solid and dashed curves are experimental and theoretical, respectively, for the disordered (fcc) alloys. The circled point is calculated for ordered Ni_3Fe.

Since our theoretical expression for T_C vs. x, Eq. (4.7) in conjunction with Eq. (7.1), has effectively three independent parameters: T_{AA}, T_{BB}, and $T_{AB}T_{BA}$ (the last of which, for simplicity, is now defined as $\bar{T}_{AB}^2$), we impose on it three conditions: $T_C = 630°K$ at $x = 0$, $(T_C)_{max} = 885°K$ and occurs at $x_{max} = 0.33$. Using the relationships,

$$(T_C)_{max} = (\bar{T}_{AB}^2 - T_{AA}T_{BB})/(2\bar{T}_{AB} - T_{AA} - T_{BB})$$
$$x_{max} = (\bar{T}_{AB} - T_{AA})/(2\bar{T}_{AB} - T_{AA} - T_{BB}), \quad (7.4)$$

derived from Eqs. (4.7) and (7.1), we obtain

$$T_{AA} = 630°K, \quad \bar{T}_{AB} = 1400°K, \quad T_{BB} = -170°K, \quad (7.5)$$

for the Ni–Ni, Ni–Fe, and Fe–Fe interactions, respectively. These interaction parameter values are then substituted back into Eqs. (4.7) and (7.1) and the curve calculated for T_C vs. x. This calculated curve, plotted in Fig. 10, agrees reasonably well with experiment, considering the restriction of the model to nearest neighbor interactions.

The T_C value calculated for $x = \frac{1}{4}$, corresponding to disordered Ni$_3$Fe, is $\sim 875°K$, and if the same interaction parameter values given in Eq. (7.5) are now inserted into Eqs. (4.9) and (6.5) it is found for ordered Ni$_3$Fe that $T_C \approx 1045°K$, as indicated in Fig. 10. Although this difference between the T_C values predicted for the ordered and disordered alloy is somewhat larger than experiment would indicate, it is not seriously wrong in magnitude and it is certainly correct in sign. Moreover, the parameter values in Eq. (7.5) allow us to understand broadly why this difference comes about. We see that $\bar{T}_{AB}$ for the Ni–Fe interactions is much larger than either T_{AA} (for Ni–Ni) or T_{BB} (for Fe–Fe), the latter indeed being negative, and since the Cu$_3$Au-type ordering of Ni$_3$Fe increases the average number of Ni–Fe nearest neighbor atom pairs at the expense of the average number of Ni–Ni and Fe–Fe pairs (the latter of which are eliminated altogether), the net effect expected is a stronger ferromagnetism reflected in a higher Curie temperature. That the same interaction parameters apply for both ordered and disordered states is probably a safe assumption because the lattice spacings are essentially unchanged [35]. It should be noted that the negative T_{BB}, indicating that the Fe–Fe interactions are antiferromagnetic, is consistent with the explanation previously proposed [8, 43] for the rapid rise in the pressure dependence of the saturation moment, as well as for the rapid drop in the saturation moment itself (see Fig. 1), which occurs in the disordered, fcc Ni–Fe alloys when the Fe concentration exceeds about 60 at. %.

XI. ATOMIC ORDER–DISORDER AND MAGNETIC PROPERTIES

With regard to the magnetocrystalline anisotropy of Ni_3Fe, magnetization [20] and torque [44] measurements at room temperature have shown that the anisotropy coefficient K_1 for an ordered crystal is about -25×10^3 ergs cm^{-3}, whereas for the same crystal disordered K_1 is very close to zero. The change in K_1 with ordering, although largest for stoichiometric Ni_3Fe, is substantial over a wide range of Ni–Fe compositions [44], as indicated in Fig. 11. For Ni_3Fe this effect on K_1 is probably the main cause for the increased coercivity and reduced permeability of the ordered phase [35].

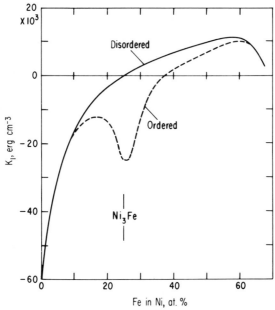

FIG. 11. Magnetocrystalline anisotropy coefficient versus composition for ordered and disordered Ni–Fe alloys.

Saturation magnetostriction data [44] for a Ni_3Fe crystal at room temperature show that with atomic ordering λ_{100} decreases from $\sim 17 \times 10^{-6}$ to $\sim 13 \times 10^{-6}$, whereas λ_{111} increases from $\sim 5 \times 10^{-6}$ to $\sim 12 \times 10^{-6}$; the effect of ordering on the saturation longitudinal magnetostriction (λ_s) of a polycrystal was found [40] to be an increase from $\sim 7 \times 10^{-6}$ to $\sim 10 \times 10^{-6}$. It has also been reported that the ΔE effect, the difference in elastic modulus between the magnetically saturated and unmagnetized states, is diminished in Ni_3Fe upon ordering [45]. Since $\Delta E \propto \mu_0 \lambda_s^2$ [46], it would appear that the drop in permeability (μ_0) more than compensates for the rise in magnetostriction, causing the change in ΔE with ordering to be a decrease.

The discovery cited earlier of irradiation-induced ordering in polycrystalline alloy samples of composition NiFe was facilitated by the fact that the magnetic field applied during irradiation with neutrons [36] or electrons [37] produces a uniaxial magnetic anisotropy that can be very large (i.e., $\sim 10^6$ erg cm^{-3}) and easily measured. A subsequent x-ray diffraction and magnetic study [47] of a neutron irradiated single crystal of NiFe confirmed the establishment of tetragonal CuAu-type order in tiny crystallites with an average linear dimension of $\sim$300 Å. The results also indicated that the majority of the crystallites had their tetragonal axes coincident with the [100] axis of the original crystal along which the field was applied during irradiation, the orientations for the other crystallites being equally distributed between the original [010] and [001] axes. Moreover, consistent with the magnetization data, the tetragonal axis of each crystallite is its single easy axis of magnetization. Thus, the ferromagnetic configuration of atomic moments in an isolated crystallite of ordered NiFe would be as shown in Fig. 12. The uniaxial anisotropy of the ordered crystallites has also been

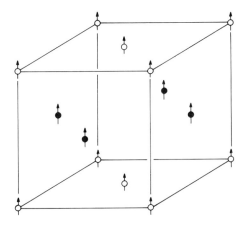

FIG. 12. Ferromagnetic configuration within an ordered NiFe crystallite. The atomic moments are oriented parallel to the tetragonal axis of a CuAu-type structure: (○) Ni; (●) Fe.

deduced from recent torque measurements [48] on a NiFe crystal that had been neutron-irradiated while in a field parallel to [111]. In this case, it is expected from symmetry that all the crystallites be equally represented among the three mutually orthogonal orientations described above, and this was borne out by the measured anisotropy of the gross crystal being cubic. The magnitude of this induced cubic anisotropy (which reaches

$K_1 \approx -2 \times 10^6$ ergs cm^{-3}, $K_2 \approx 7 \times 10^6$ ergs cm^{-3} at high irradiation energies [48]) is about two orders of magnitude larger than the cubic anisotropy of a NiFe crystal when disordered [44].

For other than the uniquely symmetrical case just described (or when the field during irradiation is zero), this field irradiation treatment of a NiFe crystal should produce a preferred orientation of the ordered crystallites and, hence, a gross anisotropy that is uniaxial. This effect therefore qualifies as a special type of *directional ordering* and as such is discussed in Chapter XII.

Nickel–Manganese

8. The nickel–manganese system provides one of the most dramatic examples of the effects of order–disorder on magnetic properties. As shown in Fig. 1 for the disordered fcc alloys of increasing Mn concentration, the saturation moment (measured and extrapolated to 0°K [9, 49]) rises from that of pure nickel, reaches a maximum of $\sim 0.8 \mu_\beta$ per atom at about 10 at. % Mn, then decreases very rapidly to about $0.1 \mu_\beta$ per atom at 25 at. % Mn. At or near the latter composition (Ni$_3$Mn) it has also been observed [9, 49–55] that an annealing treatment at about 400°C produces an extremely large increase in the low-temperature saturation moment to values as high as $1.1 \mu_\beta$ per atom [9] (corresponding to a magnetization of ~ 880 emu cm^{-3}); this also is indicated in Fig. 1. Moreover, the Curie temperature, which for the disordered (quenched) alloys decreases monotonically with increasing Mn in Ni to a value below room temperature for Ni$_3$Mn, is also increased by the annealing of Ni$_3$Mn, becoming approximately 450°C [49–55].

Figure 13 represents a composite of several sets of experimental results [50, 51, 54] for the room temperature saturation magnetization of Ni$_3$Mn, which had been slowly cooled to various temperatures and then quenched, and for the saturation magnetization of fully annealed Ni$_3$Mn measured at temperature. These results clearly indicate the onset of atomic ordering at a temperature just above the ferromagnetic Curie point of the ordered phase. That the strong ferromagnetism of annealed Ni$_3$Mn arises from a state of atomic order has been demonstrated conclusively by x-ray [56] and neutron [12, 24] diffraction evidence; the ordering is of the cubic, Cu$_3$Au type, similar to that of Ni$_3$Fe (see Fig. 4).

The room temperature neutron diffraction data [12] for ordered Ni$_3$Mn also gave $\mu_{Ni} = 0.30 \ (\pm 0.05) \mu_\beta$ and $\mu_{Mn} = 3.18 \ (\pm 0.25) \mu_\beta$ for the individual atomic moments aligned parallel to each other (an alternative ferri-

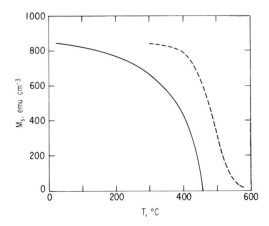

Fig. 13. Saturation magnetization of Ni$_3$Mn measured at various temperatures for highly ordered state (solid curve) and measured at room temperature for various quenching temperatures (dashed curve).

magnetic solution of these data appears highly unlikely). The question is immediately raised as to the disposition of the atomic moments responsible for the much lower saturation magnetization of Ni$_3$Mn when disordered. Since it is most improbable that the magnitudes of the atomic moments would decrease by anything like the order-of-magnitude reduction in saturation magnetization, one is forced to consider that the moments in disordered Ni$_3$Mn may not all be aligned ferromagnetically and that, instead, some of the moments may lie in opposing directions. This possibility has previously been inferred [9, 57] from the manner in which the saturation magnetization of the disordered alloys changes with composition. On this basis it was specifically proposed that the exchange interactions between Ni–Ni and Ni–Mn nearest neighbor atom pairs are ferromagnetic but that the Mn–Mn pair interactions are antiferromagnetic. This also provides a rationale for the composition dependence of the Curie point of these disordered alloys and for the increase in the Curie point of Ni$_3$Mn upon ordering (which tends to eliminate Mn–Mn nearest neighbor atom pairs). It should be noted that this interaction scheme for the Ni–Mn system is qualitatively very similar to that deduced above for Ni–Fe [see Eq. (7.5)]. Quantitatively, however, the Mn–Mn pair interactions have to be much more strongly antiferromagnetic than the Fe–Fe interactions in order to account for the larger effects of order–disorder on the magnetic properties of Ni$_3$Mn than on those of Ni$_3$Fe.

Probably the most decisive evidence that the atomic moments in disordered Ni$_3$Mn are not all aligned ferromagnetically is furnished by the unusual temperature and field dependences of the magnetization of this alloy [58–60]. As shown in Fig. 14a, the magnetization measured in a moderately high field does not increase monotonically with decreasing temperature, as does the saturation magnetization of the same alloy when ordered (see Fig. 13); instead, after reaching a maximum at about 30°K, it drops rapidly to a lower value at liquid helium temperature (4.2°K). The remanent magnetization, also represented in Fig. 14a, does not even appear until it suddenly rises to a peak near 30°K; it then decreases and again becomes vanishing small at 4.2°K. The hysteresis loop of magnetization versus field at 4.2°K, the dashed curve in Fig. 14b, shows not only the smallness of the

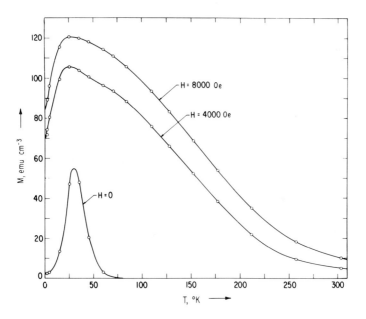

FIG. 14. (a) Disordered Ni_3Mn: magnetization at different fields, including remanence ($H = 0$), versus temperature.

remanence but also the fact that the approach to saturation at the highest fields is extremely slow. These results, having been obtained for a polycrystalline specimen, might suggest the onset of a strong magnetocrystalline anisotropy at low temperatures, but this conventional interpretation was disallowed by subsequent magnetization measurements [61] on a disordered Ni_3Mn single crystal, which gave essentially identical results for different directions of the applied field. Indeed, that a more unusual mechanism is operative in disordered Ni_3Mn had been revealed by the discovery that the same polycrystalline specimen, when cooled to 4.2°K in a magnetic field, exhibits a low-temperature hysteresis loop that is no longer symmetrical about the origin but is displaced along the field axis [59, 60]. The displaced hysteresis loop that was found is represented by the solid curve in Fig. 14b. Note that even though a large negative field reverses all of the available

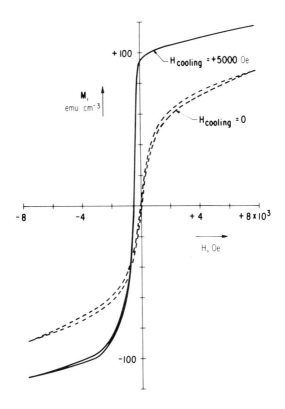

FIG. 14. (b) Disordered Ni_3Mn: hysteresis loops of magnetization versus field for alloy cooled from 300 to 4.2°K in +5000 Oe or in zero field.

magnetization, the magnetization reverts back to its positive remanent value when this field is removed; hence, the positive direction (the direction of the field applied during cooling) is the single easy direction of magnetization.

This unidirectional anisotropy in field-cooled, disordered Ni_3Mn, also revealed by torque measurement [60], is similar to that found earlier in partially oxidized cobalt particles [62], where it was attributed to an exchange coupling between ferromagnetic Co and antiferromagnetic CoO (and therefore named "exchange anisotropy"). Consequently, it was proposed [60] that disordered Ni_3Mn, though thermodynamically single phase, is composed of submicroscopic regions of ferromagnetic and antiferromagnetic order, and that this magnetic inhomogeneity arises from the local composition fluctuations existing even in a perfectly disordered alloy, in conjunction with the exchange interaction scheme described above (i.e., Ni–Ni and Ni–Mn pair interactions ferromagnetic, and Mn–Mn pair interactions antiferromagnetic). It was then shown that all the peculiar magnetic properties of disordered Ni_3Mn, including those previously mentioned, could be explained semiquantitatively in terms of this ferromagnetic–antiferromagnetic state.

Still another indication that the nearest neighbor Mn–Mn interactions are antiferromagnetic in the Ni–Mn system is provided by the arrangement of atomic moments in the ordered alloy NiMn. The ordered atomic structure of this alloy is of the tetragonal CuAu type and is therefore similar to that shown in Fig. 12 for ordered NiFe. However, in contrast to the ferromagnetic configuration of NiFe, the configuration deduced from neutron diffraction data [63] on NiMn is antiferromagnetic, with the moments of nearest neighbor Mn atoms aligned antiparallel to each other, as shown in Fig. 15.

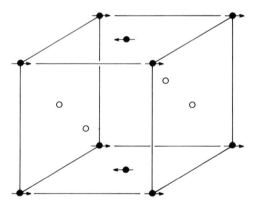

FIG. 15. Antiferromagnetic configuration for ordered NiMn. The Mn moments lie perpendicular to the tetragonal axis of a CuAu-type structure; the Ni moments are essentially zero: (○) Ni; (●) Mn.

Moreover, while the Mn moments in NiMn are each quite large ($\sim 4\mu_\beta$), the magnitude of each Ni moment is less than $0.1\mu_\beta$. Hence, it is interesting to note that the size of the Ni moment decreases steadily from $\sim 0.6\mu_\beta$ in pure nickel to $\sim 0.3\mu_\beta$ in ordered Ni_3Mn to $<0.1\mu_\beta$ in ordered NiMn, corresponding to an increase in the number of Mn nearest neighbors from

0 to 4 to 8, respectively. This variability in the Ni moment is presumably operative in the disordered Ni–Mn alloys as well.

The manner in which atomic order develops in alloy compositions near Ni_3Mn and the effects of intermediate states of order on various magnetic properties have been studied fairly extensively. In fact, it was magnetic measurement that first revealed the basic thermodynamic nature of this ordering process. Specifically, the measurement of saturation magnetization versus temperature for various annealing treatments of Ni_3Mn gave results [55] typified by the curves shown in Fig. 16. The curves for the short anneal-

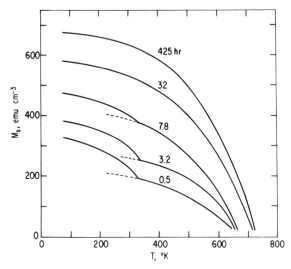

FIG. 16. Saturation magnetization versus temperature for Ni_3Mn annealed at 700°K (427°C) for various periods of time (in hours).

ing times (e.g., 3.2 hr) exhibit a distinct kink near 300°K, indicating the Curie point of a ferromagnetic phase that is coexisting with another ferromagnetic phase whose Curie point of roughly 700°K is marked by the disappearance of the magnetization. The conclusions drawn from this set of curves were that the material consists entirely of highly disordered and highly ordered regions (having low and high Curie points, respectively) and that with successive isothermal annealing the latter regions grow at the expense of the former [55]. It is clear from Fig. 16 that at 700°K (427°C) this growth process goes essentially to completion. Subsequent magnetization measurements [64] have shown that for higher annealing temperatures (e.g., 450 and 465°C) the equilibrium state itself becomes two phase, the lower Curie point approaching that of the completely disordered phase and the upper Curie point gradually decreasing from that of the perfectly ordered phase. With increasing annealing temperature these Curie point variations were interpreted respectively as a progressive loss of short-range order in the "disordered" regions and a decrease of long-range order in the "ordered" regions [64]. The coexistence of relatively disordered and ordered regions in "partially ordered" Ni_3Mn has also been deduced from neutron diffraction data [64, 65] and has been confirmed more directly by electron microscopy [66, 67]. According to the latter experiments, the ordered regions are shaped typically as prolate ellipsoids [66] or rods [67] with a short dimension of 750 Å [66] or 125–270 Å [67].

From the above description it follows that certain heat treatments of Ni_3Mn will produce a spatially inhomogeneous state in which small, highly ordered regions of large magnetization are distributed throughout a highly disordered matrix of small magnetization. Under these circumstances, if the ordered regions are of appropriate size they can be expected to act magnetically as single-domain particles. Indeed, high coercive fields [52, 55] and large magnetic aftereffects [66] (i.e., time-dependent magnetization) have been observed in partially ordered Ni_3Mn and attributed to single-domain behavior.

Magnetostriction measurements have recently been made on an ordered and disordered Ni_3Mn single crystal [68]. For the ordered state, the coefficients λ_{100} and λ_{111} were found to vary monotonically from -13.5×10^{-6} and -2.5×10^{-6} at $\sim 80°K$ to -3.7×10^{-6} and -0.5×10^{-6} at room temperature, respectively. These results were shown to be consistent with those of earlier work [54] on an ordered Ni_3Mn polycrystal for which the saturation magnetostriction was observed to change sign and become positive at 370°K. For the disordered crystal, λ_{100} and λ_{111} are much smaller in magnitude, being only -0.3×10^{-6} and -0.1×10^{-6}, respectively, at 80°K; both coefficients are zero at room temperature, which is well above the Curie point for the disordered state.

Iron–Aluminum

9. Iron–aluminum alloys of up to 50 at. % Al offer excellent opportunities for the study of magnetic properties as a function of type as well as degree of atomic order. The crystal structures of these alloys are all founded on the bcc lattice, but the arrangement of Fe and Al atoms on this lattice takes on one (or more) of three configurations, depending on the composition and thermal (or mechanical) treatment. Two of these configurations, established many years ago from x-ray diffraction evidence [69], are based on the

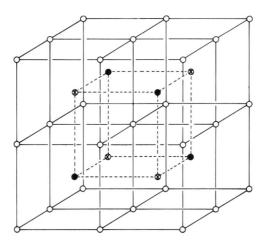

FIG. 17. Atomic sites of bcc-based structures of Fe–Al alloys: (○) α; (⊗) β_1; (●) β_2.

stoichiometries Fe$_3$Al and FeAl. With reference to Fig. 17, the α sites in both configurations are occupied by Fe atoms, while the β_1 and β_2 sites are occupied respectively by Fe and Al in Fe$_3$Al or by Al alone in FeAl; thus, FeAl has the CsCl-type structure and is isomorphous with FeCo (see Fig. 6). The occupation of these lattice sites deduced [69] for all Fe–Al compositions

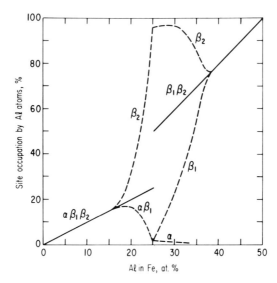

FIG. 18. Percent occupation by Al atoms of various lattice sites (shown in Fig. 17) versus composition of quenched and annealed Fe–Al alloys (solid and dashed curves, respectively).

is shown in Fig. 18 for alloys that have been either cooled very slowly or quenched from high temperature (~700°C). As indicated, the annealing of quenched alloy samples of about 20-35 at. % Al transforms their structure from disordered bcc or FeAl type to Fe_3Al type. This and the later observation [70] that cold-working of quenched samples with FeAl-type structure (~25-50 at. % Al) produces partial if not complete disordering are consistent with the obvious crystallographic fact that the sequence of structures

$$\text{bcc (disordered)} \rightarrow \text{FeAl-type} \rightarrow \text{Fe}_3\text{Al-type}$$

represents a progressive improvement in atomic order. Subsequent x-ray diffraction work [70] has shown that the information in Fig. 18 should be modified for the annealed alloys of ~18-20 at. % Al, which were found to develop a complex ordered structure of $Fe_{13}Al_3$ stoichiometry (the third of the atomic configurations mentioned earlier).

Since little is known magnetically about the ordered $Fe_{13}Al_3$ phase, our discussion of magnetic properties will be directed towards Fe-Al alloys whose atomic structure is disordered bcc, ordered FeAl type, or ordered Fe_3Al type. At or near the composition Fe_3Al all three of the latter structures are ferromagnetic at room temperature. Moreover, room temperature measurements have shown that the annealing of quenched samples of ~20-23 at. % Al results in a small decrease in saturation magnetization, whereas over the range ~23-31 at. % Al a small increase in magnetization was observed [71]. Similar results were obtained at very low temperature (4.2°K) and therefore cannot be ascribed to changes in Curie temperature (which will be discussed later). Relating these results to the structural differences between the annealed and quenched alloys of various composition (see Fig. 18), it was concluded [71, 72] that the saturation moment was highest for the disordered bcc alloys, lowest for those with FeAl-type order, and intermediate for those with Fe_3Al-type order. Thus, the saturation moment does not vary monotonically within the sequence of improved atomic order described above.

Several other basic experimental facts about the saturation moment of Fe-Al alloys are also very revealing. First, the low-temperature saturation moment of the disordered alloys decreases linearly with increasing Al concentration (up to ~23 at. %, beyond which the disordered state can no longer be quenched in), and the rate of decrease is approximately such that the moment extrapolates to zero at pure aluminum [71, 73, 74]. It is as though the Fe atoms retain the value of their moments in pure iron (i.e., ~$2.2\mu_\beta$) and are simply diluted by the nonmagnetic Al atoms, indicating that the moment value for a Fe atom is independent of the number of neighboring Al atoms, at least as long as this number is small. Secondly, neutron diffraction measurements at room temperature on ordered, stoichiometric Fe_3Al have revealed that the Fe moments on α and β_1 sites (Fig. 17) are different, being reported as $1.46 \pm 0.1\mu_\beta$ and $2.14 \pm 0.1\mu_\beta$, respectively [75] (which together give a saturation magnetization of $\sim$145 emu g^{-1}). That these room temperature results reflect an intrinsic difference between these moments at low temperature, rather than a difference in the thermal disordering of these moments [72], was subsequently demonstrated by low-temperature Mössbauer effect experiments [76]. Thus, for ordered Fe_3Al at low temperatures it can be concluded that the Fe moment on a β_1 site (whose eight nearest neighbors are all Fe atoms) is closely equal to that in pure iron, whereas an α site (with four Fe and four Al nearest neighbors) carries a Fe moment that is only about three-quarters as large. Finally, the saturation magnetization of FeAl-type ordered alloys is known to drop very rapidly as the Al concentration is increased beyond 25 at. % [70, 71, 73], which, in view of other more unusual properties of these alloys (anomalous temperature dependence of their magnetization [71] and displaced hysteresis loops when cooled to low temperatures in a magnetic field [77]), has been interpreted in terms of a gradual transition to antiferromagnetism [72], possibly involving a mixed ferromagnetic–antiferromagnetic state [77] similar to that described above for disordered Ni–Mn alloys. This interpretation extended to ordered, stoichiometric FeAl would predict a simple antiferromagnetic state. Instead, recent neutron diffraction work [78] on this alloy has revealed no magnetic ordering of any kind. It therefore appears likely that a Fe atom surrounded by eight Al nearest-neighbor atoms (as in ordered FeAl) has no local magnetic moment.

Taken together, all of the above properties suggest that in the Fe–Al system the magnitude of the moment of an individual Fe atom (μ_{Fe}) is fairly constant at $\sim 2.2\mu_\beta$ when the number of Al nearest-neighbor atoms (n_{Al}) is small, decreases to about three-quarters of this value at $n_{Al} = 4$, then decreases more rapidly and goes to zero at $n_{Al} \leq 8$. Mössbauer-effect measurements [17] on these alloys have recently shown that the nearest neighbor site occupation does have a much stronger influence on the size of an individual Fe moment than the occupation of more distant sites. Hence, we are encouraged to use the μ_{Fe} vs. n_{Al} dependence just described, in conjunction with the model discussed in the previous section, and calculate the saturation moments for the different Fe–Al structures, ordered and disordered, at various compositions. The simplest power series expression that conforms to this μ_{Fe} vs. n_{Al} dependence is

$$\mu_{Fe}(n_{Al})/\mu_{Fe}(0) = 1 - n_{Al}^2/64, \tag{9.1}$$

where $\mu_{Fe}(0) = 2.22\mu_\beta$, the atomic moment in pure iron; as shown in Fig. 19, $\mu_{Fe}(n_{Al})$ decreases at an increasing rate and reaches zero when $n_{Al} = 8$, the total number of nearest neighbors. For the disordered bcc

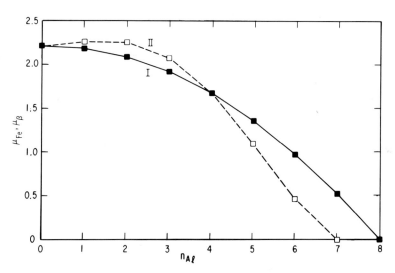

FIG. 19. Dependence of individual Fe moment on the number of nearest neighbor Al atoms according to Eq. (9.1), curve I, and Eq. (9.6), curve II.

alloys of composition $Fe_{1-x}Al_x$, we identify Eq. (9.1) with Eq. (3.8a), use the conversion to Eq. (3.9a), and obtain

$$\bar{\mu}_{Fe}(x)/\mu_{Fe}(0) = 1 - 0.125x - 0.875x^2 \tag{9.2}$$

for the average moment per Fe Atom, from which we then calculate, by means of

$$\bar{\mu}(x)_{bcc} = (1-x)\bar{\mu}_{Fe}(x), \tag{9.3}$$

the saturation moment (per average atom) of the alloy. The $\bar{\mu}(x)_{bcc}$ values for various x (the Al concentration) are plotted in Fig. 20, where we see they are fairly close to the curve for simple dilution at small x but lie increasingly below it at larger x. For the FeAl-type ordered alloys, it is clear from Figs. 17 and 18 that the Fe atoms located at random on β_1 and β_2 sites have no Al but only Fe nearest neighbors, whereas the Fe atoms on α sites have a nearest neighbor distribution which at composition $Fe_{1-x}Al_x$ is equivalent to that of an atom in a disordered bcc alloy of composition

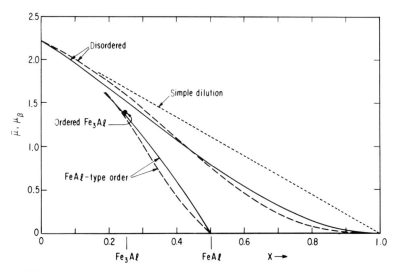

FIG. 20. Saturation moment per atom versus Fe–Al composition, calculated for disordered and FeAl-type ordered states. The solid, dashed, and dotted curves were obtained for $\mu_{Fe}(n_{Al})$ dependence I and II in Fig. 19 and for simple dilution, respectively. The moment value shown for ordered Fe$_3$Al represents both dependence I and II.

Fe$_{1-2x}$Al$_{2x}$. Thus, the saturation moment is now expressed as

$$\bar{\mu}(x)_{\text{FeAl}} = (\tfrac{1}{2} - x)\mu_{\text{Fe}}(0) + \tfrac{1}{2}\bar{\mu}_{\text{Fe}}(2x), \qquad (9.4)$$

where $\bar{\mu}_{\text{Fe}}(2x)$ is defined by Eq. (9.2), with x replaced by $2x$. As shown in Fig. 20, $\bar{\mu}(x)_{\text{FeAl}}$ calculated from Eq. (9.4) is smaller than $\bar{\mu}(x)_{\text{bcc}}$ at all compositions, decreasing rapidly to zero at $x = \tfrac{1}{2}$ (i.e., for stoichiometric FeAl). For the composition Fe$_3$Al ($x = \tfrac{1}{4}$) our calculations give $\bar{\mu}(\tfrac{1}{4})_{\text{FeAl}} = 1.35\mu_\beta$ and $\bar{\mu}(\tfrac{1}{4})_{\text{bcc}} = 1.52\mu_\beta$. The saturation moment for ordered, stoichiometric Fe$_3$Al may be expressed as

$$\bar{\mu}(\text{Fe}_3\text{Al}) = \tfrac{1}{4}\mu_{\text{Fe}}(0) + \tfrac{1}{2}\mu_{\text{Fe}}(4), \qquad (9.5)$$

which on the basis of Eq. (9.1) gives $\bar{\mu}(\text{Fe}_3\text{Al}) = 1.39\mu_\beta$. Thus, we find that $\bar{\mu}(\tfrac{1}{4})_{\text{FeAl}} < \bar{\mu}(\text{Fe}_3\text{Al}) < \bar{\mu}(\tfrac{1}{4})_{\text{bcc}}$, in qualitative accord with experiment.

Instead of Eq. (9.1), let us consider an alternative relationship between the moment of an individual Fe atom and the number of Al nearest neighbors, such that $d\mu_{\text{Fe}}/dn_{\text{Al}} = 0$ at $n_{\text{Al}} = 0$ and $\mu_{\text{Fe}}(4) = \tfrac{3}{4}\mu_{\text{Fe}}(0)$ as before, but with $\mu_{\text{Fe}} = 0$ at $n_{\text{Al}} = 7$ and 8 rather than just at $n_{\text{Al}} = 8$. The latter condition, though more stringent, lies well within the experimental limits of the μ_{Fe} vs. n_{Al} dependence discussed earlier. The power series expression of lowest order that incorporates all these conditions is

XI. ATOMIC ORDER–DISORDER AND MAGNETIC PROPERTIES

$$\mu_{Fe}(n_{Al})/\mu_{Fe}(0) = 1 + 0.0352 n_{Al}^2 - 0.0190 n_{Al}^3 + 0.0016 n_{Al}^4, \quad (9.6)$$

which is illustrated in Fig. 19. The conversion from Eq. (3.8a) to Eq. (3.9a) now gives

$$\bar{\mu}_{Fe}(x)/\mu_{Fe}(0) = 1 + 0.14x - 0.61x^2 - 3.20x^3 + 2.67x^4, \quad (9.7)$$

in place of Eq. (9.2). Using Eq. (9.7) in conjunction with eqs. (9.3) and (9.4), we again compute the saturation moment for the disordered bcc and ordered FeAl-type structures at various compositions. The results are the dashed curves in Fig. 20, which show $\bar{\mu}(x)_{bcc}$ varying more closely along the simple dilution line for small x and $\bar{\mu}(x)_{FeAl}$ dropping more rapidly to zero as x approaches $\frac{1}{2}$, compared to the corresponding solid curves based on Eqs. (9.1) and (9.2). Furthermore, at composition Fe$_3$Al, $\bar{\mu}(\frac{1}{4})_{FeAl} = 1.32\mu_\beta$ and $\bar{\mu}(\frac{1}{4})_{bcc} = 1.60\mu_\beta$, while $\bar{\mu}(Fe_3Al)$ is again $1.39\mu_\beta$. Thus, the saturation moments calculated for the three phases follow the same sequence as before, but with a wider separation of their values. In every respect, the latter calculations give better quantitative agreement with the experimental work cited earlier, indicating that Eq. (9.6) is an improvement over Eq. (9.1) as a description of the atomic moment dependence on local environment in the Fe–Al system.

If the moment of an iron atom were assumed to be independent of the neighboring atom distribution, our simple model would obviously predict that for any specified composition all the ordered and disordered Fe–Al phases would have the same saturation moment, namely, that for simple dilution. This is not so for the Curie temperature, which depends on the number of interactions between neighboring magnetic atoms as much as on the size of their moments (which affects the strength of the interactions) and will therefore change with atomic ordering, even if all the moments remain constant in magnitude. The sensitivity of the Curie temperature to ordering is especially easy to demonstrate for the Fe–Al system where only the Fe atoms are moment bearing. For a disordered bcc alloy of composition Fe$_{1-x}$Al$_x$, we obtain from Eqs. (4.6) and (4.7)

$$T_C(x)_{bbc} = \frac{j+1}{6jk} J \sum_{n_{Al}=0}^{N} (N - n_{Al}) P(n_{Al}, x) [\mu_{Fe}^{(0)}(n_{Al})]^2, \quad (9.8)$$

where J is the exchange coefficient for nearest neighbor Fe–Fe interactions. Since this expression gives for pure iron,

$$T_C(Fe) = (j + 1) N J [\mu_{Fe}^{(0)}(0)]^2 / 6jk, \quad (9.9)$$

it may be rewritten as

$$T_C(x)_{bcc}/T_C(Fe) = \sum_{n_{Al}=0}^{N} (1 - n_{Al}/N)P(n_{Al}, x)[\varrho(n_{Al})]^2, \quad (9.10)$$

where $\varrho(n_{Al}) \equiv \mu_{Fe}^{(0)}(n_{Al})/\mu_{Fe}^{(0)}(0)$. For a FeAl-type ordered alloy of the same composition, it follows from our previous description of the occupation and environment of the different lattice sites that

$$T_C(x)_{FeAl}/T_C(Fe) = \left\{ \sum_{n_{Al}=0}^{N} (1 - n_{Al}/N)P(n_{Al}, 2x)[\varrho(n_{Al})]^2 \right\}^{1/2} \quad (9.11)$$

It can similarly be shown for ordered, stoichiometric Fe_3Al that

$$T_C(Fe_3Al)/T_C(Fe) = \varrho(4)/\sqrt{2}. \quad (9.12)$$

For the case of simple dilution, where $\varrho(n_{Al}) = 1$ for all n_{Al}, the right-hand sides of Eqs. (9.10)–(9.12) reduce respectively to $1 - x$, $(1 - 2x)^{1/2}$, and $1/\sqrt{2}$, which are plotted versus x in Fig. 21. Thus, at composition Fe_3Al ($x = \frac{1}{4}$) the Curie points of both FeAl-type and Fe_3Al-type ordered alloys are predicted to be the same and somewhat lower than T_C for the disordered alloy.

For a more realistic case in which the atomic moments are not constant but vary with their local environment, let us first assume that this variation

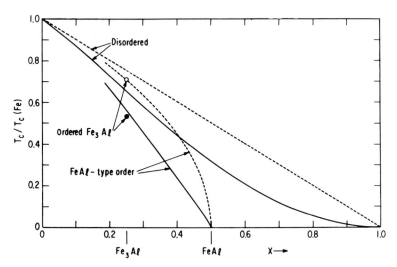

FIG. 21. Curie temperature, normalized to that of pure Fe, versus Fe–Al composition, calculated for disordered and FeAl-type ordered states and for ordered Fe_3Al. The solid curves and closed circle represent $\mu_{Fe}(n_{Al})$ dependence I in Fig. 19; the dotted curves and open circle represent simple dilution.

is given by Eq. (9.1). Inserting this relationship into Eqs. (9.10)–(9.12) and enlisting the aid of Eqs. (3.6) and (3.7), we again derive expressions for the Curie points of the three Fe–Al phases normalized to T_C for pure iron. The results are shown in Fig. 21, and it is evident that the assumed reduction in μ_{Fe} with increasing n_{Al} causes a lowering of T_C for all three phases. Furthermore, for $x = \frac{1}{4}$, calculation of the right-hand sides of Eqs. (9.10)–(9.12) now give 0.656, 0.563, and 0.530, respectively, so that $T_C(\frac{1}{4})_{bcc} > T_C(\frac{1}{4})_{FeAl} > T_C(Fe_3Al)$, whereas in the simple dilution case the latter two are the same. Indeed, this sequence in the values for T_C, which is different from the corresponding sequence for the saturation moments (see Fig. 20), agrees with the conclusions drawn from recent experimental work [79]. If we next assume that $\mu_{Fe}(n_{Al})$ follows the more abrupt variation expressed by Eq. (9.6), we find by analogous calculation that the right hand sides of Eqs. (9.10)–(9.12) give 0.721, 0.576, and 0.530, respectively. This sequence in T_C, which is the same as that above but with a wider spread in values, is in equally good agreement with experiment [79].

In the above calculations, any variation in the exchange coupling with lattice parameter has been ignored. In the Fe–Al system, the differences in lattice parameter between the various phases discussed are substantial [70] and may, in themselves, be responsible for some of the observed differences in Curie temperature.

The magnetocrystalline anisotropy of alloys of compositions near Fe_3Al has been found to decrease from positive to negative values upon annealing of a disordered (quenched) sample [80–82]. The effect of ordering is largest in stoichiometric Fe_3Al, in which K_1 at room temperature changes from $+4$ to -8 ($\times 10^4$ ergs cm^{-3}); the change at $\sim 80°K$ is from approximately the same positive value to -19×10^4 ergs cm^{-3} [82]. The effect of ordering (annealing) on the magnetostriction also is maximum at the stoichiometric Fe_3Al composition, where λ_{100} increases from 50 to 90 ($\times 10^{-6}$) and λ_{111} increases from 15 to 35 ($\times 10^{-6}$) [81]. Presumably related to this increase in magnetostriction with Fe_3Al-type ordering is the corresponding increase that has been observed in the ΔE effect (the change in elastic modulus with magnetization) [83].

Iron–Palladium and Iron–Platinum

10. Many similarities and a few important differences in the magnetic and crystallographic properties of Fe–Pd and Fe–Pt alloys make it profitable to examine the two systems together. Both systems at or near the compositions of 25 and 50 at. % Fe readily assume ordered structures of the cubic

Cu$_3$Au type and the tetragonal CuAu type, respectively [84]; these structures have been illustrated in Figs. 4 and 12. Order of the Cu$_3$Au type also forms in Fe$_3$Pt, but no ordering has been reported for the comparable Fe$_3$Pd alloy. At and between the stoichiometric compositions cited above (and extending to pure Pd or Pt) the alloys can be retained in a disordered fcc form by quenching from high temperatures, although in some cases (e.g., compositions very near FePt$_3$ [85]) severe cold work is required for appreciable disordering.

The magnetic properties of disordered Fe–Pd and Fe–Pt are analogous in two major respects. Starting from the weakly paramagnetic Pd and Pt metals, both systems become ferromagnetic (at low temperatures) at Fe concentrations less than 1 at. % [86]. The Curie point rises rapidly and exceeds room temperature above about 12 at. % Fe in Pd [87]; for the comparable Fe–Pt composition the Curie point is $\sim$200°K [88]. Furthermore, in the very dilute alloys the low-temperature saturation moment on a per-Fe-atom basis is extraordinarily large; in 1 at. % Fe in Pd(Pt) it is $\sim$8(5)μ_β [86]. Neutron diffraction measurements [89] on dilute Fe in Pd have shown that the giant moment per Fe atom resides only partly on the Fe atoms themselves, the remainder being distributed over the Pd atoms. For example, for Fe$_{0.03}$Pd$_{0.97}$ whose saturation moment $\bar{\mu} = 0.03\bar{\mu}_{Fe} + 0.97\bar{\mu}_{Pd} = 0.237\mu_\beta$ per average atom (or $\bar{\mu}/0.03 = 7.9\mu_\beta$ per Fe atom) it was deduced that $\bar{\mu}_{Fe} = 3.0\mu_\beta$ and $\bar{\mu}_{Pd} = 0.15\mu_\beta$. With increasing Fe concentration (up to 50 at. %) $\bar{\mu}_{Fe}$ remains essentially constant at $\sim 3\mu_\beta$ whereas $\bar{\mu}_{Pd}$ rises to a limiting value of $\sim 0.35\mu_\beta$. According to more detailed neutron diffraction work on Fe–Pd [90], $\bar{\mu}_{Pd}$ in alloys with more than $\sim$1 at. % Fe represents a fairly uniform moment distribution among all the Pd atoms, but in alloys with less Fe the moment on a Pd atom decreases (slowly) with increasing distance from a Fe atom. The latter observation indicates definitively that the moment on a Pd atom is induced by an exchange interaction field from nearby Fe atoms; presumably the same exchange polarization effect is also operative in the Fe–Pt system.

The second area in which the magnetic properties of disordered (fcc) Fe–Pd and Fe–Pt alloys show a marked resemblance is in the Fe-rich compositions of up to $\sim$75 at. % Fe (above which the alloys are partly or wholly bcc at room temperature). In both systems, the Curie point rises to a maximum near 50 at. % Fe and decreases steadily with further increases in Fe concentration [88, 91]. In Fe–Pt, the saturation magnetization continues to increase until it reaches a peak at $\sim$65 at. % Fe, and then it too begins to drop with increasing Fe [92]. Measurements on Fe–Pd first indicated that the saturation magnetization increases monotonically with increasing Fe concentration [91], but later work [93] revealed a magnetization peak at ~65 at. % Fe, similar to that in Fe–Pt. Thus, the magnetic behavior of Fe-rich Fe–Pd and Fe–Pt is very much like that of Fe-rich Fe–Ni. It has therefore been proposed [93] that analogous to the current interpretation of the latter (see the previous discussion of the Ni–Fe system) the interactions between Fe–Fe atom neighbors in the Fe–Pd and Fe–Pt systems are antiferromagnetic and that at sufficiently Fe-rich compositions these interactions assert themselves against the ferromagnetic coupling between Fe–Pd or Fe–Pt atom neighbors and cause an antiparallel alignment of some of the atomic moments, thus reducing the net bulk magnetization of the alloy.

With the magnetic behavior of the disordered fcc alloys as a frame of reference, we next consider the changes from this behavior caused by atomic ordering. The most remarkable are the contrasting effects of Cu_3Au-type ordering in stoichiometric $FePd_3$ and $FePt_3$. In the case of $FePd_3$, the ordered alloy has basically the same ferromagnetic characteristics as the disordered: a Curie point of 529°K compared to 497°K [94] and a low-temperature saturation moment of $\sim 1.1\mu_\beta$ per atom compared to $\sim 1.0\mu_\beta$ per atom [95]. Moreover, this small rise in the average atomic moment with ordering, according to neutron diffraction study [95], derives from increases in both the Pd moment ($0.34\mu_\beta \rightarrow 0.42\mu_\beta$) and the Fe moment ($2.98\mu_\beta \rightarrow 3.10\mu_\beta$). In the case of $FePt_3$, however, the ordered alloy is antiferromagnetic with a Néel point marked by a susceptibility maximum at $\sim 100°K$ [85], whereas the disordered alloy (produced by cold working) is ferromagnetic, its Curie point ($\sim 425°K$ [96]) and its low-temperature saturation moment ($\sim 1.0\mu_\beta$ per atom [96]) being fairly close to those of disordered (or ordered) $FePd_3$. The antiferromagnetic structure of ordered, stoichiometric $FePt_3$, as deduced from neutron diffraction data [96] is shown in Fig. 22; the Fe moments form a simple antiferromagnetic configuration and are fairly normal in magnitude ($\sim 3.3\mu_\beta$), but there appears to be essentially no moment on the Pt atoms. In connection with the latter, it should be noted that each Pt atom has four Fe nearest neighbor atoms, two of whose moments are antiparallel to those of the other two. More generally for this configuration, the Fe moments at any particular distance from a given Pt atom add vectorially to zero. Hence, the net exchange field at a Pt atom site from all the Fe moments in the system should be identically zero, and no exchange polarization effect of the type discussed earlier for the disordered alloys can be expected. On the other hand, in ordered $FePd_3$, whose Fe moments are all aligned ferromagnetically, the latter effect is presumably large and is responsible for the substantial Pd moment observed.

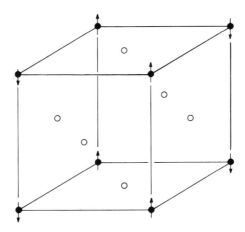

FIG. 22. Antiferromagnetic configuration for ordered FePt$_3$: (●) Fe; (○) Pt.

Atomic ordering of the tetragonal, CuAu type has little effect on the basic ferromagnetic behavior of FePd. Its Curie point is raised from ∼700 to ∼750°K, but its room temperature saturation magnetization remains at about 110 emu g^{-1}, corresponding to an average atomic moment of ∼1.5μ_β [94]. Neutron diffraction study [95] of disordered FePd has shown that the Fe and Pd moments are 2.85μ_β and 0.35μ_β, respectively; presumably, about the same moment values apply for the ordered state. In the case of FePt, the same type of atomic ordering causes an increase in Curie temperature from 530 to 750°K [92, 94] and a 15% decrease in saturation magnetization to ∼30 emu g^{-1} (∼0.7μ_β per average atom) [92]. Hence, at these equiatomic compositions also, the intrinsic magnetic properties of Fe–Pt are much more sensitive to atomic ordering than those of Fe–Pd.

Compacts of partially ordered FePt powder were recently found to have excellent permanent magnet properties [97]: a remanent induction B_R of 6200 G, a coercivity (for zero induction B) of 4600 Oe, and a maximum BH product of 7.6 × 10^6 G Oe. These values are considerably higher than those reported previously for ordered FePt [98] and are comparable to those for CoPt, the established permanent magnet alloy with the same CuAu-type structure. A large magnetocrystalline anisotropy, arising from the tetragonality of this ordered structure, is thought to be the basis for the high coercivities of these materials [97].

Atomic ordering (Cu$_3$Au type) in Fe$_3$Pt is accompanied by an increase in Curie point from 290 to 430°K [88], so that ordered Fe$_3$Pt, unlike the disordered fcc alloy, is strongly ferromagnetic at room temperature (with a saturation induction of ∼10,000 G [99]). At liquid nitrogen temperature

($\sim$77°K) the saturation induction of ordered Fe_3Pt reaches about 13,000 G, which is $\sim$10% higher than that of the disordered alloy [99]. According to a recent neutron diffraction study [100], the Fe and Pt moments in ordered Fe_3Pt are $\sim$3.3 and $\sim$1.5μ_β, respectively. The latter is the largest moment value yet reported for a Pt group or Pd group atom in a metal and presumably arises from strong exchange polarization forces imposed by the 12 Fe nearest neighbors of each Pt atom. Moreover, the increase in the Curie temperature and saturation moment of Fe_3Pt with ordering suggests that these forces counteract the tendency towards antiferromagnetic alignment among the Fe atoms that is thought to occur in the disordered Fe-rich alloys, as discussed earlier.

REFERENCES

1. W. Marshall, *Phys. Rev.* **118**, 1519 (1960); P. W. Anderson, *ibid.* **124**, 41 (1961); M. W. Klein, *ibid.* **136**, A1156 (1964); J. Kondo, *J. Appl. Phys.* **37**, 1177 (1966); S. D. Silverstein, *Phys. Rev. Letters* **16**, 466 (1966).
2. E. C. Stoner, *Rept. Progr. Phys.* **11**, 43 (1948); and references therein.
3. E. C. Stoner, *J. Phys. Radium* **12**, 372 (1951); and references therein.
4. S. A. Ahern, M. J. C. Martin, and W. Sucksmith, *Proc. Roy. Soc. (London)* **248**, 145 (1958).
5. V. Marian, *Ann. Phys.* **7**, 459 (1937).
6. P. Weiss, R. Forrer, and F. Birch, *Compt. Rend.* **189**, 789 (1929).
7. E. I. Kondorskii and L. N. Fedotov, *Izv. Akad. Nauk SSSR Ser. Fiz.* **16**, 432 (1952).
8. J. S. Kouvel and R. H. Wilson, *J. Appl. Phys.* **32**, 435 (1961).
9. G. R. Piercy and E. R. Morgan, *Can. J. Phys.* **31**, 529 (1953).
10. C. Sadron, *Ann. Phys.* **17**, 371 (1932).
11. T. Farcas, *Ann. Phys.* **8**, 146 (1937).
12. C. G. Shull and M. K. Wilkinson, *Phys. Rev.* **97**, 304 (1955).
13. G. G. Low and M. F. Collins, *J. Appl. Phys.* **34**, 1195 (1963).
14. M. F. Collins and G. G. Low, *J. Phys. (Paris)* **25**, 596 (1964).
15. M. F. Collins and G. G. Low, *Proc. Phys. Soc. (London)* **86**, 535 (1965).
16. S. Kobayashi, K. Asayama, and J. Itoh, *J. Phys. Soc. Japan* **21**, 65 (1966); R. L. Streever and G. A. Uriano, *Phys. Rev.* **149**, 295 (1966); and references therein.
17. P. A. Flinn and S. L. Ruby, *Phys. Rev.* **124**, 34 (1961); M. B. Stearns, *J. Appl. Phys.* **35**, 1095 (1964); G. K. Wertheim, V. Jaccarino, J. H. Wernick, and D. N. E. Buchanan, *Phys. Rev. Letters* **12**, 24 (1964); and references therein.
18. J. E. Goldman and R. Smoluchowski, *Phys. Rev.* **75**, 140 (1949), J. E. Goldman, *J. Appl. Phys.* **20**, 1131 (1949); R. Smoluchowski, *J. Phys. Radium* **12**, 389 (1951); *Phys. Rev.* **84**, 511 (1951).
19. H. Sato, *Proc. Conf. Magnetism Magnetic Materials, Pittsburgh, June 1955.*
19a. W. Marshall, *J. Phys. C (Proc. Phys. Soc. London)* **1**, 88 (1968).
20. E. M. Grabbe, *Phys. Rev.* **57**, 728 (1940).
21. A. Kussmann, B. Scharnow, and A. Schulze, *Z. Tech. Phys.* **13**, 449 (1932).
22. J. E. Goldman and R. Smoluchowski, *Phys. Rev.* **75**, 310 (1949).

23. W. C. Ellis and E. S. Greiner, *Trans. Am. Soc. Metals* **29**, 415 (1941).
24. C. G. Shull and S. Seigel, *Phys. Rev.* **75**, 1008 (1949).
25. S. Kaya and H. Sato, *Proc. Phys. Math. Soc. Japan* **25**, 261 (1943).
26. R. Forrer, *J. Phys. Radium* **2**, 312 (1931).
27. T. Yokoyama, *Japan. Inst. Metals* **17**, 263 (1953).
28. B. G. Lyashenko, D. F. Litvin, I. M. Puzey, and J. G. Abov, *J. Phys. Soc. Japan* **17**, Suppl. B-III, 49 (1962).
29. T. Yokoyama, *Bull. Fac. Eng. Yokohama Natl. Univ.* **7**, 27 (1958).
30. M. F. Collins and J. B. Forsyth, *Phil. Mag.* **8**, 401 (1963).
31. P. Weiss and R. Forrer, *Ann. Phys.* **12**, 279 (1929).
32. R. C. Hall, *J. Appl. Phys. Suppl.* **31**, 157S (1960).
33. R. M. Bozorth, "Ferromagnetism," pp. 196–199. Van Nostrand, Princeton, New Jersey, 1951.
34. P. Leech and C. Sykes, *Phil. Mag.* **27**, 742 (1939); F. E. Haworth, *Phys. Rev.* **56**, 289 (1939).
35. R. J. Wakelin and E. L. Yates, *Proc. Phys. Soc. (London)* **B66**, 221 (1953).
36. J. Paulevé, D. Dautreppe, J. Laugier, and L. Néel, *Compt. Rend.* **254**, 965 (1962).
37. W. Chambron, D. Dautreppe, L. Néel and J. Paulevé, *Compt. Rend.* **255**, 2037 (1962).
38. E. Kneller, private communication, 1966.
39. M. Hansen, "Constitution of Binary Alloys," 2nd ed., pp. 677–684. McGraw-Hill, New York, 1958.
40. T. Taoka and T. Ohtsuka, *J. Phys. Soc. Japan* **9**, 712 (1954).
41. E. Josso, *J. Phys. Radium* **12**, 399 (1951).
42. J. J. Went, *Physica* **17**, 596 (1951).
43. E. I. Kondorskii and V. L. Sedov, *J. Appl. Phys. Suppl.* **31**, 331S (1960).
44. R. M. Bozorth and J. G. Walker, *Phys. Rev.* **89**, 624 (1953).
45. W. Köster, *Z. Metallk.* **35**, 194 (1943).
46. R. Becker and W. Döring, "Ferromagnetismus," pp. 339–357. Springer, Berlin, 1939.
47. L. Néel, J. Paulevé, R. Pauthenet, J. Laugier, and D. Dautreppe, *J. Appl. Phys.* **35**, 873 (1964).
48. J. Paulevé, K. Krebs, A. Chamberod, and J. Laugier, *Compt. Rend.* **260**, 2439 (1965).
49. S. Kaya and A. Kussmann, *Z. Physik* **72**, 293 (1931).
50. S. Kaya and M. Nakayama, *Proc. Math. Phys. Soc. Japan* **22**, 126 (1940).
51. N. Thompson, *Proc. Phys. Soc. (London)* **52**, 217 (1940).
52. N. Volkenshtein and A. Komar, *Zh. Eksperim. i Teor. Fiz.* **11**, 723 (1941).
53. C. Guillaud, *Compt. Rend.* **219**, 614 (1944).
54. T. Taoka and T. Ohtsuka, *J. Phys. Soc. Japan* **9**, 723 (1954).
55. R. Hahn and E. Kneller, *Z. Metallk.* **49**, 426 (1958).
56. B. L. Averbach, *J. Appl. Phys.* **22**, 1088 (1951).
57. W. J. Carr, *Phys. Rev.* **85**, 590 (1952).
58. J. S. Kouvel, C. D. Graham, Jr., and J. J. Becker, *J. Appl. Phys.* **29**, 518 (1958).
59. J. S. Kouvel, C. D. Graham, Jr., and I. S. Jacobs, *J. Phys. Radium* **20**, 198 (1959).
60. J. S. Kouvel and C. D. Graham, Jr., *J. Appl. Phys. Suppl.* **30**, 312S (1959); *J. Phys. Chem. Solids* **11**, 220 (1959).
61. N. V. Volkenshtein and M. I. Turchinskaya, *Soviet Phys. JETP* **11**, 195 (1960).

62. W. H. Meiklejohn and C. P. Bean, *Phys. Rev.* **102**, 1413 (1956); **105**, 904 (1957).
63. J. S. Kasper and J. S. Kouvel, *J. Phys. Chem. Solids* **11**, 231 (1959).
64. M. J. Marcinkowski and N. Brown, *J. Appl. Phys.* **32**, 375 (1961).
65. A. Paoletti, F. P. Ricci, and L. Passari, *J. Appl. Phys.* **37**, 3236 (1966).
66. T. Taoka, *J. Phys. Soc. Japan* **11**, 537 (1956).
67. M. J. Marcinkowski and R. M. Poliak, *Phil. Mag.* **8**, 1023 (1963).
68. T. Nakamichi and M. Yamamoto, *J. Phys. Soc. Japan* **18**, 758 (1963).
69. A. J. Bradley and A. H. Jay, *Proc. Roy. Soc. (London)* **A136**, 210 (1932); *J. Iron Steel Inst. (London)* **125**, 339 (1932).
70. A. Taylor and R. M. Jones, *J. Phys. Chem. Solids* **6**, 16 (1958).
71. A. Arrott and H. Sato, *Phys. Rev.* **114**, 1420 (1959).
72. H. Sato and A. Arrott, *Phys. Rev.* **114**, 1427 (1959).
73. W. Sucksmith, *Proc. Roy. Soc. (London)* **A171**, 525 (1939).
74. W. D. Bennett, *J. Iron Steel Inst. (London)*, **171**, 372 (1952).
75. R. Nathans, M. T. Pigott, and C. G. Shull, *J. Phys. Chem. Solids* **6**, 38 (1958).
76. K. Ono, Y. Ishikawa, and A. Ito, *J. Phys. Soc. Japan* **17**, 1747 (1962).
77. J. S. Kouvel, *J. Appl. Phys. Suppl.* **30**, 313S (1959).
78. R. Nathans and S. J. Pickart, as reported by Nathans, "Magnetism" (G. T. Rado and H. Suhl, eds.), Vol. III, Chapter 5, p. 235. Academic Press, New York, 1963.
79. T. Shinohara, *J. Phys. Soc. Japan* **19**, 51 (1964).
80. H. Gengnagel, *Naturwissenschaften* **44**, 630 (1957).
81. R. C. Hall, *J. Appl. Phys.* **30**, 816 (1959).
82. I. M. Puzei, *Phys. Metals Metallog. (USSR) (English Transl.)* **9**, No. 2, 103 (1960).
83. M. Yamamoto and S. Taniguchi, *Sci. Rept. Res. Inst. Tohoku Univ.* **8A**, 193 (1956).
84. M. Hansen, "Constitution of Binary Alloys," 2nd ed., pp. 696–700. McGraw-Hill, New York, 1958.
85. J. Crangle, *J. Phys. Radium* **20**, 435 (1959).
86. R. M. Bozorth, D. D. Davis, and J. H. Wernick, *J. Phys. Soc. Japan* **17**, Suppl. B-I, 112 (1962).
87. J. Crangle, *Phil. Mag.* **5**, 335 (1960).
88. A. Kussmann and G. von Rittberg, *Z. Metallk.* **41**, 470 (1950).
89. J. W. Cable, E. O. Wollan, and W. C. Koehler, *J. Appl. Phys.* **34**, 1189 (1963).
90. G. G. Low, *Proc. Intern. Conf. Magnetism, Nottingham, Sept. 1964*, p. 133, Inst. Phys. Phys. Soc., London (1964).
91. A. Kussmann and K. Jessen, *J. Phys. Soc. Japan* **17**, Suppl. B-I, 136 (1962).
92. L. Graf and A. Kussmann, *Physik. Z.* **36**, 544 (1935).
93. H. Fujimori and H. Saitō, *J. Phys. Soc. Japan* **20**, 293 (1965).
94. M. Fallot, *Ann. Phys.* **10**, 291 (1938).
95. J. W. Cable, E. O. Wollan, and W. C. Koehler, *Phys. Rev.* **138**, A755 (1965).
96. G. E. Bacon and J. Crangle, *Proc. Roy. Soc. (London)* **A272**, 387 (1963).
97. S. Shimizu and K. Watai, *Trans. Japan Inst. Metals* **6**, 252 (1965).
98. R. M. Bozorth, "Ferromagnetism," p. 410. Van Nostrand, Princeton, New Jersey, 1951.
99. A. Kussmann, M. Auwärter, and G. von Rittberg, *Ann. Physik* **4**, 174 (1948).
100. E. Krén and P. Szabó, *Solid State Commun.* **3**, 371 (1965).

CHAPTER XII

Directional Order

S. CHIKAZUMI
The Institute for Solid State Physics
University of Tokyo
Tokyo, Japan

C. D. GRAHAM, Jr.
General Electric Research and
Development Center
Schenectady, New York

INTRODUCTION AND PHENOMENOLOGY	577
DIRECTIONAL ORDERING	580
History and Qualitative Description	580
Formal Theory of Directional Order	584
Experimental Results	590
Annealing under Stress	597
The Perminvar Loop	599
Special Related Topics	601
ROLL MAGNETIC ANISOTROPY	602
History	602
Formal Theory of Slip-Induced Anisotropy	604
Experimental Results	608
Related Topics	616
References	617

Introduction and Phenomenology

1. The phenomenon of *magnetic annealing* was first discovered by Pender and Jones [1] in 1913. In the course of measuring the magnetic properties of 3.5 and 4 wt % Si steels at elevated temperature, they found that cooling their samples from 800°C to room temperature in an applied ac field of

30 Oe resulted in a marked improvement in permeability. This improvement was stable against further annealing at 100°C.

About ten years later, the effect was rediscovered in the nickel–iron alloys (Permalloys) by Kelsall, and a series of papers published in 1934 and 1935 clarified the nature and magnitude of the effect [2–5].

Bozorth's book [6] reviews the experimental situation of about 1950 and gives complete references to the literature up to that time. Although there are some minor conflicts in the reported information, the major facts to be explained are as follows:

(1) Permanent changes in magnetic properties can be produced in many ferromagnetic materials by annealing in a magnetic field. All the alloys in the Fe–Ni–Co ternary system respond to such a magnetic annealing treatment; the pure elements Fe, Ni, and Co do not. In the alloys near Co and near 75–25% Fe–Ni, the effects of magnetic annealing are hard to detect because phase changes occur on cooling. (The high-Co alloys develop a preferred crystallographic texture when cooled through the transformation temperature in a strong field, as discussed in Chapter XIV, Sec. 16). The ferromagnetic Fe–Si and Fe–Al alloys also respond to magnetic annealing.

(2) The result of the magnetic annealing treatment is a change in the shape of the hysteresis loop of the material. If the properties are measured parallel to the direction of the field applied during cooling, the hysteresis loop is found to be more nearly rectangular; it is said to be "squared up." As a result, the maximum permeability and remanence are increased, sometimes by large factors. The coercive force is sometimes slightly improved; dc and ac losses are sometimes decreased and sometimes increased. Magnetostriction and magnetoresistance are sharply decreased.

(3) If the properties are measured perpendicular to the direction of the annealing field, the changes are opposite in sign; that is, the hysteresis loop is sheared over, remanence and permeability are decreased, etc. The effect of magnetic annealing is therefore anisotropic, and properties that are improved in one direction are degraded in the perpendicular direction.

(4) To be effective, the field must be applied below the Curie temperature of the material. There is also a lower temperature, usually near 400°C, below which the field has little effect. This suggests that atomic diffusion over short distances is required for magnetic annealing to occur. The magnitude of the applied field is not important so long as it is sufficient to reach technical saturation. Either ac or dc fields may be used.

(5) Materials that respond to magnetic annealing very commonly show a characteristic distorted hysteresis loop when they are slowly cooled in the

absence of an external field. The exact form of the loop varies, but its distinguishing feature is an abnormally low remanence, as illustrated in Fig. 1. This distorted loop is known as "Perminvar loop" (from the Fe–Ni–Co alloy in which it is most pronounced) or sometimes as a "wasp-waisted" or "constricted" loop. The constricted shape is observed only for small applied fields; if the material is driven to saturation, the loop becomes more or less normal, and the constricted shape can be restored only by another heat treatment. Materials that show the Perminvar loop also show Perminvar properties: a constant permeability with almost no losses up to a maximum induction of as much as several hundred gauss.

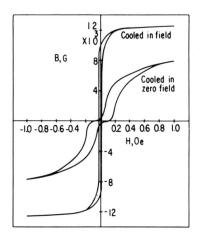

FIG. 1. Effect of magnetic annealing on 65–35% Ni–Fe alloy.

The first explanation proposed for the origin of magnetic annealing involved the relief of magnetostrictive stresses by creep deformation at elevated temperature, and the subsequent stabilization of this stress-relieved domain configuration at lower temperatures [7]. Such a theory is simple and attractive, but there are quantitative difficulties [8]. Furthermore, the response to magnetic annealing does not correlate well with magnetostriction data, and magnetic annealing is found to be effective in single crystals where there is no restraint against magnetostrictive deformation and therefore no resulting stress.

Many of the alloys that respond strongly to magnetic annealing are known to undergo long-range ordering, so it is natural to look for a connection between the two phenomena. Kaya [9] proposed that if ordering proceeds as a first-order reaction so that in the partially ordered state there are discrete regions of ordered and disordered material with different saturation magnetizations, then any departure from spherical symmetry in the shape of the ordered regions would result in magnetic shape anisotropy that could be oriented by a field applied during ordering. However, this model is incapable of explaining the fact that if a sample is reannealed with the applied field in a new direction, the anisotropy axis rotates rapidly to the new direction. Furthermore, magnetic annealing is found in alloy compositions that are not known to order [10].

Directional Ordering

History and Qualitative Description

2. The current accepted explanation of magnetic annealing lies in the phenomenon of *directional order*, or of an atomic arrangement that is anisotropic. This concept developed slowly, and was discussed independently and from various points of view by several different authors over a period of years before a more or less complete picture emerged.

J. L. Snoek, in the course of investigating magnetic aftereffects in iron containing small amounts of carbon, pointed out that there is an anisotropic distortion around each interstitial atom in a body-centered cubic structure [11a]. As shown in Fig. 2, interstitial carbon (and nitrogen) atoms in iron

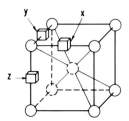

FIG. 2. Locations of interstitial sites in the bcc structure.

lie along the cube edges [11b]. Each interstitial site can therefore be associated with a direction and one can speak of x, y, or z interstitial sites. Occupation of an x site by an interstitial carbon or nitrogen atom causes a slight increase in length in the x direction. Snoek's original idea was that the spontaneous magnetostrictive distortion, which leads to an increase in length along the cube edge parallel to the local magnetization, would favor the occupation of interstitial sites along the cube edge parallel to the magnetization. In a sample magnetized to saturation at a temperature where interstitial atoms can change positions by diffusion, the x, y, and z sites will be unequally occupied. This anisotropic distribution of interstitials, which is one case of directional order, will be reflected in anisotropic magnetic properties. Snoek used this reasoning very successfully to correlate various magnetic and mechanical effects in iron containing carbon and nitrogen. Since at room temperature carbon and nitrogen can diffuse in iron at an appreciable rate, he was mostly concerned with dynamic effects connected with the motion of carbon and nitrogen to their equilibrium positions.

Néel [12] showed that the magnetostrictive interaction proposed by Snoek was too small to account for the observed effects, and postulated that an additional direct magnetic interaction exists, so that the direction of local magnetization influences the relative energies of the x, y, and z sites *independently* of any magnetostrictive distortions.

Experiments by de Vries *et al.* [13] have shown that the direct magnetic interaction is about ten times larger than the magnetostrictive interaction, and that the two effects are opposite in sign; that is, the direct magnetic interaction acts to put interstitials in sites perpendicular to the local magnetization, while the magnetostrictive interaction puts them parallel to the local magnetization.

An isolated substitutional atom in a metal crystal has no directional properties of lower symmetry than the crystal, so there is no exact analogy between the interstitial and substitutional solute atom. A *pair* of nearest neighbor substitutional solute atoms, however, defines a direction and can have directional properties. This seems to have been pointed out first by Zener [14], who was considering the effects of an applied stress, and this aspect of the directional ordering problem has since been extensively explored in connection with anelastic behavior, internal friction, etc. [15, 16].

The connection with magnetic phenomena was pointed out by Chikazumi [17] who first used the phrase "directional order." He noted that alloys near Ni_3Fe show a decrease in volume on going from the disordered to the ordered state, and he reasoned that Ni–Fe pairs must therefore be shorter than Fe–Fe or Ni–Ni pairs. A directional ordering of like-atom pairs would lead to an elongation in the direction of preferential pair alignment, and the interaction of this elongation with the magnetostrictive strain would give a coupling between the magnetization and the axis of pair alignment.

A direct magnetic interaction between the local magnetization and a solute-atom pair was suggested independently by Néel [18], by Taniguchi and Yamamoto [19], and by Chikazumi and Oomura [20].

A simple case of directional order is shown in Fig. 3. The explanation of magnetic annealing in terms of directional order is as follows. It is assumed that the energy of a pair of solute atoms depends on the angle between the

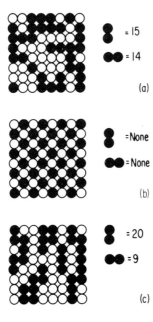

FIG. 3. Various atomic arrangements in a 50–50% alloy: (a) random solid solution; (b) perfect order; (c) directional order.

axis of the pair and the direction of local magnetization (Fig. 4). At a temperature below the Curie temperature but high enough for diffusion to occur, the pair axes will not be distributed at random, but will show some preferred direction with respect to the local magnetization. If the magnetization throughout the sample is aligned by an external magnetic field, this preferred direction of pair axes, or directional order, will extend throughout the sample. The equilibrium arrangement of pairs will develop

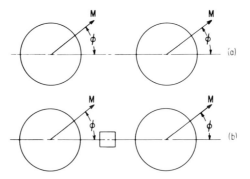

FIG. 4. The basic assumption of directional order theory is that the energy of a pair of atoms depends on the angle φ between the local magnetization M and the line joining the pair of atoms, as in (a), or the energy of an interstitial depends on the angle between M and the line joining the interstitial to its nearest neighbors, as in (b).

at a rate that depends on the details of the diffusion mechanism and on the temperature.

Now if the magnetization is rotated by an external field in a time that is short compared to the time for a pair to change its orientation, the energy of the system is increased; there now exists a magnetic anisotropy energy that favors magnetization along the previously established direction. If the temperature of the system is lowered until diffusion no longer occurs at an appreciable rate, this anisotropy is frozen into the sample and will be stable indefinitely. Because of its origin, this anisotropy is sometimes called "diffusion anisotropy."

This induced anisotropy is reflected in the magnetic properties, making magnetization easier parallel to the field applied during cooling and harder in the perpendicular direction. This accounts for the observed changes in remanence and permeability after magnetic annealing. The coercive force and losses depend primarily on the details of domain wall motion, and these are not so directly influenced by the magnetic annealing anisotropy. The magnetic annealing anisotropy does change the domain structure of the demagnetized state, acting to favor domains magnetized parallel to the field applied during cooling [21]. This leads to a decrease in magnetostriction and magnetoresistance [17] in this direction.

Note that the local magnetization is the effective agent in producing directional order, not the applied field. In the absence of an external field, directional order develops locally within each domain (and presumably within each domain wall) at elevated temperature, and at low temperature this particular domain arrangement is stabilized. This stabilized domain structure, with each domain wall at a local energy minimum, accounts for the Perminvar properties and the Perminvar loop described above. The constant permeability behavior at low inductions corresponds to domain walls moving reversibly around their minimum energy positions. At higher inductions, the walls begin to move irreversibly, but they can still remember their lowest-energy positions and will tend to return to these positions when the external field is removed.

Since these stable positions correspond to a state of small net magnetization, the remanence is found to be abnormally low. If the material is driven to saturation, the initial domain structure is destroyed, and when the field is removed, a new domain structure appears that in general does not correspond to the original minimum energy positions; the hysteresis loop after saturation is therefore normal and not constricted.

There are two general classes of phenomena associated with directional ordering. First, there are changes in properties which occur during the process of directional ordering, at temperatures where diffusion can occur. We may regard these as dynamic effects; they include the aftereffect in magnetization, the aftereffect in permeability (also known as time decay of permeability or disaccommodation), etc. These are discussed elsewhere in this book. Second, there are the phenomena observed at temperatures too low for diffusion to occur but caused by directional order produced at higher temperatures. These static effects are the primary subject of this chapter.

Formal Theory of Directional Order

3. The theory of directional order was developed independently by Néel [18] and by Taniguchi and Yamamoto [19]. We give here a somewhat modified version of Néel's treatment, since Taniguchi used a set of reference axes that make the equations rather cumbersome. Other differences between the two treatments, and some subsequent refinements, are discussed in a later section.

The fundamental assumption of directional order theory is that there exists an energy that depends on the angle between the local magnetization and the axis of the nearest neighbor pair or of an interstitial atom (Fig. 4). We write this energy as a power series in Legendre polynomials:

$$w = l(\cos^2 \varphi - 1/3) + q(\cos^4 \varphi - 6/7 \cos^2 \varphi + 3/35) + \cdots \quad (3.1)$$

in which l and q are constants characterizing a "pseudodipole" and "pseudoquadrupole" interaction, respectively, and φ is the angle between the pair axis and the local magnetization. Usually it is assumed that $q \ll l$ and only the first term of Eq. (3.1) is retained. We can write $\cos \varphi = \beta_1 \gamma_1 + \beta_2 \gamma_2 + \beta_3 \gamma_3$ where β's and γ's are the direction cosines of the magnetization and of the pair (or interstitial) axis, respectively, with respect to some set of reference axes. Note that in a single crystal only certain discrete values of the γ's are possible, since nearest neighbor pairs can only lie along certain crystallographic directions.

In a substitutional solid solution of atoms of A and B, three kinds of nearest neighbor pairs are possible: A–A, B–B, and A–B. There are accordingly three energy constants l_{AA}, l_{BB}, l_{AB} for the first term of Eq. (3.1). The total "pair energy" of an A–B alloy is then

$$E = \sum_i (N_{AAi}l_{AA} + N_{BBi}l_{BB} + N_{ABi}l_{AB})(\cos^2 \varphi_i - \tfrac{1}{3}) \quad (3.2)$$

where the sum is over all possible nearest-neighbor pair direction (i specifies the pair directions), and the second term of Eq. (3.1) has been neglected.

The values of N_{AAi}, N_{BBi}, and N_{ABi} are not independent; a change of atomic position that creates a new A–A pair also creates a new B–B pair and destroys two A–B pairs. The arrangement of *all* the pairs can therefore be specified by the arrangement of any *one* kind of pair. It seems conceptually easier to consider like-atom pairs (Fig. 3). Equation (3.2) then becomes

$$E = l \sum_i N_{BBi}(\cos^2 \varphi_i - \tfrac{1}{3}), \qquad (3.3)$$

where

$$l = l_{AA} + l_{BB} - 2l_{AB}. \qquad (3.4)$$

For further details, see Eqs. (4.1)–(4.5) in Slonczewski [22], or Eqs. (17.3)–(17.5) in Chikazumi [23]. Note that l will be a function of temperature, going to zero at the Curie temperature.

First we determine the equilibrium number of like-atom pairs (or interstitials) aligned in any direction at temperature T_1. This is given by

$$N_i = N \frac{\exp[-l_1(\cos^2 \varphi_i - \tfrac{1}{3})/kT_1]}{\sum_i \exp[-l_1(\cos^2 \varphi_i - \tfrac{1}{3})/kT_1]} \qquad (3.5)$$

where N is the total number of pairs (or interstitials) present, l_1 is the value of l at T_1, k is the Boltzmann constant, and the sum is again over all possible pair orientations.

It seems reasonable to assume that l_1 is small compared to kT_1 in all common cases. The strongest reason for this assumption is that directional order does not seem to develop very strongly, or else it should be easily observable in lattice parameter measurements, elastic constants, etc. If $l_1 \ll kT_1$, we can write

$$\exp[-l_1(\cos^2 \varphi - \tfrac{1}{3})/kT_1] \cong 1 - l_1(\cos^2\varphi - \tfrac{1}{3})/kT_1$$

and

$$\sum_i \exp[-l_1(\cos^2 \varphi_i - \tfrac{1}{3})/kT_1] \cong z$$

where z is the number of possible nearest neighbor directions (or of possible interstitial sites). Equation (3.3) then becomes

$$N_i = (N/z)\{1 - [l_1(\cos^2 \varphi_i - \tfrac{1}{3})/kT_1]\} \qquad (3.6)$$

This equation gives only the equilibrium number N_i and says nothing about the rate at which the equilibrium is established.

If a sample is held at temperature T_1 long enough for the equilibrium of Eq. (3.6) to develop, and is then cooled rapidly to temperature T_2, where diffusion cannot occur, a permanent anisotropy exists, which is given by

$$E_u = \sum_i N_i l_2 (\cos^2 \theta_i - \tfrac{1}{3}), \qquad (3.7)$$

where l_2 is the value of l at T_2, and θ_i is the angle between the local magnetization and the ith pair axis at T_2. We can write

$$\cos \theta = \alpha_1 \gamma_1 + \alpha_2 \gamma_2 + \alpha_3 \gamma_3,$$

where the α's are the direction cosines of the magnetization at T_2, and the γ's, as before, are the direction cosines of the pair (or interstitial) axes.

We have then for a cubic or isotropic distribution of pair axes

$$\begin{aligned} E_u &= \sum_i (N/z)\{1 - [l_1(\cos^2 \varphi_i - \tfrac{1}{3})/kT_1]\} l_2(\cos^2 \theta_i - \tfrac{1}{3}) \\ &= -(N/z)(l_1 l_2/kT_1) \sum_i \cos^2 \varphi_i \cos^2 \theta_i + \text{const} \\ &= -(N/z)(l_1 l_2/kT_1) \sum_i (\beta_1 \gamma_{1i} + \beta_2 \gamma_{2i} + \beta_3 \gamma_{3i})^2 (\alpha_1 \gamma_{1i} + \alpha_2 \gamma_{2i} + \alpha_3 \gamma_{3i})^2 \\ &\quad + \text{const}. \end{aligned} \qquad (3.8)$$

This is the general form of the directional order anisotropy (often called the induced anisotropy), under the various simplifying assumptions made. The value of z and the permitted values of γ depend on the crystal structure, as summarized by the following equation and Table I. The reference axes are taken along the (cubic) crystallographic axes.

$$\begin{aligned} E_u &= (-N l_1 l_2/kT_1)[k_1(\alpha_1^2 \beta_1^2 + \alpha_2^2 \beta_2^2 + \alpha_3^2 \beta_3^2) \\ &\quad + k_2(\alpha_1 \alpha_2 \beta_1 \beta_2 + \alpha_2 \alpha_3 \beta_2 \beta_3 + \alpha_1 \alpha_3 \beta_1 \beta_3)] \\ &= -(N l_1 l_2/kT_1)(k_1 \sum_i \alpha_i^2 \beta_i^2 + k_2 \sum_{i>j} \alpha_i \alpha_j \beta_i \beta_j). \end{aligned} \qquad (3.9)$$

The constants k_1 and k_2 are the directional order anisotropy coefficients. The quantities $N l_1 l_2 k_1/kT_1$ and $N l_1 l_2 k_2/kT_1$, which are determined experimentally, are called F and G by some authors.

We can make a number of remarks about the physical significance of Eq. (3.9) and Table I. For the case of a random polycrystal, the angular-dependent part of Eq. (3.9) becomes

$$\sum_i \alpha_i^2 \beta_i^2 + 2 \sum_{i>j} \alpha_i \alpha_j \beta_i \beta_j = \sum_i{}^2 \alpha_i \beta_i = \cos^2(\theta - \theta_t), \qquad (3.10)$$

TABLE I
VALUES OF k_1 AND k_2 IN EQ. (3.9)

	k_1	k_2
Random polycrystal	2/15	4/15
Interstitials in bcc	1/3	0
Pairs in simple cubic	1/3	0
Pairs in fcc	1/12	4/12
Pairs in bcc	0	4/9

where θ_t specifies the direction of the magnetization during the formation of the directional order, and θ specifies the direction of the magnetization during the measurement of the anisotropy. Usually one applies sufficiently large fields so that the magnetization is very nearly parallel to the applied field at all times, and considers that the measured angle from the applied field to the reference axis in the sample is equal to θ.

The cases of interstitial atoms in the bcc structure and of pairs in the simple cubic structure are equivalent, in the sense that both involve sources of anisotropy with axes along $\langle 100 \rangle$ type directions. Since the three $\langle 100 \rangle$ axes make equal angles with $\langle 111 \rangle$ directions, a field applied along a $\langle 111 \rangle$ direction does not favor any single anisotropy axis. This accounts for the zero value of k_2. In the case of pairs in the bcc structure, the situation is in a sense reversed: the pairs lie along $\langle 111 \rangle$ directions, and therefore annealing with the field along $\langle 100 \rangle$ does not favor occupancy of any one set of $\langle 111 \rangle$ directions. This means that k_1 must be zero. In the fcc structure, nearest neighbor pairs lie along $\langle 110 \rangle$ directions, there is a preferential occupation of pair orientations for any direction of magnetization, and both k_1 and k_2 are nonzero.

For any of these cases, Eq. (3.9) can be represented by

$$E_u = -K_u \cos^2(\theta - \theta_0),$$

in which K_u is the *induced uniaxial anisotropy constant*, and θ and θ_0 are the angles from the magnetization and from the easy axis of the induced anisotropy to a fixed reference axis. For random polycrystalline samples, θ_0 (the easy axis) coincides with the direction of the field during annealing. For single crystals, this is true only if the field during annealing is along a principal crystallographic direction: $\langle 100 \rangle$, $\langle 110 \rangle$, or $\langle 111 \rangle$. In all other cases, the easy axis deviates from the direction of the applied field during

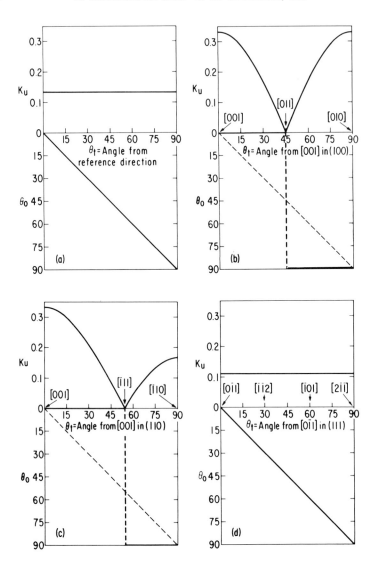

FIG. 5. Predictions of directional order theory [Eq. (3.9)] for various special cases: K_u is the uniaxial anisotropy constant in the equation $E = K_u \sin^2 \theta$; K_u is in units of $Nl_1 l_2/kT_1$; θ_t is the angle at which the field is applied during the magnetic anneal; θ_0 is the angle at which the easy axis is found. (a) Prediction for a random polycrystal. (b)–(d) prediction for the case of substitutional atoms in a simple cubic structure or of interstitials in a bcc structure; (e)–(g) for substitutionals in a fcc structure; (h)–(j) for substitutionals in a bcc structure. The cases shown are sample disks cut in (100), (110), and (111) planes, with the angles measured from the [001], [001], and [011] directions, respectively. The light dotted line indicates the condition $\theta_0 = \theta_t$.

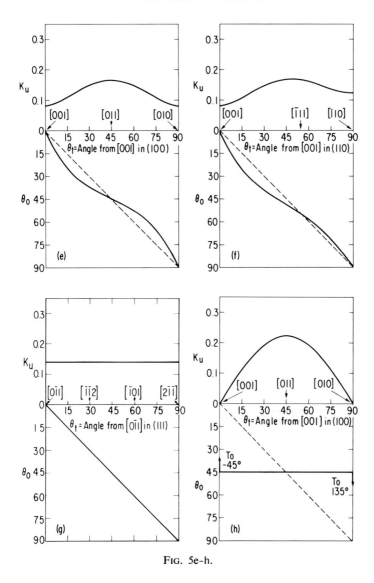

Fig. 5e–h.

annealing (usually denoted as θ_t). Furthermore, the value expected for K_u depends on θ_t. The most interesting cases in the cubic systems are measurements in a {100}, {110}, or {111} plane. The predictions of Eq. (3.9) for these cases are shown in Fig. 5, together with the result for a random polycrystal. Slonczewski has pointed out that all the geometric properties of Eq. (3.9) follow from the assumed cubic symmetry, and are not specific properties of the directional order model [22].

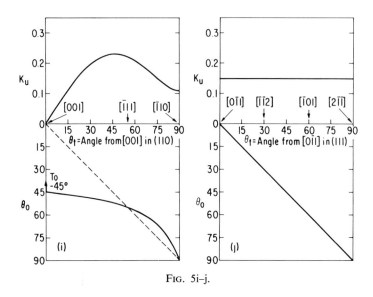

Fig. 5i–j.

Experimental Results

4. There have been many investigations of the effect of magnetic annealing on the technical magnetic properties of polycrystalline materials. There is no satisfactory way to summarize data of this kind; for a list of references, classified by material, see the review article by Graham [24].

In a relatively small number of cases, the induced anisotropy has been determined directly, either from magnetization curves or from torque data. In some cases, the measurements have been made on single crystals.

Reference to Eq. (3.9) shows the variables which can be investigated: K_u depends on the number of pairs N, on the interaction constants l_1 and l_2 (the subscripts denote the temperature of annealing and of measurement), on the annealing temperature T_1, and on the geometrical factors k_1 and k_2. Theory provides only approximate values for l_1 and l_2. Néel [18] relates the values of l to the magnetostriction constants, and deduces a value of the order of 10^{-15} erg for l at room temperature in fcc Ni–Fe alloys. The temperature dependence of l, which gives the high-temperature value l_1, is then related to the temperature dependence of the magnetostriction. Taniguchi and Yamamoto [19] give a similar estimate for the room temperature value of l, but make the simpler assumption that the temperature dependence will follow the square of the spontaneous magnetization. Both authors thus conclude that induced anisotropies K_u should be in the range 10^3–10^5 ergs cm^{-3}, as measured at room temperature for Fe–Ni and similar alloys with large (25 at. %) solute concentrations. This agrees with experiment.

The experimental variables which have been investigated include composition, annealing temperature (and annealing treatment generally), measuring temperature, and (in single crystals) crystallographic direction. Chemical composition is perhaps the most commonly studied parameter; its influence has been investigated in the Co–Ni system by Yamamoto et al. [25a,b]; in Fe–Co by Marechal [26]; in Fe–Ni by Chikazumi and Oomura [20] and Ferguson [27]; in Fe–Al by Gengnagel and Wagner [28], Birkenbeil and Cahn [29], and Chikazumi and Wakiyama [30, 31]; and in Fe–Si by Sixtus [32]. The results depend sensitively on the details of heat treatment, and the agreement between different investigators is not very good, but two general results are clear: the induced anisotropy vanishes as the concentration of solute becomes small and also vanishes at a composition and heat treatment corresponding to complete long-range order. The maximum values of the induced anisotropy K_u range up to about 10^4 ergs cm^{-3} in Fe–Co, Co–Ni, and Fe–Al; to about 4×10^3 for Fe–Ni; and to about 10^3 in Fe–Si.

The value of K_u depends on N, the number of like-atom pairs. At low concentrations, N is simply proportional to the square of the solute concentration, but at higher concentrations this is no longer true. If the alloy behaves as an ideal solid solution, the pair concentration N is proportional to $n_A{}^2 n_B{}^2$, where n_A and n_B are concentrations of A and B. Figure 6 compares experimental values of K_u in the Ni–Fe system with the curve calculated from $n_A{}^2 n_B{}^2$; the agreement is qualitatively satisfactory. One possible reason

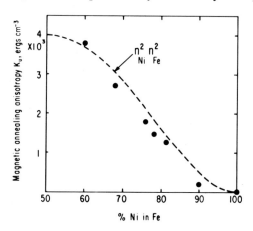

FIG. 6. Magnetic annealing anisotropy K_u versus composition in the Ni–Fe system. Dotted curve is proportional to the number of like-atom pairs in an ideal solid solution (Data from Chikazumi and Oomura [20].)

for departures from the simple curve is a departure of the system from ideal behavior, that is, a tendency to ordering or clustering. Calculations of the effect of departure from ideality were first made by Néel [12], and later in much more detail by Iwata [33].

In the simplest interpretation of the theory, the coefficient l should be independent of composition. However, l may in fact be composition dependent through variations in lattice spacing or of atomic moment with composition. A further difficulty is that in the usual experiment where the sample is slowly cooled in a field, the value of the annealing temperature T_1 is not well known. Even if the samples are quenched from a fixed temperature to define T_1, the ratio of T_1 to the Curie temperature T_C varies with composition, and the experimental conditions are not strictly comparable from one composition to another.

The effect of varying the annealing temperature T_1 at a fixed composition has been investigated by Ferguson [34] and by Yamamoto et al. [25a, 25b]. The latter authors also measured the variation of K_u with the temperature of measurement. Their results are illustrated in Figs. 7 and 8. For the cases

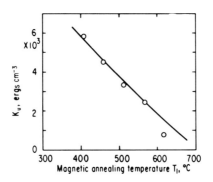

FIG. 7. Magnetic annealing anisotropy K_u versus magnetic annealing temperature T_1, for 30.8–69.2% Co–Ni. Curve is proportional to M_T^2/T_1. (From Yamamoto et al. [25b].)

examined, K_u varies with measuring temperature in the same way that M_s^2 varies, and with annealing temperature as M_s^2/T_1. Both of these results are in agreement with Taniguchi's prediction. However, there is evidence that this is not true for all compositions, especially where long-range ordering is known to occur [35].

The value of the interaction constant l and its temperature dependence are probably the most significant quantities to be determined experimentally. However, in only two cases has a direct measurement of l been obtained.

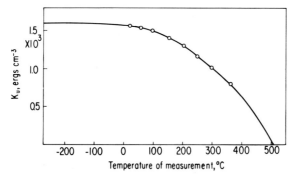

FIG. 8. Magnetic annealing anisotropy K_u versus temperature of measurement, for a single crystal of 12–88% Co–Ni. Curve is proportional to M_T^2. (From Yamamoto et al. [25b].)

In a very careful experiment, de Vries et al. [13] determined l for interstitial solid solutions of carbon and nitrogen in polycrystalline iron. For interstitial carbon and nitrogen, the number of pairs N is replaced by the number of interstitials, which can be determined directly. By choosing a temperature at which the relaxation time for diffusion was about 1 hr, they were able to make the annealing temperature the same as the measuring temperatures so that $l_1 = l_2$ and the measurement gave l^2 at the temperature of the experiment. In addition, it was necessary to make a careful analysis of the contribution of the magnetostrictive energy associated with the interstitials, and a measurement of the magnetostriction associated with a change in the direction of the induced anisotropy. The final result is that $l = 8.5 \times 10^{-16}$ erg for carbon at 239°K and $l = 5.5 \times 10^{-16}$ erg for nitrogen at both 223.5 and 227.5°K. A consequence of the magnetostriction measurement is that the direct magnetic interaction is *opposite* in sign to the magnetostrictive interaction originally postulated by Snoek; that is, the magnetic interaction acts to put the interstitial carbon or nitrogen atoms on cube edges *perpendicular* to the local magnetization.

There is no reason why analogous experiments to determine l cannot be done on substitutional alloys, except for the uncertainty in N.

A more limited but more direct test of Eq. (3.9) can be made by comparing the magnitude of K_u obtained in a single-crystal sample by identical magnetic annealing treatments with the field applied in different crystallographic directions. Such experiments were first carried out by Chikazumi [36]; the principal experimental difficulty is to separate the induced anisotropy from the crystal anisotropy, which may be large. Chikazumi's method, which has been used by all subsequent investigators, is illustrated in Fig. 9.

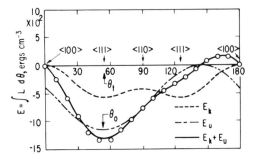

FIG. 9. Separation of induced anisotropy from crystal anisotropy in a single crystal of Ni_3Fe. The experimental torque curve, L vs. θ, of a magnetically annealed crystal (in this case cut in a {110} plane) is integrated graphically in steps to give values of $E = \int_0^\theta L\, d\theta$. These are plotted as open circles. Fourier analysis of this curve permits separation of the uniaxial component of anisotropy $E_u = -K_u \cos^2(\theta - \theta_0)$, shown as dot-dash line, from the crystal anisotropy $E_K = K_1[\sin^2\theta \cos^2\theta + (\sin^4\theta/4)] + K_2(\sin^4\theta \cos^2\theta/4)$, shown as a dotted line. The minimum point of the E_u curve is the easy axis resulting from the magnetic annealing (θ_0), and the magnitude of E_u at θ_0 is numerically equal to the uniaxial anisotropy constant K_u. The solid curve is the sum of $E_u + E_K$; the match with the data points is good. (From Chikazumi [36].)

Directional order theory makes clear predictions about the effect of applying the magnetic annealing field in different crystallographic directions; these are contained in Eq. (3.9) and Table I and illustrated in Fig. 5 for some special cases. It is customary to compare theory and experiments in terms of the ratio k_2/k_1. In the simple theory, this ratio is equal to 4 for the

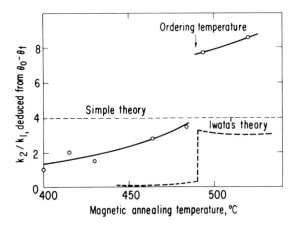

FIG. 10. Experimental values of k_2/k_1 (determined from $\theta_0 - \theta_t$) for Ni_3Fe, compared with simple directional order theory and with Iwata's modification of the theory. (From Takahashi et al. [38a].)

XII. DIRECTIONAL ORDER

fcc structure. Chikazumi found that a value of 8.5 gave a good fit to his data on a Ni_3Fe crystal [36]. He suggested that this anomalous value of k_2/k_1 might be related to the occurrence of nearest neighbor B–B–B triplets in a triangular array. Subsequent experiments on various Ni–Fe and Co–Ni alloys [25, 37, 38a], however, indicate that values of $k_2/k_1 > 4$ are characteristic only of alloys near Ni_3Fe, treated at $T > 485°C$, which is the ordering temperature for this alloy. The reason for this anomalously large value of k_2/k_1 is quite unclear. The other alloys investigated, and Ni_3Fe if treated below 485°C, have $k_2/k_1 < 4$. It is also found that k_2/k_1 depends on the details of the heat treatment, such as the cooling rate [36]. Some light is cast on these complications by the calculations of Iwata [33], who considered in detail the effects of a tendency toward ordering or clustering on the induced anisotropy. His results for the ordering alloy Ni_3Fe are compared with experiment in Fig. 10. Iwata correctly predicts a sudden large increase in k_2/k_1 at the ordering temperature, but does not predict the correct values for k_2/k_1. In particular, he cannot account for values of k_2/k_1 greater than 4 in an ordering alloy.

There are also difficulties in the interpretation of the Co–Ni results. Here the single-crystal k_2/k_1 values are less than 4, corresponding to an ordering tendency, but the composition variation of K_u, corrected for changes in Curie temperature, shows a more rapid increase of K_u with concentration than does an ideal solid solution, which indicates a clustering or precipitation tendency. The Ni–Co alloys do not show any precipitation reaction, and neutron diffraction measurements have shown no evidence of ordering [38b].

In the bcc structure, one expects that $k_1 = 0$, so that there should be no induced anisotropy along the $\langle 100 \rangle$ directions in a substitutional alloy. However, significant effects are found along $\langle 100 \rangle$ [28, 30, 31, 39] in Fe–Al alloys and probably also in Fe–Si alloys [32]. This could be caused by a contribution from second-nearest neighbor pairs, since in the bcc structure the spacing of second nearest neighbors is only 1.15 times that of nearest neighbors. Figure 11 shows some of the experimental results obtained by Wakiyama for the Fe–Al system [31]. The value of F, which is proportional to k_1, is clearly not zero. The variation of both F and G with Al content agrees reasonably well with the concentration of Al–Al pairs except for the comparison range 20–25 at. % Al, where the values of F and G show sharp maxima. One possible explanation is that the magnetic moment of the iron atoms may depend on the number of surrounding Al atoms (as has been shown for ordered Fe_3Al by Nathans *et al.* using neutron diffraction [41]), so that the alloy effectively consists of more than two kinds of atoms, and various kinds of atom pairs can be aligned. At Fe_3Al, the superlattice is formed so rapidly that directional order is completely suppressed. In the composition range 20–22.5 at. % Al it has been found [30, 31] that the induced anisotropy includes not only a simple uniaxial term ($\cos^2 \theta$) but also cubic and higher-order terms ($\cos^4 \theta$, etc.).* Wakiyama examined the possible contributions of the higher order terms in Eq. (3.1) to the fourfold anisotropy and found that the magnitude of the cubic anisotropy should

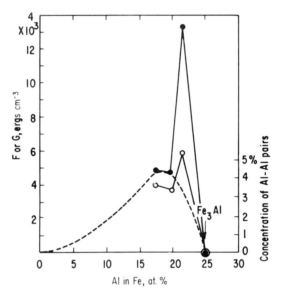

FIG. 11. Composition dependence of the induced anisotropy constants F and G (proportional to k_1 and k_2) in the Fe–Al system: open circles, F; filled circles, G. Dotted line gives the concentration of Al–Al nearest neighbor pairs as determined by the x-ray analysis of Bradley and Jay [40]. (From Wakiyama [31].)

be independent of the direction of annealing field in the {100} plane. The observed fourfold anisotropy, however, depends upon the direction of the annealing field in the {100} plane as well as in the {110} plane. Perhaps the heat treatments with the magnetic field applied in different directions suppress the normal superlattice formation to different degrees. Since the magnetocrystalline cubic anisotropy depends upon the superlattice formation, such a difference in normal ordering will result in a cubic anisotropy.†

* Birkenbeil and Cahn [29] also found a cubic term in the induced anisotropy of a sample of 9.6 at. % Al–Fe annealed under stress.
† It is sometimes suggested that directional order may compete with normal long-range order; the idea is usually that the presence of a field during cooling may influence the

The values of G/F or k_2/k_1 for Fe–Al alloys depend upon the composition as shown in Fig. 12; they increase rapidly as the composition approaches Fe$_3$Al. The reason for this is unknown.

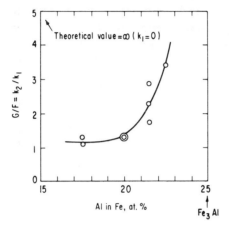

FIG. 12. Composition dependence of $G/F = k_2/k_1$ in the Fe–Al system. (From Wakiyama [31].)

We may conclude that directional order theory as proposed by Néel and by Taniguchi, and extended by Iwata, provides an adequate qualitative explanation of induced anisotropy and its variation with composition, heat treatment, etc. However, the detailed predictions of the theory, especially those applying to single crystals, are not very satisfactory. Probably the trouble lies in the nature of the assumptions made in the theory, namely, that all the effects can be attributed to nearest neighbor interactions in a cubic lattice, with a single interaction constant which does not depend on the local configuration of atoms.

Annealing under Stress

5. The related problem of magnetic anisotropy produced by annealing under an applied stress has not been studied in much detail, either theoretically or experimentally. The physical situation in this case is more complica-

degree of order developed. However, an applied field will have a very small effect on the *magnitude* of the local magnization, since each domain is locally magnetized to saturation, even in the absence of an applied field. It follows that the influence of an applied field on the degree of order should be very small. The applied field will, however, turn the local magnetization out of the easy direction, and it is perhaps possible that this *rotation* of the magnetization could influence the degree of order.

ted. In general, an interstitial atom or a substitutional pair will be associated with an anisotropic local strain. We expect this to be an extension along the direction defined by the interstitial, but it can be either an extension or a contraction along the axis of the substitutional pair. If a crystal is subject to a uniform strain, the energy of an interstitial or a pair will depend on its orientation with respect to the strain. Therefore, if any solid solution alloy (not necessarily magnetic) is subjected to an applied stress, the resulting strain will tend to cause an anisotropic arrangement of the atom pairs.

In a ferromagnetic alloy below its Curie temperature, there is always a local uniform strain due to the spontaneous magnetostriction. This strain will tend to produce pair alignment, but the axis of this strain-induced alignment need not be the same as the axis produced by the interaction of the pairs with the local magnetization. In fact, in the two cases that have been examined with sufficient care, namely, carbon and nitrogen interstitials in iron, the two effects are opposite in sign, with the magnetic interaction about 10 times larger than the magnetostrictive strain interaction (see the preceding section).

If uniform strain is imposed on a ferromagnetic material by an applied stress, two things will occur. First, the interaction of the stress with the magnetostriction will tend to rearrange the domain configuration so that the magnetization is either parallel or perpendicular to the applied stress, depending on the signs of the magnetostriction constants and of the stress. This resulting magnetization direction will tend to produce a particular anisotropic pair arrangement. Second, the direct interaction of the applied strain with the pairs will tend to produce a particular pair arrangement just as in the case of a nonmagnetic material. Again, the two arrangements may or may not be the same. The magnetic interaction has a limiting value, which is reached when the domain rearrangement is complete; the direct stress interaction increases with increasing stress until plastic flow begins. De Vries has analyzed the effects of magnetization and tension on interstitials in the bcc structure [42].

Experiments on the stress-induced anisotropy must be done with care to avoid creep deformation during the high-temperature treatment. Data have been reported only for 50–50% Ni–Co [43], 45–55% Ni–Fe [44], and some Al–Fe alloys [29], all polycrystalline. In Ni–Co and Ni–Fe, the anisotropy developed under stresses up to 2.5 kg mm^{-2} is less than 10^3 ergs cm^{-3}; in Ni–Co the easy axis of the anisotropy is parallel to the tension axis, and in Ni–Fe, perpendicular. The Fe–Al alloys give values of K_u up to 17×10^3 ergs cm^{-3} when annealed under 14 kg mm^{-2}; the easy axis is parallel to the tension axis.

Uchikoshi and Chikazumi [45] examined the effect of tension, applied to a cylindrical rod of 21.5–78.5% Fe–Ni during annealing at 490°C, on the shape of the magnetization curve at room temperature. The magnetization was aligned during annealing by the solenoidal field of an electric current flowing in the specimen rod itself and was therefore everywhere perpendicular to the applied stress. No effect was observed up to a stress of 0.5 kg mm^{-2}. This stress produced a strain of about 25×10^{-6}, which is over 10 times larger than the magnetostrictive strain of 2×10^{-6}. This means that the magnetic or pseudodipolar interaction is much larger than the strain interaction in this alloy for strains below the elastic limit. A tension of 1.5 kg mm^{-2} caused a slight change in the induced anisotropy.

The Perminvar Loop

6. If a material which responds to magnetic annealing is cooled in the absence of a magnetic field, directional order will be developed independently in each domain and presumably within each domain wall. As a result, the local magnetization directions which existed during the cooling become stabilized, and the response of the material to small applied fields is altered. The details of this problem have been treated by Néel [46] for the case of interstitials in iron, and more generally by Taniguchi [47]. The calculations give the restoring force on a domain wall as it moves away from the minimum energy position created by the local directional order. Some of Taniguchi's results are shown in Fig. 13; the general result is that 90° domain walls are bound to their sites by long-range forces, while 180° walls are held only by relatively short-range forces. The physical reason for this result is not hard to see: as a 90° wall moves through the crystal, it leaves behind it a region whose magnetization is turned at right angles to its original direction. But this original direction was the direction favored by the local induced anisotropy, and the 90° rotation of the magnetization requires an energy K_u per unit unit volume rotated. This energy must be supplied by the force on the moving wall, so that the force remains constant, even for large displacements of the wall. The 180° wall, however, merely reverses the direction of the magnetization as it moves, and since the induced anisotropy is uniaxial, this does not result in any increase in energy. Therefore, the 180° wall experiences a restoring force only over a distance comparable to its own thickness.

In real materials, the distinction between the behavior of 90 and 180° walls is of less significance, since all the domain walls are in a kind of connected network, so that no domain wall can move independently of its

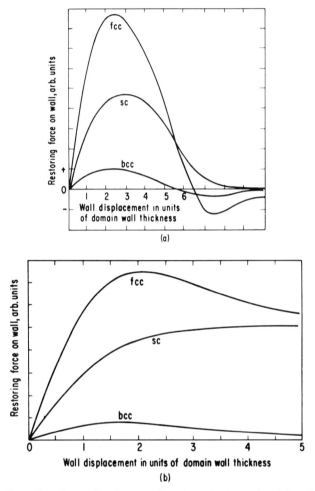

FIG. 13. Examples of restoring force on domain walls due to local directional order: (a) (100) 180° wall, $K_1 > 0$; (b) (100) 90° wall, $K_1 > 0$. (From Taniguchi [47].)

neighbors. Qualitatively, the constricted hysteresis loops and Perminvar properties of typical magnetic annealing alloys can be explained well in terms of domain walls pinned in their sites by local directional order. Small fields will lead to wall displacements proportional to the field, giving constant-permeability behavior. Larger applied fields will lead to irreversible domain wall displacements of 180° walls. Removal of the field allows each domain wall to seek a low-energy site, which in general will correspond to its position during cooling; this will give the typical low-remanence hysteresis curve of Perminvar (see Fig. 1). Finally, saturating the sample with a large field will wipe out the domain structure, and the new structure formed on lowering the field will be unable to find its way back to the original arrangement, and the Perminvar loop will be destroyed.

Special Related Topics

7. Effect of Oxygen. A marked effect of very small quantities of oxygen on the response to magnetic annealing of certain Ni–Fe and Fe–Ni–Co alloys has been reported by Heidenreich et al. [48]. A crystal of 43–34–23% Ni–Fe–Co, for example, responded to magnetic annealing when its oxygen content was 0.0014% (14 ppm) but did not when the oxygen content was reduced to 0.0001% (1 ppm). Electron diffraction work indicates that the oxygen is present in layers on {111} planes when it influences the magnetic annealing. A later paper [49] suggests that small amounts of oxygen inhibit the formation of long-range order, and therefore allow directional order to persist through heat treatments that in a purer material would give only long-range order.

Heidenreich and Nesbitt report that there was considerable difficulty in their experiments with nonreproducible sample behavior, and chemical analysis for oxygen in the range of a few parts per million pushes present techniques to their limit. Probably for these reasons, the oxygen effect has not been studied further.

8. Effect of Irradiation. The requirement for the appearance of directional order is that atomic rearrangement be able to take place at a temperature below the Curie temperature. Usually, such rearrangement is possible only by thermal diffusion, and only above some minimum temperature. However, other rearrangement mechanisms are conceivable. Néel suggested that plastic flow might cause local diffusion [18], although it now appears that another mechanism is responsible for the anisotropy developed during cold-rolling (see the next section). Irradiation of various kinds can replace heat as the source of atomic mobility [50]. Paulevé et al. [51] have found a large anisotropy resulting from neutron irradiation (>1 MeV) of 50–50% Ni–Fe in a magnetic field. Room temperature irradiation seems to give a normal directional order anisotropy, rather larger in magnitude than usual since the temperature of the "anneal" is lower. In this case, irradiation simply permits the equilibrium state of directional order to be attained or approached in reasonably short times at low temperatures. However, irradiation at 200–320°C gives exceedingly large anisotropy values, up to 2×10^6 ergs cm^{-3} in polycrystalline samples, which must be the result of some new process. This is found to be long-range ordering to a tetragonal Au–Cu type structure consisting of alternate layers of iron and nickel atoms. This structure was previously unknown in the Fe–Ni system. Such a long-range directional arrangement of atoms has been called an "oriented superstructure" by Néel [52] to distinguish it from the short-range "directional order" we have discussed in this report. The radiation-induced structure has been shown to consist of crystallites of the Au–Cu structure, about 300 Å in diameter, with their tetragonal axes along the cube directions of the parent crystal. If the irradiation is carried out in a magnetic field, the three kinds of crystallites are no longer present in equal numbers, and magnetic anisotropy results [53]. Irradiation with 1 MeV electrons has an effect similar to but smaller than irradiation with neutrons [54].

9. *Other Related Topics.* We list here some topics that are related to directional order, either conceptually or experimentally, with some primary references.

(1) Aftereffect and disaccommodation are the kinetic effects associated with directional ordering. They are discussed in Chapter XVI.

(2) Directional ordering effects have been extensively studied in ferrites (magnetic oxides). The subject is discussed and references are given in the books by Smit and Wijn [55] and by Chikazumi [23].

(3) Anisotropy resulting from cooling certain materials in a magnetic field can result from the phenomenon of "exchange anisotropy" (see Meiklejohn [56]).

(4) Permanent magnet alloys of the Alnico type show a large anisotropy as a result of heat treatment in a magnetic field. This is fundamentally a shape anisotropy effect, and is discussed in this book in connection with permanent magnet materials.

(5) Under certain special conditions, heat treatment in a magnetic field can produce a preferred crystallographic texture in a sample, with a resulting anisotropy which is experimentally similar to a directional order anisotropy. This is discussed in Chapter XV on textured magnetic materials.

Roll Magnetic Anisotropy

History

10. Roll magnetic anisotropy is the term used to describe a magnetic anisotropy induced by plastic deformation of ferromagnetic alloys. The phenomenon was first discovered in 1934 by Six *et al.* [57], who used the phenomenon to make the magnetic material called Isoperm. This is a 50–50% Fe–Ni alloy that is heavily cold-rolled, recrystallized to the cube texture (001) [100] by annealing, and finally cold-rolled again by about 50% reduction. The final rolling produces a magnetic anisotropy whose easy axis is perpendicular to the rolling direction. When a field is applied parallel to the rolling direction, the magnetization process occurs exclusively by rotation, and the resulting magnetization curve is linear (Fig. 14). This gives a material that has constant permeability over a range of fields, and which is therefore useful as a core material in high-quality transformers or loading coils.

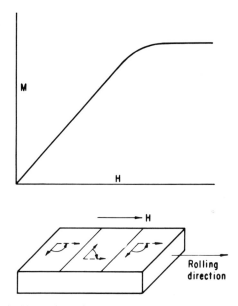

FIG. 14. Schematic illustration of the magnetization curve and domain structure in Isoperm.

Detailed investigations of this phenomenon were made by Conradt et al. [58] in 1940, and by Rathenau and Snoek [59] in 1941. It was concluded that the roll magnetic anisotropy cannot be explained in terms of internal residual stresses which might induce a magnetic anisotropy by interaction with the magnetostriction. Néel [18, 60] and Taniguchi and Yamamoto [19] extended their directional order theory to this phenomenon. They suggested that directional order could develop during the process of plastic deformation so as to favor directed pairs in the direction of local magnetization. An entirely different interpretation was proposed by Chikazumi et al.

[61a–c], who proposed that new atom pairs will be formed across the atomic planes on which slip occurs and that these atom pairs will show directional order unless the slip deformation occurs with cubic symmetry. This theory has been successful in explaining various magnetic anisotropies observed in rolled single crystals, as explained in the following section. A similar interpretation has been proposed by Bunge and Müller [62]. The magnetic anisotropy which results from this mechanism is sometimes called "slip-induced anisotropy."

There is little doubt that roll magnetic anisotropy is caused by directional order, because the magnitude of the induced anisotropy varies with composition very much as does the magnetic annealing anisotropy.

Formal Theory of Slip-Induced Anisotropy

11. Here we discuss the mechanism of slip-induced magnetic anisotropy in various ordered alloys.

Figure 15 shows a perfectly ordered fcc lattice of the A_3B type which is sheared by the motion of a single dislocation in the $[01\bar{1}]$ direction along the (111) plane. There are no BB pairs in the undeformed portion that is still in the perfectly ordered state, while a number of BB pairs are created across the slipped plane. In Fig. 15, all the BB pairs are parallel to the [011] direction, as indicated by double lines. The number of BB pairs thus induced is given by $1/\sqrt{3}a^2$ per unit area of the slipped plane, where a is the lattice

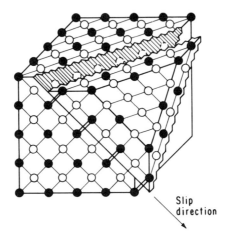

FIG. 15. Production of BB pairs by unit slip on the (111) plane in the $[01\bar{1}]$ direction in an A_3B type long-range ordered structure. BB pairs indicated by double lines. (From Chikazumi *et al.* [61b].)

constant. When the long-range parameter is S, this number is $S^2/\sqrt{3}a^2$. Now we define the slip density s as the average number of dislocations that have passed through a given plane in the crystal. Since there are $\sqrt{3}/a$ atom layers in a unit length perpendicular to the slip plane, the number of planes on which slip occurs is $\sqrt{3}s/a$, provided each dislocation travels along a separate atomic plane. Actually, however, a number of successive dislocations may travel along a single atomic plane. In such a case the second dislocation cancels the previously induced BB pairs. Since a pair of dislocations traveling together does not create any region of disorder, the two dislocations tend to approach to each other, as pointed out by Fisher [63]. We introduce the probability of having an isolated dislocation, p_0, and the probability of creating dislocations in a new atomic plane, p'. Then the number of effective slipped planes becomes $\sqrt{3p_0p's/a}$, so that the number of BB pairs induced in a unit volume is given by

$$N_{BBi} = (S^2/\sqrt{3}a^2)(\sqrt{3p_0p'}/2a)|s_i| = \tfrac{1}{8}NpS^2|s_i| \qquad (11.1)$$

where $p = p_0p'$, and N is the number of atoms per unit volume. The factor 2 in the denominator of Eq. (11.1) is put in because, even if all dislocations are isolated from one another, the probability of having a disordered plane is still $\tfrac{1}{2}$. Putting Eq. (11.1) into Eq. (3.3), we have the anisotropy energy,

$$E_a = \tfrac{1}{8}NlpS^2 \sum_i |s_i|(\alpha_1\gamma_{1i} + \alpha_2\gamma_{2i} + \alpha_3\gamma_{3i})^2$$
$$= \tfrac{1}{8}NlpS^2 \sum_i |s_i| f_i(\alpha_1, \alpha_2, \alpha_3) \qquad (11.2)$$

where

$$f_i(\alpha_1, \alpha_2, \alpha_3) = \gamma_{1i}^2\alpha_1^2 + \gamma_{2i}^2\alpha_2^2 + \gamma_{3i}^2\alpha_3^2 + 2\gamma_{2i}\gamma_{3i}\alpha_2\alpha_3$$
$$+ 2\gamma_{3i}\gamma_{1i}\alpha_3\alpha_1 + 2\gamma_{1i}\gamma_{2i}\alpha_1\alpha_2. \qquad (11.3)$$

Table II gives the direction cosines of the induced BB pairs ($\gamma_{1i}, \gamma_{2i}, \gamma_{3i}$), and the coefficients of the $\alpha_i\alpha_j$'s of each term in Eq. (11.3) for 12 slip systems. This kind of induced anisotropy is expected for an alloy with long-range order such that the slip distance is smaller than the size of the ordered antiphase domains. In this sense, we call this anisotropy the Long-range order, Fine slip type (LF type).

Another type of anisotropy can be induced when a part of the crystal on one side of the slipped plane is brought in contact with another out-of-step, ordered domain. Then the direction of BB pairs can be [011], [110], or [101] for the (111) slip plane. On the other hand, when the crystal has

TABLE II

Direction Cosines of BB Pairs and Coefficients of Each Term in $f_1(\alpha_1, \alpha_2, \alpha_3)$ in Eq. (11.3) for Long-Range Order Fine Slip Roll Anisotropy[a]

Slip system	Slip plane	Slip direction	Direction cosines of BB pairs			Coefficients in $f(\alpha_1, \alpha_2, \alpha_3)$					
			β_1	β_2	β_3	α_1^2	α_2^2	α_3^2	$\alpha_2\alpha_3$	$\alpha_3\alpha_1$	$\alpha_1\alpha_2$
1	(111)	[01$\bar{1}$]	0	$1/\sqrt{2}$	$1/\sqrt{2}$	0	$\frac{1}{2}$	$\frac{1}{2}$	1	0	0
2		[10$\bar{1}$]	$1/\sqrt{2}$	0	$1/\sqrt{2}$	$\frac{1}{2}$	0	$\frac{1}{2}$	0	1	0
3		[1$\bar{1}$0]	$1/\sqrt{2}$	$1/\sqrt{2}$	0	$\frac{1}{2}$	$\frac{1}{2}$	0	0	0	−1
4	(11$\bar{1}$)	[101]	$1/\sqrt{2}$	0	$-1/\sqrt{2}$	$\frac{1}{2}$	0	$\frac{1}{2}$	0	−1	0
5		[011]	0	$1/\sqrt{2}$	$-1/\sqrt{2}$	0	$\frac{1}{2}$	$\frac{1}{2}$	−1	0	0
6		[1$\bar{1}$0]	$1/\sqrt{2}$	$1/\sqrt{2}$	0	$\frac{1}{2}$	$\frac{1}{2}$	0	0	0	−1
7	(1$\bar{1}$1)	[110]	$1/\sqrt{2}$	$1/\sqrt{2}$	0	$\frac{1}{2}$	$\frac{1}{2}$	0	0	0	1
8		[10$\bar{1}$]	$1/\sqrt{2}$	0	$1/\sqrt{2}$	$\frac{1}{2}$	0	$\frac{1}{2}$	0	1	0
9		[011]	0	$1/\sqrt{2}$	$-1/\sqrt{2}$	0	$\frac{1}{2}$	$\frac{1}{2}$	−1	0	0
10	($\bar{1}$11)	[01$\bar{1}$]	0	$1/\sqrt{2}$	$1/\sqrt{2}$	0	$\frac{1}{2}$	$\frac{1}{2}$	1	0	0
11		[101]	$1/\sqrt{2}$	0	$-1/\sqrt{2}$	$\frac{1}{2}$	0	$\frac{1}{2}$	0	−1	0
12		[110]	$1/\sqrt{2}$	$-1/\sqrt{2}$	0	$\frac{1}{2}$	$\frac{1}{2}$	0	0	0	−1

[a] From Chikazumi and Suzuki [61a–c].

only short-range order, slip along the (111) plane causes an increase in the number of BB pairs across the slipped plane. The directions of BB pairs are again [011], [110], and [101]. In any case, the average anisotropy of these three BB pairs (a, b, c) is given by

$$w = (l/3)(\cos^2 \varphi_a + \cos^2 \varphi_b + \cos^2 \varphi_c)$$
$$= (l/3) \sum_{k=a,b,c} (\gamma_{1k}\alpha_1 + \gamma_{2k}\alpha_2 + \gamma_{3k}\alpha_3)^2$$
$$= l(n_{1i}n_{2i}\gamma_1\gamma_2 + n_{1i}n_{3i}\gamma_1\gamma_3 + n_{2i}n_{3i}\gamma_2\gamma_3) \quad (11.4)$$

where (n_{1i}, n_{2i}, n_{3i}) are the direction cosines of the normal to the slip plane. Equation (11.4) describes a uniaxial anisotropy with the hard axis normal to the slip plane, provided $l > 0$. The number of BB pairs produced per unit area of the slipped plane is

$$N_{BB} = \tfrac{1}{16} N p' \sigma \, | s_i |, \quad (11.5)$$

where σ is the short-range order parameter. The anisotropy energy is then expressed as

$$E_a = \tfrac{1}{16} N l p' \sigma \sum_i | s_i | \, g_i(\alpha_1, \alpha_2, \alpha_3), \quad (11.6)$$

where

$$g_i(\alpha_1, \alpha_2, \alpha_3) = n_{1i}n_{2i}\alpha_1\alpha_2 + n_{1i}n_{3i}\alpha_1\alpha_3 + n_{2i}n_{3i}\alpha_2\alpha_3. \quad (11.7)$$

The values of the (n_{1i}, n_{2i}, n_{3i}) and the coefficients of the $\alpha_i\alpha_j$'s in each term of Eq. (11.7) are given in Table III for 12 slip systems. This anisotropy we call the Short-range order, Coarse slip type (SC type).

TABLE III

DIRECTION COSINES OF THE NORMAL TO THE SLIP PLANE AND THE COEFFICIENT OF EACH TERM IN EQ. (11.7) FOR SHORT-RANGE ORDER, COARSE SLIP ROLL ANISOTROPY[a]

Slip system	Direction cosines of normal to slip plane			Coefficients in g $(\alpha_1\alpha_2\alpha_3)$		
	n_1	n_2	n_3	$\alpha_2\alpha_3$	$\alpha_3\alpha_1$	$\alpha_1\alpha_2$
1, 2, 3	$1/\sqrt{3}$	$1/\sqrt{3}$	$1/\sqrt{3}$	$\tfrac{1}{3}$	$\tfrac{1}{3}$	$\tfrac{1}{3}$
4, 5, 6	$1/\sqrt{3}$	$1/\sqrt{3}$	$-1/\sqrt{3}$	$-\tfrac{1}{3}$	$-\tfrac{1}{3}$	$\tfrac{1}{3}$
7, 8, 9	$1/\sqrt{3}$	$-1/\sqrt{3}$	$1/\sqrt{3}$	$-\tfrac{1}{3}$	$\tfrac{1}{3}$	$-\tfrac{1}{3}$
10, 11, 12	$-1/\sqrt{3}$	$1/\sqrt{3}$	$1/\sqrt{3}$	$\tfrac{1}{3}$	$-\tfrac{1}{3}$	$-\tfrac{1}{3}$

[a] From Chikazumi and Suzuki [61a–c].

608   S. CHIKAZUMI AND C. D. GRAHAM, JR.

In the next subsection, various experimental results for Ni_3Fe single crystals will be explained in terms of these two types of anisotropy.

Experimental Results

12. One of the mysteries of roll magnetic anisotropy is that the induced anisotropy in polycrystals is entirely different from that in single crystals. We shall describe first the experimental results on single crystals of Ni_3Fe, and then will discuss the case of polycrystals of the same alloy.

The roll magnetic anisotropy depends strongly on the crystallographic planes and directions of rolling. For (110) [001] rolling, the easy axis of the anisotropy is in the rolling plane and perpendicular to the rolling direction. Although no distinct domain patterns can be observed on unrolled Ni_3Fe crystals, well-defined domains with their magnetizations lying parallel to the easy axis appear after the crystal is rolled (Fig. 16). The

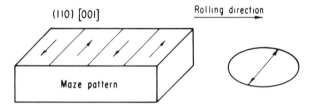

FIG. 16. Domain structure and average easy axis in the rolling plane of a Ni_3Fe crystal rolled in the (110) plane, [001] direction. (From Chikazumi *et al.* [61b].)

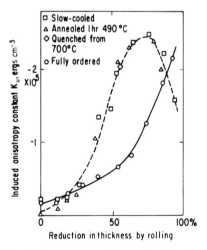

FIG. 17. Induced roll anisotropy versus rolling reduction: Ni_3Fe crystal, (110) [001] rolling, previous heat treatment as indicated. (From Chikazumi *et al.* [61b].)

XII. DIRECTIONAL ORDER

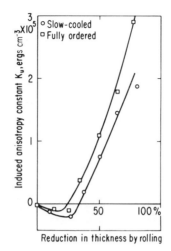

FIG. 18. Induced roll anisotropy versus rolling reduction: Ni₃Fe crystal, (001) [110] rolling, previous heat treatment as indicated. (From Chikazumi et al. [61b].)

magnitude of the induced anisotropy increases with the degree of rolling deformation (Fig. 17), but drops off for reductions greater than 75%. It is interesting to note that a perfectly ordered crystal develops the anisotropy rather slowly. This is presumably because the deformation occurs largely by the motion of pairs of dislocations, with little change in the distribution of atom pairs. In contrast, a quenched specimen, which is usually believed to be disordered, exhibits an anisotropy as large as that of the partially ordered crystals. This fact is explained by assuming that some degree of order may exist even after quenching, and such a partial order may be easily destroyed by any kind of dislocations (paired or unpaired).

For (001) [110] rolling, the easy direction of the anisotropy measured in the rolling plane is perpendicular to the rolling direction in the early stage of rolling, but later it changes to parallel to the rolling direction. The anisotropy constant is plotted against reduction in Fig. 18; there is a

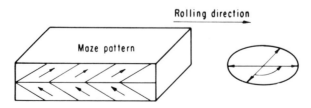

FIG. 19. Domain structure and average easy axis in the rolling plane of an Ni₃Fe crystal rolled in the (001) plane, [110] direction. (From Chikazumi et al. [61b].)

change in sign at about 20–30% reduction. Domain observations, however, indicate that the magnetization actually lies at an angle to the rolling plane (Fig. 19). The angle decreases with additional rolling.

In the case of (001) [100], the easy direction of the anisotropy as measured by torque magnetometer is perpendicular to the rolling direction in the rolling plane. Only maze domain patterns are observed on the top and side

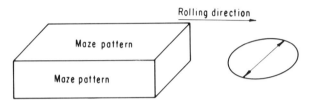

FIG. 20. Domain structure and average easy axis in the rolling plane of an Ni$_3$Fe crystal rolled in the (001) plane, [001] direction. (From Chikazumi et al. [61b].)

surfaces of the rolled crystal. This means the local easy axis is parallel neither to the rolling plane nor to the side surface (Fig. 20). The anisotropy constant, K_u, is 0.5×10^5 ergs cm^{-3} for 52% reduction.

These experiments can all be explained in terms of slip-induced directional order. For (110) [001] rolling, four slip systems, as shown in Fig. 21, should

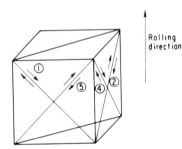

FIG. 21. Slip systems contributing to deformation in (110) [001] rolling. (From Chikazumi et al. [61b].)

take part in the deformation. The slip densities of these slip systems, 1, 2, 4, and 5, are expressed in terms of r, the fractional reduction of the thickness, as

$$s_1 = s_4 = -r/2, \qquad s_2 = s_5 = r/2. \tag{12.1}$$

Referring to the form of $f_i(\alpha_1, \alpha_2, \alpha_3)$ for these slip systems as given in Table II and also to Eq. (12.1), we have the anisotropy energy

$$E_a = \tfrac{1}{16} N l p S^2 r \alpha_3^2 \tag{12.2}$$

for the LF type. Since l is positive for the Fe–Ni alloys system, Eq. (12.2) makes the magnetization stable in the (001) plane. Similarly, referring to the form of $g_i(\alpha_1, \alpha_2, \alpha_3)$ for the four slip systems, we have

$$E_a = (1/24)Nlp'\sigma r\alpha_1\alpha_2 \qquad (12.3)$$

for the SC type. This indicates that the easy axis is parallel to [110], which is perpendicular to the rolling direction in the rolling plane. The superposition of these two anisotropies makes [110] the easy axis, thus explaining the experimental result.

In (001) [110] rolling, the same four slip systems are expected to contribute to the slip deformation. Actually, however, only two systems, 1 and 2 (or 4 and 5) occur in the upper half of the crystal plate, with the other two systems operating in the lower half, as determined by observation of slip bands (Fig. 22). The reason why easy glide along the (111) plane (common

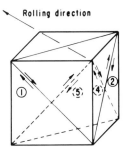

FIG. 22. Slip systems contributing to deformation in (001) [110] rolling. (From Chikazumi *et al.* [61b].)

for 1 and 2 systems) occurs only in (001) [110] rolling, and not in (110) [001] rolling, can be understood as follows. In the case of (001) [110] rolling, the glide plane (111) for the systems 1 and 2 makes an angle 54.5° with the rolling plane (Fig. 23), and, therefore, it rotates during rolling to approach the plane of maximum shearing stress—that is, the plane which lies 45° from the rolling plane. Thus, the original glide plane becomes more favorable than the other possible glide plane (111). On the contrary, in the case of (110) [001] rolling, the two glide planes make an angle of 35.5° with the roll plane; thus, slip along one glide plane rotates the other glide plane toward the plane of maximum shear stress, making this more favorable than the original glide plane. This means that both glide planes should contribute to the deformation.

The slip densities for (001) [110] rolling are, therefore, given by

$$s_1 = r, \qquad s_2 = -r. \qquad (12.4)$$

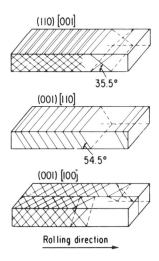

Fig. 23. Observed slip in rolling of Ni₃Fe single crystals of various orientations. (From Chikazumi et al. [61b].)

These two slip systems give rise to anisotropy of the LF type

$$E_a = \tfrac{1}{16}NlpS^2 r\alpha_3(2\alpha_1 + 2\alpha_2 + \alpha_3), \tag{12.5}$$

and of the SC type

$$E_a = (1/24)Nlp'\sigma r(\alpha_1\alpha_2 + \alpha_2\alpha_3 + \alpha_3\alpha_1). \tag{12.6}$$

Since $\alpha_3 = 0$ in the rolling plane, the LF type does not contribute to the anisotropy in the rolling plane, while the SC type makes the easy direction perpendicular to the rolling direction in the rolling plane. Stereographic consideration of Eqs. (12.5) and (12.6) reveals that the former has its easy axis parallel to [111] while the latter has an easy plane parallel to the (111) plane. In either case we can expect that the easy direction lies 35.5 or 55.5° from the rolling plane if we observe it on the edge surface of the rolled crystal. This agrees with the experiment (Fig. 19). As the rolling proceeds, easy glide causes rotation of the crystal and makes the easy axis rotate away from the rolling plane. This is the reason why the easy direction in the rolling plane rotates from perpendicular to parallel to the roll direction.

In (001) [100] rolling, a pair of slip systems 2 and 11 (or 4 and 8) is expected to occur (Fig. 24), because the other four systems (3, 6, 7, and 12) do not contribute to reducing the thickness of the plate and the remaining four systems lead to widening of the crystal plate. The slip densities are given by

$$s_2 = -r, \quad s_{11} = r, \tag{12.7}$$

which gives for the LF type anisotropy

$$E_a = -\tfrac{1}{8}NlS^2 r\alpha_2^2, \tag{12.8}$$

and for the SC type

$$E_a = (1/24)Nlp'\sigma\alpha_2\alpha_3. \tag{12.9}$$

The former has its easy axis perpendicular to the rolling direction in the rolling plane while the latter does not contribute to the anisotropy in the

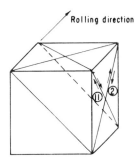

Fig. 24. Slip systems contributing to deformation in (001) [100] rolling. (From Chikazumi et al. [61b].)

rolling plane ($\alpha_3 = 0$). This agrees with experiment. The latter type, however, has its easy axis parallel to the [011] direction causing the resultant easy axis to make a glancing angle with the rolling plane, as was observed experimentally.

The magnitude of the induced anisotropies can also be explained quantitatively by Eqs. (12.2)–(12.9) by assuming values of pS^2 and $p'\sigma$ ranging from 0.2 to 5.9%.

The roll magnetic anisotropy of polycrystalline Ni_3Fe has its easy axis parallel to the rolling direction. Domain observation reveals that the easy axis (the direction of domain magnetization) differs in different crystallites (Fig. 25). One of the striking features of the polycrystalline material is that the induced anisotropy is larger for well-ordered material than for less ordered (Fig. 26). This fact may be explained as follows. The deformation

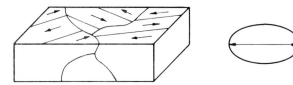

Fig. 25. Domain structure and average easy axis in rolled polycrystalline Ni_3Fe. (From Chikazumi et al. [61b].)

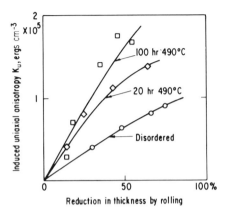

FIG. 26. Anisotropy induced by rolling polycrystalline Ni$_3$Fe after various heat treatments. (From Chikazumi et al. [61b].)

of each crystallite is constrained to match the deformation of the surrounding crystallites, so that more than two slip systems must operate simultaneously. If a pair of dislocations travels along a slip plane in the ordered crystal, and another dislocation which is traveling on an intersecting slip

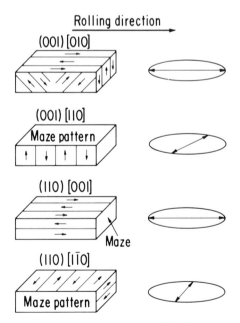

FIG. 27. Domain structure and easy axes of rolled Fe$_3$Al crystals of various orientations. (From Chikazumi et al. [64].)

plane happens to pass through between the paired dislocations, these two dislocations are no longer on the same atomic plane and thus become effective in destroying the ordered arrangement of atoms. Such an event is probably rare in the perfectly ordered crystal, because it will increase the ordering energy. It was, in fact, observed experimentally that a perfectly ordered sample cracked at grain boundaries after only 10% reduction by rolling.

The investigation of roll anisotropy has been extended to Fe_3Al, which has the bcc structure [64]. The domain structures and the easy directions in the rolling plane are shown in Fig. 27. The slip system in a bcc crystal is commonly accepted to be (110) $\langle 111 \rangle$, as shown in Fig. 28. In most cases

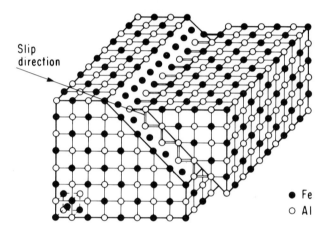

FIG. 28. Directional order produced by unit slip on the (110) plane in the [1$\bar{1}$1] direction in ordered bcc Fe_3Al. Like-atom pairs are indicated by double lines. (From Chikazumi et al. [64].)

the nearest neighbor pairs do not contribute to the roll magnetic anisotropy. The observed values of the induced anisotropy and the calculated formulas are listed in Table IV, where l' represents the value of l for the second-nearest neighbor pairs.

From magnetostriction and magnetic annealing data, we can assume $Nl' = -30 \times 10^8$ ergs cm^{-3}, from which we have $p\sigma = 13.3\%$. This value is slightly larger than the values of PS^2 or $p'\sigma$ for Ni_3Fe, for reasons that are still unclear. The investigation of roll magnetic anisotropy on single-crystal samples has been extended to Ni–Co [65] and Ni–Mn [66a] alloys. In both cases, the magnitude of the induced anisotropy was small, about 10^3 ergs cm^{-3}. For Ni–Co alloys, fourth-order anisotropy was observed.

TABLE IV

COMPARISON BETWEEN EXPERIMENT AND CALCULATED VALUES OF THE COEFFICIENT OF ROLL MAGNETIC ANISOTROPY IN Fe_3Al[a]

Rolling plane	Rolling direction	Observed		Calculated K_u	$P\sigma Nl'$ (ergs cm^{-3})
		K_u (erg cm^{-3})	Easy direction		
(001)	[010]	3.70×10^5 $r = 10.3\%$	[010]	$-\dfrac{p\sigma Nl_2}{32} r$	-1.15×10^8
	[110]	1.76×10^5 $r = 10\%$	27° from roll direction	0	—
(110)	[001]	7.06×10^5 $r = 12\%$	[001]	$-\dfrac{p\sigma Nl_2}{64} r$	-3.76×10^8
	[1$\bar{1}$0]	-4.34×10^5 $r = 8.7\%$	[001]	$\dfrac{3p\sigma l_2}{256} r$	-4.25×10^8

[a] Bunge and Müller [62].

Calculations based on the theory of the slip-induced directional order have been extended by Chin [66b] to the case of wire drawing, to rolling in other crystallographic orientations, and to other alloy systems [66c].

Related Topics

13. Roll Magnetic Anisotropy of Precipitation Alloys. Roll magnetic anisotropy has been investigated in dilute precipitation alloys, such as Co–Cu [67, 68], Fe–Cu [69], and Co–Au [70]. The induced anisotropy in these alloys has been attributed to a change in the shape anisotropy of ferromagnetic particles and also to a crystal transformation of the precipitated cobalt from fcc to hexagonal. Single crystals of Co–Cu have also been investigated [71].

14. Experiments on Single Crystals of Ni_3Mn. It has been found that the superlattice of Ni_3Mn is destroyed by rolling so that the ferromagnetic properties are lost as rolling progresses. Hahn and Kneller [72] and Sato [66a] discovered that the saturation magnetization is first increased in the initial stage of rolling, and then decreased with further rolling. One possible explanation for the original increase is that the width of a pair of dislocations may change so as to increase the number of AB pairs.

ACKNOWLEDGMENTS

We are grateful to T. Wakiyama and J. J. Becker for critically reviewing the manuscript.

REFERENCES

1. H. Pender and R. L. Jones, *Phys. Rev.* **1**, 259 (1913).
2. G. A. Kelsall, *Physics* **5**, 169 (1934).
3. J. F. Dillinger and R. M. Bozorth, *Physics* **6**, 279 (1935).
4. R. M. Bozorth and J. F. Dillinger, *Physics* **6**, 285 (1935).
5. O. Dahl and F. Pawlek, *Z. Physik* **94**, 504 (1935).
6. R. M. Bozorth, "Ferromagnetism," pp. 117 ff and 171 ff. Van Nostrand, Princeton, New Jersey, 1951.
7. R. M. Bozorth, *Phys. Rev.* **46**, 232 (1934).
8. R. Becker and W. Döring, "Ferromagnetismus," pp. 418 ff. Springer, Berlin, 1939.
9. S. Kaya, *Rev. Mod. Phys.* **25**, 49 (1953).
10. G. W. Rathenau, discussion of Kaya [9].
11a. J. L. Snoek, *Physica* **8**, 711 (1941).
11b. D. N. Beshers, *J. Appl. Phys.* **36**, 290 (1965).
12. L. Néel, *J. Phys. Radium* **12**, 339 (1952); **13**, 249 (1952).
13. G. de Vries, D. W. Van Geest, R. Gersdorf, and G. W. Rathenau, *Physica* **25**, 1131 (1959).
14. C. Zener, *Phys. Rev.* **71**, 34 (1947).
15. A. D. LeClaire and W. M. Lomer, *Acta Met.* **2**, 731 (1954).
16. A. S. Nowick and D. P. Seraphim, *Acta Met.* **9**, 40 (1961).
17. S. Chikazumi, *J. Phys. Soc. Japan* **5**, 327, 333 (1950).
18. L. Néel, *J. Phys. Radium* **15**, 225 (1954).
19. S. Taniguchi and M. Yamamoto, *Sci. Rept. Res. Inst. Tohoku Univ.* **A6**, 330 (1954); S. Taniguchi, *ibid.* **A7**, 269 (1955).
20. S. Chikazumi and T. Oomura, *J. Phys. Soc. Japan* **10**, 842 (1955).
21. M. Yamamoto, S. Taniguchi, and K. Aoyagi, *Phys. Rev.* **102**, 1295 (1956).
22. J. C. Slonczewski, Magnetic Annealing, *in* "Magnetism" (G. T. Rado and H. Suhl, eds.), Vol. I, Chapter 5. Academic Press, New York, 1963.
23. S. Chikazumi, "Physics of Magnetism," Wiley, New York, 1964.
24. C. D. Graham, Jr., *in* "Magnetic Properties of Metals and Alloys," pp. 288 ff. Am. Soc. Metals, Cleveland, Ohio, 1959.
25a. M. Yamamoto, S. Taniguchi, and K. Aoyagi, *Sci. Rept. Res. Inst. Tohoku Univ.* **A13**, 117 (1961).
25b. M. Yamamoto, S. Taniguchi, and K. Aoyagi, *J. Phys. Soc. Japan Suppl. B-I* **17**, 328 (1962).
26. J. Marechal, *J. Phys. Radium* **16**, 122S (1955).
27. E. T. Ferguson, *J. Appl. Phys.* **29**, 252 (1958).
28. H. Gengnagel and H. Wagner, *Z. Angew. Phys.* **13**, 174 (1961).

29. H. J. Birkenbeil and R. W. Cahn, *Proc. Phys. Soc.* (*London*), **79**, 831 (1962); *J. Appl. Phys.* **16**, 362S (1961).
30. S. Chikazumi and T. Wakiyama, *J. Phys. Soc. Japan Suppl.* B-I **17**, 325 (1962).
31. T. Wakiyama, thesis, University of Tokyo, 1965; *Tech. Rept. Inst Solid State Phys. Univ. Tokyo Ser. A*, p. 147 (1965).
32. K. J. Sixtus, *Z. Angew. Phys.* **14**, 241 (1962).
33. T. Iwata, *Sci. Rept. Res. Inst. Tohoku Univ.* **A10**, 34 (1958); *Trans. Japan. Inst. Metals* **2**, 86, 92 (1961).
34. E. T. Ferguson, *Intern. Colloq. Magnétisme, Grenoble*, p. 187 (1958); *J. Appl. Phys.* **29**, 252 (1958).
35. M. Yamamoto, *Bull. Electrotech. Lab.* (*in Japanese*) **24**, 161 (1960); quoted by T. Iwata [33]; H. Wagner and H. Gengnagel, *Monatsber. Deut. Akad. Wiss.* (*Berlin*) **5**, 94 (1963).
36. S. Chikazumi, *J. Phys. Soc. Japan* **11**, 551 (1956).
37. K. Aoyagi, *Sci. Rept. Res. Inst. Tohoku Univ.* **A13**, 137 (1961).
38a. M. Takahashi, T. Sasakawa, and H. Fujimori, *J. Phys. Soc. Japan* **17**, 585 (1962); **18**, 734 (1963); *J. Appl. Phys.* **35**, 869 (1964).
38b. M. F. Collins and D. A. Wheeler, *Proc. Phys. Soc.* (*London*) **82**, 633 (1963).
39. K. Suzuki, *J. Phys. Soc. Japan* **13**, 756 (1958).
40. A. J. Bradley and A. H. Jay, *Proc. Roy. Soc.* (*London*) **A136**, 210 (1932).
41. R. Nathans, M. T. Pigott, and C. G. Shull, *J. Phys. Chem. Solids* **6**, 38 (1958).
42. G. de Vries, *Physica* **25**, 1211 (1959).
43. R. Vergne, *Compt. Rend.* **252**, 82 (1961).
44. R. Vergne, *Intern. Colloq. Magnétisme, Grenoble*, p. 190 (1958).
45. H. Uchikoshi and S. Chikazumi, unpublished data, 1953.
46. L. Néel, *J. Phys. Radium* **13**, 249 (1952).
47. S. Taniguchi, *Sci. Rept. Res. Inst. Tohoku Univ.* **A8**, 173 (1956).
48. R. D. Heidenreich, E. A. Nesbitt, and R. D. Burbank, *J. Appl. Phys.* **30**, 995 (1959); E. A. Nesbitt and R. D. Heidenreich, *ibid.* **30**, 1000 (1959).
49. E. A. Nesbitt, B. W. Batterman, L. D. Fullerton, and A. J. Williams, *J. Appl. Phys.* **36**, 1235 (1965).
50. A. I. Schindler and E. I. Salkovitz, *J. Appl. Phys.* **31**, 245S (1960); A. I. Schindler and R. H. Kernohan, *J. Phys. Soc. Japan Suppl.* B-III **18**, 314 (1963).
51. J. Paulevé, D. Dautreppe, J. Laugier, and L. Néel, *Compt. Rend.* **254**, 965 (1962); *J. Phys. Radium* **23**, 841 (1962).
52. L. Néel, *J. Appl. Phys.* **30**, 3S (1959).
53. L. Néel, J. Paulevé, R. Pauthenet, J. Laugier, and D. Dautreppe, *J. Appl. Phys.* **35**, 873 (1964).
54. W. Chambron, D. Dautreppe, L. Néel, and J. Paulevé, *Compt. Rend.* **255**, 2037 (1962).
55. J. Smit and H. P. J. Wijn, "Ferrites." Wiley, New York, 1959.
56. W. H. Meiklejohn, *J. Appl. Phys.* **33**, 1328 (1962).
57. W. Six, J. L. Snoek, and W. G. Burgers, *Ingenieur* (*Utrecht*) **49**, E195 (1934).
58. H. W. Conradt, O. Dahl. and K. J. Sixtus, *Z. Metallk.* **32**, 231 (1940).
59. G. W. Rathenau and J. L. Snoek, *Physica* **8**, 555 (1941).
60. L. Néel, *Compt. Rend.* **238**, 305 (1954).
61a. S. Chikazumi and K. Suzuki, *Phys. Rev.* **98**, 1130 (1955).
61b. S. Chikazumi, K. Suzuki, and H. Iwata, *J. Phys. Soc. Japan* **12**, 1259 (1957).

XII. DIRECTIONAL ORDER

61c. S. Chikazumi, *J. Appl. Phys.* **29**, 346 (1958).
62. H. J. Bunge and J. G. Müller, *Wiss. Z. Hochsch. Verk. Dresden* **5**, 327 (1957); H. J. Bunge, *Z. Metallk.* **49**, 40 (1958).
63. J. C. Fisher, *Acta Met.* **2**, 9 (1954).
64. S. Chikazumi, K. Suzuki, and H. Iwata, *J. Phys. Soc. Japan* **15**, 250 (1960); S. Chikazumi, *J. Appl. Phys.* **31**, 158S (1960).
65. N. Tamagawa, Y. Nakagawa, and S. Chikazumi, *J. Phys. Soc. Japan* **17**, 1256 (1962).
66a. T. Sato, thesis, Gakushuin Univ., Tokyo, 1964; *J. Phys. Soc. Japan* **21**, 1892 (1966).
66b. G. Y. Chin, *J. Appl. Phys.* **36**, 2915 (1965); *Mater. Sci. Eng.* **1**, 77 (1966).
66c. A. R. Von Neida, G. Y. Chin, and A. T. English, *J. Appl. Phys.* **39**, 610 (1968).
67. J. D. Livingston and J. J. Becker, *Trans. AIME* **212**, 316 (1958).
68. T. Mitui, M. Sato, and T. Sambongi, *J. Phys. Soc. Japan* **17**, 639 (1962); T. Mitui, *J. Phys. Soc. Japan* **15**, 929 (1960).
69. N. Tamagawa and T. Mitui, *J. Phys. Soc. Japan* **17**, 718 (1962).
70. T. Mitui and M. Sato, *J. Phys. Soc. Japan* **18**, 740 (1963).
71. N. Tamagawa and T. Mitui, *J. Phys. Soc. Japan* **20**, 1988 (1965).
72. R. Hahn and E. Kneller, *Z. Metallk.* **49**, 480 (1958).

CHAPTER XIII

The Influence of Crystal Defects on Magnetization Processes in Ferromagnetic Single Crystals

HERMANN TRAUBLE*
Max Planck Institute for Metals Research
Institute of Physics
Stuttgart, Germany

INTRODUCTION	622
DISLOCATIONS IN SINGLE CRYSTALS	625
Plastic Deformation	625
Recovery	628
INTERACTIONS BETWEEN DISLOCATIONS AND MAGNETIZATION	628
Survey	628
Interactions between Plane 180° Bloch Walls and Single Dislocations Parallel to the Wall Planes	630
Interactions between 180° Bloch Walls and Configurations of Several Dislocations	635
Interactions between Dislocations and Nearly Uniform Magnetization	641
DOMAIN STRUCTURE	646
Magnetically Uniaxial Crystals (Cobalt)	647
Magnetically Multiaxial Crystals (Silicon–Iron)	647
THEORY OF THE MAGNETIZATION CURVE	650
General Theory	650
Bloch Wall Displacement in Real Crystals	652
Coercive Field, H_c	655
Initial Susceptibility, χ_i	658
The Influence of the Dislocation Arrangement on Coercive Field and Initial Susceptibility	661
Approach to Saturation	666

* Present address: Max-Planck Institut für Physikalische Chemie, Göttingen, Germany.

EXPERIMENTAL RESULTS	667
General	667
Coercive Field, H_c	668
Initial Susceptibility, χ_i	677
Reversible Susceptibility, χ_{rev}	682
Approach to Saturation	683
References	685

Introduction

1. Many of the properties of large ferromagnetic crystals* depend on structural defects. As examples, magnetic hysteresis and ferromagnetic aftereffects are phenomena which are caused primarily by interactions between the magnetization and lattice imperfections. Table I shows the various types of structural defects in pure crystals, the most important methods for producing them, and some of the related magnetic effects.

What, then, are the causes of an interaction between a lattice defect and the magnetization? In many cases the principal contribution originates from the magnetostrictive coupling, which is determined by the magnetostriction constants and the stress field of the structural defects in the crystal. In this connection, therefore, the dislocations, which are the most important sources of internal stresses, play a dominant role. Another contribution arises from the fact that in the neighborhood of lattice defects the so-called crystal properties, such as spontaneous magnetization, magnetocrystalline anisotropy, etc., may be changed. Due to the magnetoelastic coupling and the change in crystal properties, the magnetization distribution may be changed in the vicinity of lattice defects.

As is well known, the elementary processes of magnetization changes are Bloch wall displacement and magnetization rotation. In the simplest case, i.e., in single crystals, the contributions of these processes to the overall magnetization change are uniquely determined by the external field. In the case of magnetically multiaxial monocrystals (e.g., iron), magnetization changes in small fields are mainly due to Bloch wall displacement. At higher fields both Bloch wall displacement and magnetization rotation

---

* The term signifies crystals which, in a demagnetized state, are subdivided into magnetic domains.

† The fundamentally different situation in magnetically uniaxial crystals is discussed in Section 24.

XIII. CRYSTAL DEFECTS IN FERROMAGNETIC SINGLE CRYSTALS

TABLE I

THE MOST IMPORTANT STRUCTURAL DEFECTS IN PURE SINGLE CRYSTALS, THE COMMON WAYS IN WHICH THEY MAY BE PRODUCED, AND THEIR EFFECTS ON MAGNETIC PROPERTIES

Type of defect	Methods for producing the defects	Magnetic parameters which are influenced by the defects
Dislocations (line defects, most important sources of internal stresses)	Plastic deformation	Shape and characteristic quantities (χ_i, χ_{rev}, χ_s, α, H_c, M_R) of the magnetization curve; magnetic hysteresis; Barkhausen spectrum
Stacking faults	Plastic deformation	—
Point defects (vacancies, interstitial atoms)	Quenching, irradiation with electrons, plastic deformation	Time dependence of all parameters that are connected with Bloch wall movements; ferromagnetic lag
Multiple vacancies and other agglomerates of point defects	Plastic deformation, quenching, irradiation with electrons or neutrons	Shape and characteristic quantities of the magnetization curve; magnetic hysteresis
Defect zones [1-4]	Irradiation with neutrons	Shape and characteristic quantities of the magnetization curve; magnetic hysteresis

play a role, until finally only rotation will occur.[†] This behavior may be qualitatively understood as follows. In a crystal without lattice defects Bloch walls are almost completely free to move. Hence, in weak external fields, the magnetization changes are brought about essentially by Bloch wall displacement, and a very rapid increase in the overall magnetization is observed. In contrast, when homogeneous rotation occurs, it is necessary to overcome the generally high magnetocrystalline anisotropy. Thus, homogeneous rotation occurs only in high fields, and the slope of the corresponding portion of the magnetization curve is small. Therefore, the motion and the mobility of Bloch walls are rather sensitive to lattice defects, whereas the rotation process is not, especially when the crystalline anisotropy is sufficiently large. In fact, magnetic hysteresis occurs mainly in the steep portion of the magnetization curve, where Bloch wall displacement predominates. Also, ferromagnetic lag phenomena are connected with the existence of Bloch walls.

Although the strong influence of structural defects in a crystal on the magnetization processes has long been known [5], it became possible only recently to understand quantitatively the physical background. This is due, in the first place, to the fact that the nature and properties of various structural defects in crystals have been investigated only recently to such an extent that it became possible to treat systematically their effects on magnetization processes. More specifically, we are now able to alter, in a controlled manner, the density and geometrical arrangements of the various lattice defects by means of plastic deformation, quenching, recovery, and irradiation, and thus can assess quantitatively their magnetic effects. More recently, by application of the micromagnetic concept methods have been developed which make it possible to calculate the interactions between lattice defects and magnetization, and particularly between Bloch walls and dislocations. However, quantitative comparison of experiment and theory are yet restricted to single crystals, where our knowledge of structural defects is most advanced. In addition, single crystals have a relatively simple magnetic domain structure, and the magnetization processes with increasing field strength follow simple laws. We shall therefore limit the following discussion to single crystals. In polycrystals, the conditions are basically more complex, mainly because there is a strong magnetostatic coupling between the grains which is difficult to treat mathematically.

2. In this chapter we shall primarily be concerned with the influence of dislocations, on the static magnetization curves of pure metal single crystals. Particular attention will be paid to the dependence on temperature and deformation of coercive field and initial susceptibility. The influence of point defects on magnetization processes has been explained elsewhere in a study of lag phenomena [6, 87] (see also Chapter XIV).

Our first step towards an interpretation of the magnetization curve of real crystals is the investigation of the interaction between individual dislocations and the magnetization. In this connection, we shall also discuss the effect of specific, frequently occurring dislocation arrangements on the displacement of a Bloch wall. We shall then turn to the question of how the domain structure and the contributions to magnetization changes of Bloch wall motion and rotation depend on field strength (Secs. 19–23 and 24). Sections 25–30 are devoted to the basic problem of a theory of the magnetization curve, namely to the study of the displacement of a Bloch wall in a crystal with lattice defects; this requires statistical considerations of domain and defect structures. The comparison of theory with experimental results on cobalt, iron, and nickel single crystals is given in Secs. 34–40. It will show to what extent our present theoretical concepts are validated, and where unsolved problems still remain. Besides the problem of the effects of lattice defects on magnetization processes, considerable interest is taken in the complementary question, namely, to what extent can magnetic measurements be used to determine the nature, the density, and the geometrical arrangement of structural defects. In this respect, measurements of the initial susceptibility, the coercive field, and, in particular, the approach to magnetic saturation give valuable information [6, 87]. We shall go into more details of this question where appropriate.

Dislocations in Single Crystals

3. The most important lattice defects with respect to the study of the magnetization process in pure metal single crystals are the dislocations. An individual dislocation causes stresses whose magnitude varies inversely with the distance r from the dislocation line, i.e., as $1/r$. The density i.e., the number of dislocations per square centimeter or their total length per cubic centimeter and the arrangement of the dislocations can be changed in a controlled manner: (1) by plastic deformation and (2) by annealing the crystals.

Plastic Deformation

4. The mechanism of plastic deformation is the movement of dislocations. The elastic interactions between the dislocations give rise to the work hardening of the crystal. This phenomenon is described by the work hardening curve $\tau(a)$, which shows the relationship between the shear stress τ, and the shear strain a.

For face-centered cubic single crystals with initially intermediate orientations (e.g., nickel) the work hardening curve (Fig. 1) exhibits three regions, I, II, and III, each having a different slope. The dislocation processes occurring in each of these three regions were studied in detail by electron microscopy, x-rays, and other methods [7–11, 15, 59, 88]. These investigations have resulted in the following concepts and relations between the total dislocation density, ϱ, and the shear stress, τ.

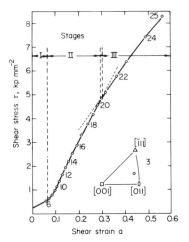

FIG. 1. Work hardening curve of an intermediately oriented nickel single crystal (tensile stress experiment at room temperature). The magnetic measurements reproduced in Figs. 23 and 31 were obtained after removing the external load at the points 1–25.

5. Region I. Here, only a single slip system, the so-called primary system, is being activated. With respect to work hardening, the dislocations act as individual sources of stress. Dislocations in closely neighboring slip planes may form dislocation dipoles [12–15, 59]. The density per unit volume, N, of dislocation sources remains constant, whereas the number of dislocations, n, produced per source increases linearly with shear stress, following the relation [16]

$$n = [2\pi/bG(L_1 N/L_2^2)^{1/4}](\tau - \tau_0), \tag{5.1}$$

where τ is the resolved shear stress, τ_0 is the critical shear stress, G is the shear modulus, b is the magnitude of the Burgers vector, and L_1 and L_2 are the slip line lengths of edge and screw dislocations, respectively; $L_1 \simeq L_2$. If we simplify the picture by considering the dislocations produced by the sources as dislocation rings with diameter $d \simeq L_1 \simeq L_2$, the dislocation density ϱ is given by

$$\varrho = \pi d N n, \tag{5.2}$$

where n is given by Eq. (5.1). Thus, region I is characterized by the proportionality

$$\varrho \propto (\tau - \tau_0). \tag{5.3}$$

For example, hexagonal cobalt shows up to very high amounts of slip

($a \simeq 1.0$–1.5), a behavior that is characteristic of region I, with a linear work hardening curve [17, 59]

$$\tau - \tau_0 = \vartheta a, \tag{5.4}$$

where ϑ is the work hardening coefficient (approximately equal to 1 kp mm^{-2} at room temperature). According to electron microscope investigations by Thieringer [18, 59] the dislocations formed in cobalt during plastic deformation are fairly uniformly distributed. However, large numbers of screw dislocations are frequently observed to lie near each other in the same slip plane. In cobalt the dislocation density ϱ_0 of the undeformed crystals is relatively high. This is due to the phase transformation at 417°C from the fcc lattice into the hexagonal structure. Let ϱ_v represent the density of the dislocations that originate during deformation. ϱ_v increases as the deformation increases, according to Eqs. (5.1) and (5.2). We then have the total dislocation density,

$$\varrho_\sigma = \varrho_0 + \varrho_v. \tag{5.5}$$

According to electron microscope investigations [17, 18] the total dislocation density ϱ_σ in intermediately oriented cobalt crystals increases with increasing shear strain a according to the equation

$$\varrho_\sigma = \varrho_0 + \varrho_v = 3.5 \times 10^9 (0.0286 + a) \quad \text{cm}^{-2} \tag{5.6}$$

6. Region II. In this region, in addition to the primary system secondary slip systems are being activated. The interactions between dislocations in the various slip systems give rise to the formation of obstacles in the primary slip plane; these strongly impede dislocation motion. Hence, the dislocation pattern becomes far more complicated than it is in region I. There develop dislocation networks, deformation bands, dislocation pileups with long-range stress fields, etc. [15, 19, 21, 88]. In a simplified fashion, the processes in region II may be described as follows [7, 22]. The number of the dislocation sources, N, increases with increasing slip, with each source producing a constant number of dislocations, n. The total dislocation density increases with deformation as

$$\varrho = (16\pi^2/G^2 b^2 n^2)\tau^2. \tag{6.1}$$

Region II is therefore characterized by

$$\varrho \propto \tau^2. \tag{6.2}$$

The grown-in dislocations are no longer of importance.

7. *Region III.* Region III is characterized by a gradual reduction of the work hardening coefficient. This decrease of slope is due to the fact that at high enough stresses thermally activated cross slip of screw dislocations begins which opposes work hardening [22]. Equation (6.2) holds also in region III of the work hardening curve.

For intermediately oriented, bcc single crystals, such as iron, for example, the work hardening curve, like that of the fcc single crystals, exhibits three regions, and it is likely that Eqs. (5.3) and (6.1) are valid there, too [24]. However, in contrast to the fcc crystals, there are generally several slip systems contributing to plastic slip, such that in detail the process may become quite involved.

Recovery

8. The lattice defects that have been originated during plastic deformation, quenching, or irradiation may in part be eliminated by annealing the crystals at sufficiently high temperatures. In general, recovery occurs in several steps at different temperatures. In the ideal case, at each step only one type of defect will be eliminated totally or partially [23, 26, 27]. Internal stresses can be reduced in a particularly effective way by changing the density and arrangement of dislocations. However, the elementary processes of recovery have not yet been investigated as extensively as those of plastic deformation. When a crystal is annealed at sufficiently high temperatures where self-diffusion is possible (recovery stage V), the dislocations can move over large distances, and depending on sign, may mutually annihilate each other. One then observes a large decrease of the dislocation density and hence of the internal stresses. Some reduction of internal stresses can also occur at considerably lower temperatures, namely, at the temperature of vacancy migration (recovery stage IV). In this case, vacancies condense at dislocations and thus cause movements of the dislocations. Internal stresses are then relieved by the climbing of dislocations, by their rearrangement into more stable configurations (dipoles, dislocation walls), or by the decomposition of dislocation groups. In contrast to the annealing in Stage V, the dislocation density does not decrease substantially in this case.

Interactions between Dislocations and Magnetization

Survey

9. The magnetostrictive interaction between the magnetization and dislocations is described by the magnetoelastic coupling energy Φ_{el}. For crystals with cubic symmetry Φ_{el} is given by [5]

$$\Phi_{\text{el}} = -\tfrac{3}{2}\lambda_{100}(\gamma_1{}^2\sigma_{11} + \gamma_2{}^2\sigma_{22} + \gamma_3{}^2\sigma_{33}) - 3\lambda_{111}(\gamma_1\gamma_2\sigma_{12} + \gamma_2\gamma_3\sigma_{23} + \gamma_3\gamma_1\sigma_{31}). \tag{9.1}$$

The σ_{ij} are the components of the stress tensor of the dislocation, λ_{100} and λ_{111} are the magnetostriction constants, and the γ_i, are the direction cosines of the magnetization with respect to the cube axes. Vicena [28] was the first one to calculate the magnetoelastic coupling energy between a Bloch wall and a dislocation; he considered a plane 180° Bloch wall and a straight dislocation running parallel to the plane of the wall. However, much simpler is the complementary procedure, in which one first calculates the force, p^v, which a dislocation experiences in the stress field of the Bloch wall. One then obtains the force on the Bloch wall, $p^\omega = -p^v$. According to Peach and Koehler [29]* p^ω is given by

$$p^\omega = -\int (d\mathbf{f} \times \nabla) \times \sigma^M \mathbf{b}. \tag{9.2}$$

Using Stoke's theorem, this integral may be transformed to give

$$p^\omega = \int d\mathbf{l} \times \sigma^M \mathbf{b}. \tag{9.3}$$

Here $\mathbf{b}$ is the Burgers vector, the σ^M are the components of the extra stress tensor of the Bloch wall, $d\mathbf{l}$ is a line element of the dislocation, and $d\mathbf{f}$ is a surface element. In Eq. (9.2) one has to integrate over the surfaces bordered by the dislocation, and in Eq. (9.3) the integral must be taken over the entire length of the dislocation line.

Rieder [31] has calculated the extra magnetostrictive stresses for the most important types of plane Bloch walls in iron, nickel, and cobalt. In Cartesian coordinates with the z axis perpendicular to the plane of the Bloch wall, the interaction of the wall with any particular source of internal stresses is determined by the components σ^M_{11}, σ^M_{12}, σ^M_{22} of the extra stress tensor.

For a single, straight dislocation parallel to the plane of the Bloch wall, with the line vector, $\mathbf{l}$ (with components l_x, l_y, and l_z), and with the Burgers vector, $\mathbf{b}$ (with components b_x, b_y, b_z), Eq. (9.3) gives the force p_z exerted by the dislocation on the Bloch wall in the direction of its normal,

$$p_z = \sigma^M_{11} l_y b_x - \sigma^M_{22} l_x b_y + \sigma^M_{12}(l_y b_y - l_x b_x) \tag{9.4}$$

Accordingly, this force only depends on the projections of the Burgers vector and the line vector of the dislocation on the plane of the Bloch wall; it therefore vanishes when the dislocation or its Burgers vector are perpendicular to the plane of the Bloch wall.

* As to the choice of sign for this equation, refer to a note by De Wit [30].

10. For the understanding of the magnetization processes in real crystals it is important to distinguish between type-I and type-II Bloch walls.

Type-I Bloch walls produce long-range internal stresses which are constant in the neighboring regions of a wall. The magnitudes of the stresses depend, aside from the elastic constants, only on the magnetostriction constants in the easy directions. Examples are the (110) 90° and the (111) 90° Bloch walls in iron, and the (100) 71°, (110) 71°, (110) 109°, and (111) 109° Bloch walls in nickel.

On the other hand, the internal stresses of *type-II Bloch walls* are confined essentially to the wall itself, if one disregards the extremely small residual stresses in the adjacent domains. This type includes all of the Bloch walls for which the projections of the magnetization vectors $\mathbf{M}_1$ and $\mathbf{M}_2$ on either side of the wall fall on a common line in the plane of the Bloch wall. The extra stresses accompanying the inhomogeneous magnetization distribution within the Bloch walls are responsible for the interaction between the walls and the dislocations. Hence, type-II walls interact magnetostrictively only with lattice defects within the wall whereas type-I walls interact with all of the defects in the crystal. Thus, type-II walls are far more mobile than type-I walls. For example, in deformed silicon–iron single crystals at room temperature, the type-II walls are some 50 times more mobile than type-I walls [32].

This leads to important consequences for the understanding of the magnetization curve. In particular, magnetization changes in small fields, are primarily due to displacements of type-II walls, particularly 180° walls (see also Secs. 19–23); whereas the displacement of type-I Bloch walls requires a higher external field strength.

Interactions between Plane 180° Bloch Walls and Single Dislocations Parallel to the Wall Planes

11. Iron. Consider the interaction between a (110) 180° Bloch wall and a straight dislocation parallel to the plane of the Bloch wall with a $\langle 111 \rangle$ Burgers vector parallel to the plane of the wall (Fig. 2). Let $\mathbf{l} = l\,(i \cos \alpha + j \sin \alpha)$ be the line vector of the dislocation, where α represents the angle between $\mathbf{l}$ and the positive x axis. The preferred magnetic directions are $\langle 100 \rangle$ directions; the magnetization on either side of the wall is parallel to the x axis. The spin configuration in the Bloch wall is represented by $\phi(z)$, where ϕ is the angle between the magnetization direction and the positive x axis.

As the Bloch wall moves across such a dislocation, the force p_z, from

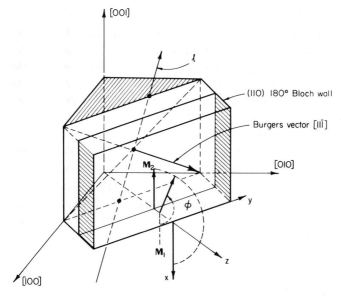

FIG. 2. Illustration referring to the interaction between a (110) 180° Bloch wall and a dislocation line in a {110} plane with a ⟨111⟩ Burgers vector: l is the direction of the dislocation line.

Eq. (9.4) is

$$p_z = \sigma_{11}^M b_x l \sin\alpha - \sigma_{22}^M b_y l \cos\alpha + \sigma_{12}^M (b_y l \sin\alpha - b_x l \cos\alpha). \quad (11.1)$$

The components σ_{11}^M, σ_{22}^M, and σ_{12}^M of the extra stress tensor of a (110) 180° Bloch wall can be written as follows, according to Rieder [31]:

$$\sigma_{11}^M = -[\tfrac{3}{2}\lambda_{100}(-c'_{11} - c'_{12}/2) + \tfrac{3}{4}\lambda_{111}c'_{12}]\sin^2\phi,$$
$$\sigma_{22}^M = -[\tfrac{3}{2}\lambda_{100}(-c'_{12} + c'_{22}/2) + \tfrac{3}{4}\lambda_{111}c'_{22}]\sin^2\phi,$$
$$\sigma_{12}^M = -\tfrac{3}{2}\lambda_{111}c_{44}\sin 2\phi. \quad (11.2)$$

The c'_{ij} are the elasticity constants as transformed into the Bloch wall system; they are connected with the conventional Voigt's elastic constants c_{ik} by

$$c'_{11} = c_{11} - \frac{2c_{12}^2}{c_{11} + c_{12} + 2c_{44}}; \quad c'_{22} = \frac{4c_{44}(c_{11} + c_{12})}{c_{11} + c_{12} + 2c_{44}};$$
$$c'_{12} = \frac{4c_{12}c_{44}}{c_{11} + c_{12} + 2c_{44}}. \quad (11.3)$$

According to Radeloff [33] at room temperature $\lambda_{100} = 22.2 \times 10^{-6}$, λ_{111}

$= -20.7 \times 10^{-6}$, and according to Lee [34] $c_{44} = 1.12 \times 10^{12}$ dyn cm^{-2}, $c_{11} = 2.41 \times 10^{12}$ dyn cm^{-2}, and $c_{12} = 1.46 \times 10^{12}$ dyn cm^{-2}.

We now calculate the force p_z for edge and for screw dislocations with Burgers vectors $\pm \langle \bar{1}1\bar{1} \rangle$. For edge dislocations

$$p_z = \pm \{\tfrac{1}{3} bl\sigma_{11}^M + \tfrac{2}{3} bl\sigma_{22}^M + (2\sqrt{2}/3) bl\sigma_{12}^M \}. \tag{11.4}$$

From Eqs. (11.2)–(11.4) and the data given above, we obtain the force exerted on the Bloch wall along the z axis per unit length of dislocation

$$p_z(\phi) = \pm \{4.03 \sin^2 \phi + 3.28 \sin 2\phi\} \times 10^7 b \quad \text{dyn}. \tag{11.5}$$

The sign depends on the directions of the Burgers vector and the line vector; for the configuration shown in Fig. 2, the correct sign is $(+)$.

Similarly, we obtain for screw dislocations

$$p_z = \pm \{\tfrac{2}{3} bl\sigma_{11}^M - \sqrt{2/3}\, bl\sigma_{22}^M + \tfrac{1}{3} bl\sigma_{12}^M \}, \tag{11.6}$$

which, at room temperature, gives

$$p_z = \pm \{6.86 \sin^2 \phi + 1.16 \sin 2\phi\} \times 10^7 b \quad \text{dyn}. \tag{11.7}$$

The curves $p_z(\phi)$ as given by Eqs. (11.5) and (11.6) are shown in Fig. 3.

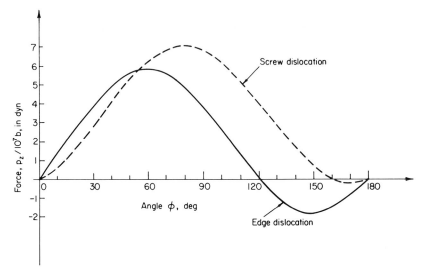

FIG. 3. Force, p_z, at room temperature as a function of the displacement of a (100) 180° Bloch wall in iron going through respectively an edge or screw dislocation, both parallel to {110} planes (see Fig. 2).

From these curves one finds the $p_z(z)$ dependence by taking into account the spin configuration $\phi(z)$ in the (110) 180° Bloch wall [35].

12. Cobalt. For $T < 518°K$, cobalt has hexagonal structure and is magnetically uniaxial (c axis is the preferred magnetic axis). Hence, there exist only 180° Bloch walls, all of which are perpendicular to the basal plane. The dislocations that are responsible for plastic deformation lie in the basal plane and have Burgers vectors of the type $\frac{1}{3}\langle\bar{2}110\rangle$. We calculate the force experienced by a 180° Bloch wall that moves across straight dislocation lines that are parallel to the plane of the wall (see Fig. 4). We consider (1)

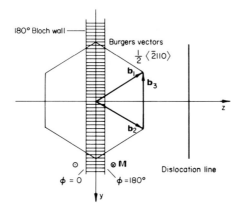

Fig. 4. Cobalt: interaction between a 180° Bloch wall with dislocations in the basal plane. The section is parallel to the basal plane.

the case of a pure screw dislocation (Burgers vector $\mathbf{b}_3$) and (2) a dislocation with a 60° edge component (Burgers vectors $\mathbf{b}_1$ and $\mathbf{b}_2$). Since, in each case, only the l_y and b_y components of, respectively, the dislocation line vector and the Burgers vector do not vanish (see Fig. 4), Eq. (9.4) reduces to

$$p_z = \sigma_{12}^M l_y b_y. \tag{12.1}$$

According to Rieder [31]

$$\sigma_{12}^M = -\lambda_{44} c_{44} \sin\phi \cos\phi,$$

where λ_{44} is a magnetostriction constant and c_{44} is a Voigt elasticity constant. As above, ϕ represents the angle between the magnetization direction and the positive x direction. Hence, from Eq. (12.1)

$$p_z(\phi) = \pm c \sin 2\phi, \tag{12.2}$$

where
$$c = (b/2)\lambda_{44}c_{44} \tag{12.2a}$$
for a screw dislocation, and
$$c = (b/4)\lambda_{44}c_{44} \tag{12.2b}$$
for mixed dislocations.

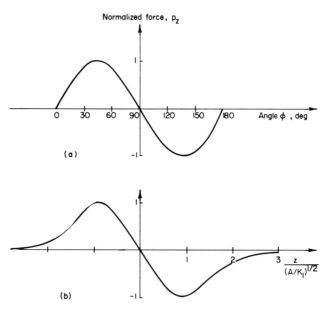

FIG. 5. Force, p_z, as a function of the displacement of a 180° Bloch wall in cobalt going through respectively an edge or screw dislocation in the basal plane, corresponding to Fig. 4: (a) the variable is the angle ϕ between the magnetization in the Bloch wall and the x axis; (b) the magnetization configuration $\phi(z)$ in the Bloch wall is taken into account.

The quantity $p_z(\phi)$ is shown in Fig. 5a for both cases. The magnetization configuration $\phi(z)$ in the Bloch wall is given by [36]

$$dz/d\phi = A^{1/2}(K_1 + K_2 \sin^2 \phi)^{-1/2}(\sin \phi)^{-1}, \tag{12.3}$$

where A is the exchange constant, and K_1 and K_2 are the constants of crystalline anisotropy. Neglecting K_2, we obtain the function $p_z(z)$ as shown in Fig. 5b. In many other cases, as, for example, in the case of the interaction between a $(\bar{1}\bar{1}2)$ 180° Bloch wall in nickel with a screw dislocation in the [110] direction [37], one also obtains a $p_z(z)$ curve which is point symmetrical with respect to the center of the Bloch wall, i.e., to $\phi = 90°$.

Interactions between 180° Bloch Walls and Configurations of Several Dislocations

13. Electron microscopy investigations [15, 19–21, 59] have shown that dislocations are by no means arranged in a completely random fashion, particularly in deformed crystals. In particular, certain energetically favorable configurations, such as *dislocation dipoles* and *tripoles, dislocation walls*, etc. occur frequently. Furthermore, the piling up of dislocations at some obstacles may result in the formation of *groups of dislocations* of the same kind. When a crystal contains a large number of such dislocation configurations, and when the mean spacing between dislocations in the direction of the normal to the Bloch wall is smaller than the thickness of the wall, then the walls will sense the existence of such specific dislocation configurations. In the following we shall calculate the resultant interaction force, p_z, between a 180° Bloch wall and (1) dislocation dipoles and (2) dislocation groups. Any possible influence of the dislocations on the magnetization distribution in the Bloch wall will be neglected; a solution of this problem has recently been given by Pfeffer [89].

14. Dislocation Dipoles. Figure 6 presents a dilatation dislocation dipole composed of two edge dislocations; its parameters are the distance between the slip planes, h (length of the dipole), and the angle of orientation, α, which, in the absence of an external stress, is equal to $\pi/4$ or to $3\pi/4$. The

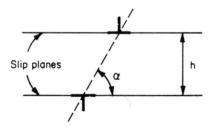

Fig. 6. Sketch of a dilation dislocation dipole as formed by two edge dislocations: h is the length of the dipole.

interaction between a Bloch wall and such a dipole depends on the nature of the Bloch wall and the dislocations, and further on the spacing, d, between the individual dislocations along the normal to the Bloch wall.

Let us first consider the force function when a Bloch wall is shifted across an arrangement of two dislocations parallel to the plane of the Bloch wall, with the spacing d perpendicular to the Bloch wall (Fig. 7a). According to Sec. 11, we take the force of the individual dislocations as proportional to sin 2ϕ (see Fig. 5). This is true, for example, for magnetically uniaxial cobalt where 180° Bloch walls are perpendicular to the slip planes, and where one has screw dislocations and 60° dislocations in the slip plane.

Consider the two cases given in Fig. 7b.

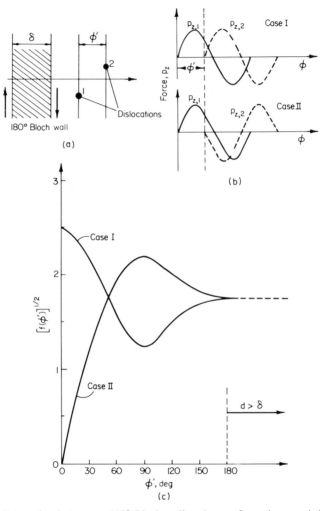

FIG. 7. Interaction between a 180° Bloch wall and a configuration consisting of two dislocations. The quantities ϕ' and d characterize the dislocation spacing perpendicular to the Bloch wall. (a) Geometric relation; (b) force, p_z, corresponding to two individual dislocations; (c) variation of the function $[f(\phi')]^{1/2}$ with dislocation spacing, where $\phi' = \pi d/\delta \sim$ distance of the dislocations in the z direction [see Eq. (14.4)]. δ is the thickness of the Bloch wall.

XIII. CRYSTAL DEFECTS IN FERROMAGNETIC SINGLE CRYSTALS

Case 1. Both dislocations have the same sign. In that case

$$p_{z,1}(\phi) = c \sin 2\phi \quad \text{and} \quad p_{z,2}(\phi) = c \sin 2(\phi - \phi'),$$

where $\phi(z)$ gives the magnetization distribution within the Bloch wall, and ϕ' characterizes the spacing between the dislocations along the z direction. If we assume a linear variation of the magnetization direction in the wall with thickness δ,

$$\phi = (\pi/\delta)z,$$

then

$$\phi' = (\pi/\delta)d. \tag{14.1}$$

Case 2. The two dislocations have opposite sign. This gives

$$p_{z,1}(\phi) = c \sin 2\phi \quad \text{and} \quad p_{z,2}(\phi) = -c \sin 2(\phi - \phi').$$

The resultant force function p_z is obtained by adding up the individual forces. With regard to the discussions of the coercive field and the initial susceptibility (Secs. 24–40), we consider the average values $M(p_z)$ and $M(p_z^2)$ as functions of the dislocation spacing, d, or ϕ', respectively. $M(p_z)$ is obviously equal to zero. $M(p_z^2)$ may be calculated from

$$M(p_z^2) = (d + \delta)^{-1} \int p_z^2(z)\,dz, \tag{14.2}$$

where $d + \delta$ characterizes the range of the $p_z(z)$ function. For a linear variation of spin direction within the Bloch wall [see Eq. (14.1)] we have

$$M(p_z^2) = (\pi + \phi')^{-1} \int p_z^2(\phi)\,d\phi. \tag{14.3}$$

We may therefore write $M(p_z^2)$ in the form

$$M(p_z^2) = c^2/(\pi + \phi') \left\{ \int_0^{\phi'} \sin^2 2\phi\, d\phi \right.$$
$$+ \int_{\phi'}^{\pi} [\sin 2\phi \pm \sin 2(\phi - \phi')]^2\, d\phi$$
$$\left. + \int_{\pi}^{\pi+\phi'} \sin^2 2(\phi - \phi')\, d\phi \right\},$$

using the plus sign in the first case, and the negative sign in the second. The evaluation yields

$$M(p^2) = [c^2/(\pi+\phi')]\{\pi + [(\pi - \phi')\cos 2\phi' + \tfrac{1}{4}\cos 2\phi' \sin 4\phi' + \tfrac{1}{2}\sin^3 2\phi']\}$$
$$= [c^2/(\pi + \phi')]f(\phi'), \tag{14.4}$$

where the factor c (for cobalt) is given by Eq. (12.2). As will be seen later, the parameters of the magnetization curve, coercive field, and initial susceptibility are proportional to $f^{1/2}(\phi')$ and $1/f^{1/2}(\phi')$, respectively, where $f(\phi')$ is defined by Eq. (14.3). The function $f^{1/2}(\phi')$ is shown in Fig. 7c.

As a second example, we shall discuss the interaction between a (110) 180° Bloch wall in iron and a dislocation dipole which consists of two edge dislocations with opposite signs, as shown in Fig. 8 (see Sec. 11). As above, we shall assume the spin variation in the Bloch wall to be linear and shall characterize the dislocation spacing along the normal to the Bloch wall (along the z axis) by the angle $\phi' = \pi d/\delta$.

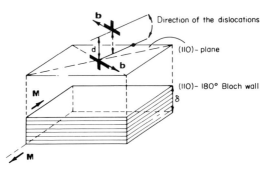

FIG. 8. Iron: interaction between a (110) 180° Bloch wall and an edge dislocation dipole (see Fig. 2). The dislocations lie in {110} planes.

The resultant force function is obtained by adding of the individual forces $p_z = p_{z,1} + p_{z,2}$ with

$$p_{z,1} = 10^7 b(4.03 \sin^2 2\phi + 3.28 \sin 2\phi) \quad \text{dyn cm}^{-1},$$
$$p_{z,2} = -10^7 b\{(4.03 \sin^2 2(\phi - \phi') + 3.28 \sin 2(\phi' - \phi')\} \quad \text{dyn cm}^{-1},$$

According to Sec. 11 again, the mean value $M(p_z)$ vanishes, while $M(p_z^2)$ must be calculated according to Eqs. (14.2) or (14.3).

With the previously given room temperature values of the constants λ and c one obtains

$$\begin{aligned} M(p_z^2) &= [10^{14} \pi b^2/(\pi + \phi')]\{22.92 - (2/\pi)[(\pi - \phi')(11.47\beta^2 - 3.35\alpha^2) \\ &\quad + 18.17\alpha^3\beta + 11.47\alpha\beta^3 + 26.40\alpha^2\beta^4 + 6.70\alpha^3\beta^3 - 6.70\alpha\beta^5 \\ &\quad - 13.40\alpha^5\beta + 13.20\alpha^6 - 13.20\alpha^4\beta^4 + 13.20\alpha^4]\} \\ &= [10^{14} \pi b^2/(\pi + \phi')] f(\phi') \end{aligned} \qquad (14.5)$$

for the range $0 \leq \phi' \leq \pi$.

In the above equation the substitutions $\beta = \cos \phi'$ and $\alpha = \sin \phi'$ were used. Figure 9 shows the variation of the function $f(\phi')^{1/2}$ with dislocation spacing.

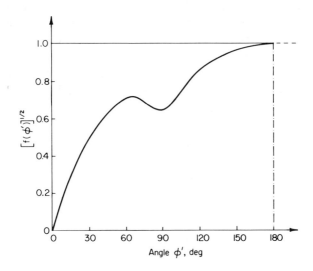

FIG. 9. Interaction between a (110) 180° Bloch wall and an edge dislocation dipole in iron corresponding to Fig. 8. Variation of the function $[f(\phi')]^{1/2}$ [see Eq. (14.5)] with dislocation spacing ϕ' or d, respectively, where $\phi' = \pi d/\delta \sim$ spacing between the dislocation in the direction of the normal to the Bloch wall and $[f(\phi')]^{1/2}$ is normalized to the maximum value 1.

15. Interactions between Bloch Walls and Dislocation Groups. As has already been mentioned, plastic deformation of crystals may lead to the formation of dislocation groups. The arrangements of dislocations in such groups have also been calculated by Eshelby *et al.* [38] and by Liebfried [39] for the case in which the dislocations are pushed by shear stresses against an obstacle. In such cases the dislocation density is especially high at the head of the group, i.e., near the obstacle. Actually such dislocation groups have only rarely been observed in deformed unloaded crystals.* According to electron micrographs it appears to be more realistic to assume the dislocations within a group to be equally spaced. The configuration that will be studied in the following is shown in Fig. 10a. A plane perpendicular to a 180° Bloch wall contains a group of n equally spaced dislocations with

---

* Recently such dislocation groups have been observed by Mughrabi [88] in deformed copper single crystals using transmission electron microscopy. In this paper the dislocation arrangement of the deformed crystals had been fixed by neutron bombardment before unloading the crystals.

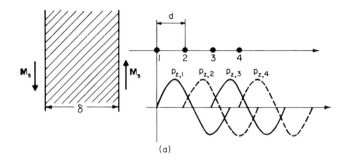
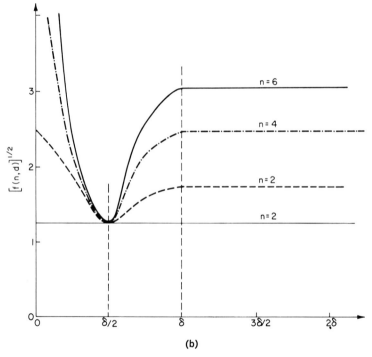

Fig. 10. Interaction between a 180° Bloch wall and a group of n equally spaced dislocations in a plane perpendicular to the plane of the Bloch wall. The force curve p_z for the individual dislocations has been taken according to Fig. 5 (point symmetrical): (a) geometric relations; (b) variation of the function $f(n, d)$ as defined by Eq. (15.1b), where m is the number of dislocations per group and d is the spacing between dislocations in the group.

their line vectors parallel to the Bloch wall. The resultant force function p_z is obtained by adding up the individual forces,

$$p_z = p_{z,1} + p_{z,2} + \cdots + p_{z,n}.$$

Let the functions $p_{z,i}$ be point symmetrical with respect to the center of the Bloch wall. Accordingly, and with respect to Sec. 11 we write

$$\begin{aligned}
p_{z,1} &= c \sin 2\phi, & 0 &\leq \phi \leq \pi \\
p_{z,2} &= c \sin 2(\phi - \phi'), & \phi' &\leq \phi \leq \phi' + \pi \\
p_{z,n} &= c \sin 2(\phi - (n-1)\phi'), & (n-1)\phi' &\leq \phi \leq (n-1)\phi' + \pi.
\end{aligned}$$

It is obvious that the mean value $M(p_z)$ vanishes. For a linear change of spin direction in the Bloch wall, $M(p_z^2)$ may be calculated from

$$M(p_z^2) = [\pi + (n-1)\phi']^{-1} \int p_z^2(\phi)\, d\phi. \tag{15.1}$$

The calculation yields

$$\{M(p_z^2)[\pi+(n-1)\phi']\}/c^2 = \begin{array}{l} (\pi/2)n^2, \quad \text{for} \quad \phi' = 0 \\ n\pi/2, \quad \text{for} \quad \phi' \leq \pi \\ n\pi/2 + (n-1)v(\phi'), \\ \quad \text{for} \quad \pi/2 \leq \phi' \leq \pi \\ n\pi/2 + (n-1)v(\phi') + (n-2)v(2\phi'), \\ \quad \text{for} \quad \pi/4 \leq \phi' \leq \pi/2. \end{array} \tag{15.1a}$$

where $v(\phi')$ is defined by Eq. (14.4).

In connection with applications the expression

$$(M(p_z^2)[\pi + (n-1)\phi'/c^2])^{1/2} = [f(n,d)]^{1/2} \tag{15.1b}$$

will be of interest. Its dependence on the dislocation spacing and on the size of the group are indicated in Fig. 10b.

It is possible, in very much the same way, to study dislocation groups that obey other distribution laws and are composed of other kinds of dislocations. A study of the interaction between correlated dislocation arrangements and Bloch walls taking into account the spin rearrangement in the walls has recently been published by Pfeffer [89].

Interactions between Dislocations and Nearly Uniform Magnetization

16. General. In ideal crystals, it is a good approximation to assume that the magnetization is uniform in zero field within each individual magnetic domain, and in a sufficiently high external field, within the entire crystal. For $H = 0$, the spontaneous magnetization is directed along one of the so-called "preferred" or "easy" directions (directions in which the magnetocrystalline anisotropy energy has a minimum), whereas at $H > 0$, the magnetization directions are determined by the potential energy, the magnetocrystalline anisotropy, and the magnetostatic energy. When an increasing external field is applied, the magnetization in the individual domains is rotated uniformly.

On the other hand, in real crystals, the homogeneous magnetization pattern is distorted locally in the vicinity of dislocations, due to the magnetoelastic coupling. This in turn has an influence on the process of magnetization rotation. In addition to being dependent on the field intensity and the magnetic and elastic constants, the magnetization is also sensitive to the spatial arrangement of the dislocations.

The theoretical treatment of this problem belongs to the field of micromagnetism, the magnetization continuum theory developed by W. F. Brown. The starting point in this theory is the minimization problem

$$\delta(\Phi_K + \Phi_H + \Phi_A + \Phi_S + \Phi_{\text{el}}) = 0, \qquad (16.1)$$

where the Φ's are local free energy densities, namely Φ_K the magnetocrystalline anisotropy energy, Φ_H the magnetostatic energy, Φ_A the exchange energy, Φ_S, the magnetostatic field energy, and Φ_{el} the magnetoelastic energy. From Eq. (16.1) we may derive the basic micromagnetic equations, or the so-called Brown's equations, which are a system of nonlinear partial differential equations [40, 41]. These may be solved rigorously in special cases only, for example, when the magnetic inhomogeneities that are caused by lattice defects such as dislocations are small deviations about a mean magnetization direction (Brown's approximation). This situation is found, for example, (1) in the region of approach to magnetic saturation, where the magnetization is nearly parallel to the external field throughout the entire crystal and deviates from that direction only slightly, and only locally, due to interactions with dislocations and (2) within the magnetic domains in cases where the magnetocrystalline anisotropy energy is sufficiently large.

The influence of dislocations on rotation processes was first studied by W. F. Brown [42, 43] for the region of approach to saturation. The theory was generalized by Seeger and Kronmüller [44, 45] by using the Kröner continuum theory of internal stresses [46], in such a way that it became possible to calculate the approach to saturation susceptibility χ_S for any given arrangement of dislocations. By introducing modifications into this scheme, Kronmüller [47, 48] was later able to treat theoretically the effect of dislocations on the magnetization pattern in magnetic domains.

In order to understand the physical nature of the results, it is useful to introduce the concepts of the exchange lengths of the stray field, l_S, the magnetocrystalline energy, l_K, and the field, l_H. These quantities characterize the extent of magnetization inhomogeneities in the vicinity of defects, depending on which energy determines preferentially the mean magnetization direction. They are given by

$$l_S \equiv 1/\varkappa_S = (A/2\pi M_S^2)^{1/2}, \qquad (16.2)$$

$$l_K \equiv 1/\varkappa_K = (A/K_1)^{1/2}, \qquad (16.3)$$

$$l_H \equiv 1/\varkappa_H = (2A/HM_S)^{1/2}. \qquad (16.4)$$

The room temperature values of these exchange lengths for iron, cobalt, and nickel at $H = 1000$ Oe are given in Table II. Only within the region of approach to saturation does the exchange length l_H play a role at room temperature, whereas the magnetization pattern in the magnetic domains is, in small fields, determined by an exchange length that is composed of l_K and l_S.

TABLE II

EXCHANGE LENGTHS OF IRON, COBALT, AND NICKEL AT ROOM TEMPERATURE (295°K) IN A MAGNETIC FIELD $H = 1000$ Oe[a]

Element	l_S (Å)	l_K (Å)	l_H (Å)
Fe	33	209	153
Co	32	55	136
Ni	76.5	453	189

[a] Kronmüller [48].

17. High Fields. In this case, the magnetization is locally deflected from its mean direction along the field direction by local stress sources such as dislocations; the deflections depend on the stress field and the magnetostriction constants. Because of exchange coupling, these inhomogeneities extend into the crystal over distances that are characterized by the exchange length l_H. In case of a localized platelike stress source (Fig. 11), the exchange length gives the distance from the stress source where the deflection, γ, of the magnetization from its mean direction is reduced to e^{-1} of its value at the stress source. The corresponding magnetization distri-

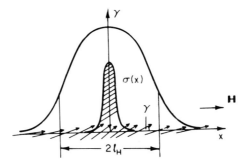

FIG. 11. Schematic deflection of the magnetization from direction of the field **H** in the case of a localized laminar stress $\sigma(x)$, where $l_H = (2A/HM_s)^{1/2}$ [48].

bution in the vicinity of an edge dislocation in nickel at $\mathbf{H} = 1000$ Oe is shown in Fig. 12. As noted above, the magnetization pattern in the neighborhood of dislocations is very sensitive to the particular arrangement of the dislocations. When the dislocation spacing R_{ij} is less than l_H, the magnetization does not completely reflect the stress fields of the individual dislocations. There is a strong magnetic interaction between the dislocations. In the limiting case where the spacing between dislocation is very small, the magnetization in the neighborhood of a dislocation group behaves as if there were only one single dislocation with a correspondingly higher strength, a so-called superdislocation. In the other extreme case characterized by $R_{ij} \gg l_H$, each individual dislocation is completely reflected in the magnetization pattern. Both extreme cases of dislocation distributions are represented schematically in Fig. 13. Since l_H can be varied with the field

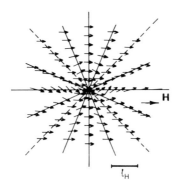

FIG. 12. Magnetization pattern in the vicinity of an edge dislocation in nickel at a field $\mathbf{H} = 1000$ Oe parallel to the x axis; $l_H = (2A/HM_s)^{1/2}$. For clarity, the deflection has been enlarged in the drawing by a factor of 5 [48].

according to Eq. (16.4), we have an elegant method by means of magnetic measurements to study the dislocation distribution in large samples. For further details on this subject, see Brown [48], and Secs. 34 and 40.

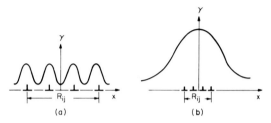

FIG. 13. Magnetization pattern in the vicinity of dislocation groups for the two extreme cases: (a) $R_{ij} \gg l_H$, very large spacing between dislocations; (b) $R_{ij} \ll l_H$, very small spacing between dislocations (superdislocation case) [48].

18. Low Fields. The magnetization inhomogeneities in the interior of ferromagnetic domains may be treated by the same methods as those used for very strong fields, provided the magnetocrystalline energy is large enough to permit the application of Brown's approximation in solving the micromagnetic equations. The mean magnetization direction will now be determined by the external field, the demagnetizing field, and the crystal field. Instead of the exchange length l_H, we have an exchange length composed of l_S and l_K. For nickel [48] we have, with $K_1 < 0$ and $H = 0$,

$$l_K' = (3A/2 \,|\, K_1 |)^{1/2}. \tag{18.1}$$

In very weak fields where the deflection γ of the mean magnetization direction from the easy directions is very small, the change in the free energy due to the presence of dislocations may be written in the form

$$\phi = c(\varrho G^2 b^2/\pi \,|\, K_1 |) \sin^2 \gamma. \tag{18.2}$$

Here ϱ is the dislocation density per cubic centimeter, G is the shear modulus, b is the Burgers vector, K_1 is the fourth-order magnetocrystalline anisotropy constant, and c is a constant which is a function of the magnetostriction constants and the dislocation distribution. For nickel at room temperatures, the order of magnitude of c is about 10^{-8} [48]. The influence of dislocations on rotation processes may then be taken into account by introducing an effective anisotropy constant.

$$K_{\text{eff}} = K_1 + c\varrho Gb^2/K_1. \tag{18.3}$$

Domain Structure

19. The influence of structural defects on the motion of a Bloch wall decidedly depends on whether the wall is type I or type II (see Sec. 10). The interpretation of the magnetization curve depends essentially on the following questions:

(1) Which types of Bloch walls are responsible for the magnetization changes in the various regions of the magnetization curve?
(2) How does a plastic deformation affect the domain structure?

These questions can be answered by Bloch wall observations using, for example, the Bitter technique (Bitter [49] and Williams *et al.* [50]). For this purpose, we studied the domain structure of undeformed and deformed single crystals of cobalt, nickel, and Fe–Si along the virgin curve, and along the hysteresis loop [32, 51].

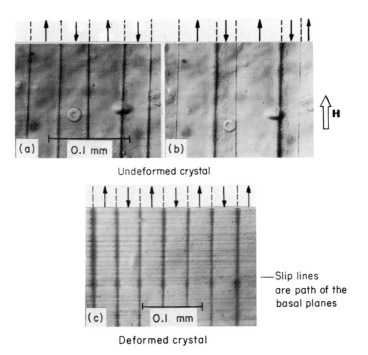

FIG. 14. Cobalt: domain structure on planes parallel to the preferred magnetic axis (Bitter technique, room temperature): (a and b) behavior of the domain structure of an undeformed crystal under the effects of an increasing field (a is the demagnetized crystal); (c) deformed crystal, demagnetized (slip lines are parallel to basal planes).

Magnetically Uniaxial Crystals (Cobalt)

20. In hexagonal magnetically uniaxial cobalt the only Bloch walls present are 180° Bloch walls which lie perpendicular to the basal plane. The shape of the Bloch walls and the magnetic domains are very sensitive to the dimensions of the crystal, as well as the demagnetization procedure. Thus, after demagnetization, the domain structures of rod-shaped, intermediately oriented, single crystals are labyrinthine [51]. In a increasing field, the energetically favorably oriented domains grow at the expense of the unfavorably oriented domains by displacements of the 180° Bloch walls, as shown in Fig. 14a and b. In magnetically uniaxial crystals, rotational processes occur to a considerable extent along the entire magnetization curve, depending upon the orientation of the crystals (see Sec. 24). The character of the Bloch wall changes in such a way that the angle between the magnetization vectors M_1 and M_2 continually decreases on either side of the Bloch wall. As is seen in Fig. 14b, the domain structure (and its behavior during magnetization changes) is not greatly influenced by plastic deformation of the crystal. This is essentially due to the fact that the crystal anisotropy of cobalt at room temperature and lower temperatures is so high as to mask perturbations in the anisotropy resulting from internal stresses. The temperature dependence of the domain structure of cobalt has been studied by Hubert [90].

Magnetically Multiaxial Crystals (Silicon–Iron)

21. Undeformed Crystals. As an example of magnetically multiaxial crystals we shall discuss the results for Fe–Si single crystals. The shapes and orientations of the crystals investigated are given in Fig. 15. The plane of observation is a {100} plane; i.e., the traces of the 180° Bloch walls are parallel to ⟨100⟩ directions, while those of the 90° walls are at an angle of 45° to these directions (dashed lines in Fig. 15). This choice of crystal shape and orientation has two advantages:

(1) The preferred directions perpendicular to the observation surface (and to the external field) play a minor role; this simplifies the problem to an extent that it may be treated theoretically.

(2) Since there is no preferred direction parallel to the rod axis or to the external field, the results may be generalized.

The most important results are presented in Fig. 16. They are understood most readily by considering the magnetization processes in a decreasing field (the initial stage is saturation). Let us first consider the undeformed

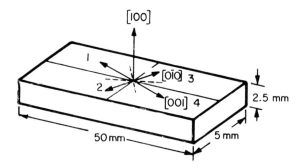

FIG. 15. Shape and orientation of an iron–silicon single crystal on which the Bloch wall studies in Fig. 16 were carried out (external field and tensile stress were applied parallel to the specimen axis).

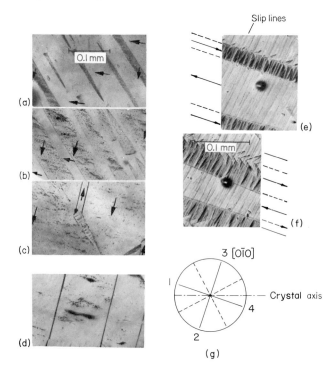

FIG. 16. Field dependence of the domain structure of an iron–silicon single crystal (3.5 wt % Si) with shape and orientation as shown in Fig. 15. Observations on a (100) surface (Bitter technique, room temperature). (a–d) Undeformed crystal: transition from saturation to zero magnetization: (a) $H = 44$ Oe; (b) $H = 40$ Oe; (c) $H = 7$ Oe; (d) $H = 0$. (e and f) Deformed crystal: change of the domain structure with decreasing field. (g) Positions of the $\langle 100 \rangle$ preferred directions in the (100) observation plane corresponding to the photographs.

crystal (Fig. 16a–d). At about 44 Oe we first observe, in Fig. 16a, the 1 → 2 phase transition that is characterized by formation and displacement of 90° Bloch walls. The nuclei form preferentially on the surface of the sample. In this process the magnetization changes from the preferred direction 1 nearest the field direction into the next favorable direction, 2. Then, as the field is decreased, the crystal becomes preferentially magnetized parallel to direction 2 (Fig. 16b). At the same time, the transverse demagnetizing field increases. At about 7 Oe, 180° wall nuclei develop, as shown in Fig. 16c; the growth of these initiates the 2 → 3 phase transition. Further changes in magnetization occur primarily by 180° wall displacements. In the demagnetized state (Fig. 16c) the domain structure mainly consists of antiparallel magnetized domains with magnetization directions along 2 and 3. This structure will also be found after ac field demagnetization of the crystal. When, conversely, we start with the demagnetized state, then the processes described occur in reverse sequence, i.e., in a weak field it is primarily the 180° Bloch walls that are displaced, and only after the field has become relatively high do 90° wall and rotation processes occur. 2 → 1 or 3 → 1 transitions have never been observed in weak fields because the mobility of 90° Bloch walls is low since these walls are representatives of type-I Bloch walls.* Starting from the H_c point on the hysteresis loop and following up to saturation, virtually the same sequence of processes is found as along the virgin curve.

22. Deformed Crystals. After a plastic deformation of the crystals the domain structure has a fundamentally different appearance (Fig. 16e and f). Starting from saturation, the 1 → 2 transition, which is the first to appear in an undeformed crystal, is severely inhibited, since it is difficult to move the type I 90° Bloch walls in a deformed crystal. Only at much lower fields does the 180° type 1 → 4 transition set in (Fig. 16e), a process which does not occur at all in undeformed crystals. The demagnetized state is reached in decreasing field as a result of lateral displacements of these 180° walls. As may be seen from a comparison of Fig. 16d and f, the resulting domain structure is fundamentally different from that in the undeformed crystal. Generalizing, we may say that in plastically deformed crystals the contribution of type-I Bloch walls is greatly reduced.

23. Summary. The most important results of these studies may be summarized as follows:

---

* Exceptions are (110) 90° Bloch walls. However, in real crystals, deviations from this orientation will always be present.

(1) In the case of low fields, i.e., at the beginning of the virgin curve, and in the vicinity of the H_c point, magnetization changes occur by displacements of type-II Bloch walls, predominantly 180° Bloch walls.

(2) Along the virgin magnetization curve and along the corresponding section of the hysteresis loop, the sequence of magnetization processes is by and large the same.

(3) The domain structure of the demagnetized crystal can be understood only after an analysis of the processes *during demagnetization*; these depend on the mobility of the Bloch walls (stresses), the orientation of the crystal, and the conditions for nucleation.

(4) The domain structure is quite different after a sufficiently strong plastic deformation; in particular, the contribution to magnetization processes of type-I Bloch walls is reduced.

As has been shown elsewhere [32, 51], it is possible to interpret the above processes qualitatively, and, in part, quantitatively in terms of the so-called phase theory (see Sec. 24). It is necessary for this purpose to expand the phase theory as originally conceived for ideal crystals by Néel [25] and by Lawton and Stewart [52], by taking into account the mobilities of the Bloch walls and the conditions for nucleation.

Theory of the Magnetization Curve

General Theory

24. Generally, the changes in magnetization occur simultaneously by Bloch wall movements and by rotations. When these processes are reversible, the state of magnetization of a crystal may be calculated for any given external field from the variation principle, $\delta\Phi_G = 0$ [53], where Φ_G is the free enthalpy of the crystal. Following Néel [25] and Lawton and Stewart [52], the subdivision into magnetic domains can then be taken into account by assigning all domains with the same magnetization direction (characterized by the direction cosines, $\gamma_j^{(i)}$, of the magnetization, referred to any arbitrary Cartesian coordinate system) to a common phase with a volume fraction v_i. The free enthalpy is then minimized with respect to the phase volumes v_i and the magnetization directions, $\gamma_j^{(i)}$. This leads to the following equations:

$$\partial\Phi_G/\partial v_i = 0, \qquad \partial\Phi_G/\partial\gamma_j^{(i)} = 0, \qquad (24.1)$$

with the supplementary conditions

$$\sum_i v_i = 1, \qquad \sum_j (\gamma_j^{(i)})^2 = 1. \qquad (24.2)$$

XIII. CRYSTAL DEFECTS IN FERROMAGNETIC SINGLE CRYSTALS

The free enthalpy is composed of the following terms:

$$\Phi_G = \Phi_K + \Phi_S + \Phi_H + \Phi_W + \Phi_\lambda + \Phi_{el}$$

where Φ_K is the magnetocrystalline energy, Φ_S is the stray field energy, Φ_H is the potential energy in the external field, Φ_W is the Bloch wall energy, Φ_λ is the energy of the magnetostrictive distortions, and Φ_{el} is the magnetoelastic coupling energy. The term Φ_{el} may be split into two components $\Phi_{el,W}$ and $\Phi_{el,R}$ where $\Phi_{el,W}$ takes into account the interactions between Bloch walls and the dislocations, and $\Phi_{el,R}$ those of the magnetization in the domains with dislocations (rotations). In the case of an ideal crystal the terms Φ_{el} and Φ_W are very small.

With this procedure it becomes possible to calculate ideal hysteresis-free magnetization curves, which are in good agreement with measurements on annealed single crystals [25, 53, 54]. Some generally important results of these calculations should be emphasized.

(1) In ideal magnetically multiaxial single crystals such as iron or nickel, the change in magnetization in the steep portion of the magnetization curve is entirely due to Bloch wall motion (at vanishing internal field). For real crystals with lattice defects and demagnetizing effects this is only approximately true.

(2) In magnetically uniaxial crystals, on the other hand (e.g., cobalt), a considerable amount of magnetization rotation will occur from the beginning, except for the case in which the field is parallel to the preferred axis [52, 53]. Furthermore, Eqs. (18.3) and (24.1) make it possible to calculate those parameters of the real crystal which describe reversible processes, e.g., initial susceptibility and reversible susceptibility. Here, the terms $\Phi_{el,W}$ and $\Phi_{el,R}$ are particularly important.

For the case of small reversible magnetization changes, $\Phi_{el,W}$ may be written as

$$\Phi_{el,W} = \tfrac{1}{2} \sum_i R_i (\Delta v_i)^2, \qquad (24.3)$$

where Δv_i is the change in phase volume for small field changes, and R_i is the local function of the Bloch wall energy for small wall deflections. In Sec. 30 we shall calculate the force constants, R_i, as used here, for the 180° Bloch walls.

At a sufficiently large magnetocrystalline anisotropy, the interaction between the magnetization within the domains and the dislocations can, according to Sec. 16, be taken into account by an effective anisotropy constant, $K_{1,\text{eff}}$, in Eq. (18.3).

It has already been stated in the introduction that Bloch wall movements are impeded far more by lattice defects than the rotations when the magnetocrystalline anisotropy is high. Our next step is therefore to investigate the movement of a Bloch wall in a crystal with many lattice defects. We may then calculate the constants R_i in Eq. (24.3), as well as that portion of the magnetization curve in which the change in magnetization occurs by Bloch wall movement.

Bloch Wall Displacement in Real Crystals

25. Force Function. According to Sec. 19, the most important mechanism of magnetization change is the displacement of type-II Bloch walls. We shall now study the influence of a large number of statistically distributed lattice defects on the displacement of such a Bloch wall, and shall assume for purposes of simplification that the Bloch wall is not coupled with other Bloch walls, and that no rotational processes occur.

The resultant force, $\Sigma(z)$, which acts on a Bloch wall at some position, z, within the crystal as a result of the interactions with lattice defects, depends *inter alia* on density and distribution of the lattice defects, and hence on the position of the Bloch wall within the crystal. The spatial dependence of $\Sigma(z)$ has two reasons:

(1) The number m_j of lattice defects of some type j, which interact with the Bloch wall, varies as the Bloch wall moves. At a statistical distribution of lattice defects the m_j's will vary according to normal distributions.

(2) The force p_z between a lattice defect of some type j and the Bloch wall depends on the position of the defect relative to the Bloch wall, and, in the simplest case, on the distance $(z_0 - z)$ between the lattice defect and the center of the Bloch wall. We therefore think of the p_j's as being random variables, the corresponding probability distributions being determined by the spatial variations of the interaction forces, p_j, and perhaps also by other parameters.

If we therefore consider the m_j's and the p_j's to be random variables, then the resultant force $\Sigma(z)$ experienced by the Bloch wall at some position z may be written as

$$\Sigma(z) = \sum_j m_j p_j(z_0 - z). \tag{25.1}$$

This expression simply states that $\Sigma(z)$ is the sum of all individual forces, to be taken in accordance with the specific arrangement of the lattice defects.

XIII. CRYSTAL DEFECTS IN FERROMAGNETIC SINGLE CRYSTALS

The Bloch wall will be in equilibrium when the force exerted by a magnetic field H on the Bloch wall and the force $\Sigma(z)$ resulting from the lattice defects are equal and opposite in sign. Using the internal field, H_i, the equilibrium condition for a plane 180° Bloch wall is given by

$$2H_i M_s F \cos \varphi = \Sigma(z). \tag{25.2}$$

Here, φ is the angle between the field direction and the magnetization vector, F is the area of the Bloch wall, and M_s is the spontaneous magnetization.

Figure 17 shows, as an example, a rather arbitrarily assumed function $\Sigma(z)$. The dashed part of the curve indicates the displacement of a Bloch wall under the action of a field that first increases and then decreases. The corresponding microhysteresis loop is given in Fig. 17b. It may be characterized by such parameters as coercive field, remanence, initial susceptibility, etc. As is seen from a comparison of Figs. 17b and 17a, the coercive field H_c is determined by the maximum amplitude, $\Sigma_{\max}$, of the $\Sigma(z)$ curve that

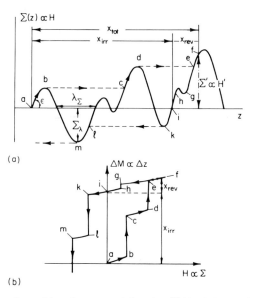

FIG. 17. (a) A rather arbitrarily assumed function $\Sigma(z)$ of the resultant force exerted on a Bloch wall by lattice imperfections; z is the coordinate perpendicular to the Bloch wall plane. The dashed lines are the paths corresponding to Bloch wall displacements in increasing and decreasing fields. The zero point slopes of the $\Sigma(z)$ curve are characterized by $\tan \varepsilon \equiv \alpha$. (b) Microscopic magnetization curve corresponding to the dashed Bloch wall path. The total path, x_{tot} traversed by a Bloch wall when a field H' is applied can be decomposed into a reversible part x_{rev} and an irreversible part x_{irr} (remanence); these two parts can be calculated separately.

must be overcome when the Bloch wall is driven back from its remanence position (i) into its initial position (a). The initial susceptibility, χ_i, on the other hand, is connected with the zero-point slope ($\alpha = \tan \epsilon$) of the $\Sigma(z)$ curve. The resultant hysteresis loop of a large crystal with many Bloch walls is obtained by superimposing the corresponding microhysteresis loops of the kind shown in Fig. 17b. Its parameters are obtained by averaging over the corresponding parameters of the individual $\Sigma(z)$ diagrams.

26. Hysteresis Loop. To calculate the hysteresis loop, one has to express the displacements of the Bloch walls as a function of the field. As may be seen from Fig. 17, the position of a Bloch wall (its z coordinate) is determined for any given $\Sigma(z)$ curve by the initial position of the Bloch wall, and by the applied field as a function of time. This means that the following steps must be made in a statistical calculation of the hysteresis loop:

(1) First, an average and perhaps simplified $\Sigma(z)$ curve must be characterized with sufficient accuracy by means of statistical parameters. These parameters must then be related both to the physical properties of the sample, i.e., the nature, density, and distribution of the lattice defects, and to the properties of the Bloch wall.

(2) The second step is to describe mathematically in a $\Sigma(z)$ diagram, characterized as above, the displacement of a Bloch wall under the influence of a field that varies with time.

We must confine ourselves here to the sketching of the fundamental aspects of the solution of these problems (for details, see Träuble [51]). We consider the most important parameter of the $\Sigma(z)$ curve to be the probability distribution $f(\Sigma)$ of the amplitudes. According to the central limit theorem of statistics, the distribution $f(\Sigma)$ converges to the normal distribution as the number of lattice defects, i.e., the number of random variables in Eq. (25.1) increases. Its parameters (mean and standard deviation) can however be expressed in terms of the well known random variables m_j and p_j (see Träuble [51]). The $\Sigma(z)$ curve can be further characterized by means of the probability distributions of the zero-point distances $f(\lambda)$, the maxima of the $\Sigma(z)$ curve, $f(\hat{\Sigma})$, and the zero-point slopes, $f(\alpha)$. In order to simplify their calculation [50], the $\Sigma(z)$ curve was considered to be the resultant of statistically independent quantities. It is an essential condition for the validity of this approach that the Bloch walls remain rigid and do not deform at lattice defects. This condition is generally fulfilled because deformations of Bloch walls (with some exceptions [55]) cause high stray fields, and this is energetically unfavorable.

We shall return to this point in the discussion of the experiments. After the parameters of the $\Sigma(z)$ curve have been determined in this way, the displacements of the Bloch walls under the action of a field can be described by introducing a hysteresis function, which characterizes the distribution of the Bloch walls along the $\Sigma(z)$ function as a function of the applied field. Here, the initial distribution of the Bloch walls (demagnetization process) enters into the calculation [51].

We shall confine ourselves in the following to a discussion of the parameters coercive field and initial susceptibility. A comparison of the theoretical results with experimental results will provide sufficient evidence as to extent to which our model corresponds to the actual situation.

In the final results [Eqs. (27.2) and (28.1)], lattice defects will be characterized by means of the following parameters: (1) their volume density ϱ_j or, respectively, the average number $\bar{m}_j$ of defects which interact with the Bloch walls, and (2), parameters which describe the interactions of the defects with the Bloch walls; the latter parameters are essentially the mean values $M(p_j)$ and the standard deviations $D(p_j)$ of the interaction functions p_j. In case the interaction function depends only on the distance between the lattice defect and the center of the Bloch wall, we have

$$M(p_j) = d^{-1} \int p_j(z)\, dz, \tag{26.1}$$

$$D(p_j) = d^{-1} \int z^2 p_j(z)\, dz = M(p_j^2) - M^2(p_j), \tag{26.2}$$

where d characterizes the range of the interaction functions p_j in the direction perpendicular to the Bloch wall. In many cases d will be approximately equal to the Bloch wall thickness, δ. If the interaction functions p_j depend also on other parameters, then in calculating $M(p_j)$ and $D(p_j)$ one has to average over these as well.

Coercive Field, H_c

27. Magnetization Changes by Bloch Wall Displacement. If magnetization changes in the neighborhood of H_c are preferentially due to Bloch wall displacements, as is the case for magnetically multiaxial crystals with sufficiently high crystal anisotropy, then the coercive field is approximately equal to the field required to drive Bloch walls through the crystal over distances which are equal to the mean Bloch wall spacing L, i.e., the coercive field is determined by the expectation value $|\Sigma_{\max}|$ of the maximum amplitude of the $\Sigma(z)$ function over a distance L (see Sec. 25). In the case of magnetization changes by 180° Bloch wall movements (see Sec. 20), one may define

$$H_c = (2M_sF \cos \varphi)^{-1} \langle | \Sigma_{\max} | \rangle_{\mathrm{av}} \qquad (27.1)$$

where Σ is given by Eq. (25.1). The average in this case must be taken over all of the Bloch walls present in the crystal. M_s is the spontaneous magnetization, F is the mean area of the individual Bloch walls, and φ is the angle between the field and the magnetization vector.

In the next step an astimate is made of the expectation value $\langle | \Sigma_{\max} | \rangle_{\mathrm{av}}$ corresponding to the movement of a Bloch wall over a distance L. If we want to apply the theory of statistically independent processes, the distance L must be assigned a number of events, v. As in Section 25, we again consider the $\Sigma(z)$ curve to be a set of statistically independent values, and take the smallest distance between statistically independent Σ values as being approximately equal to the Bloch wall thickness, δ. We then have $v \simeq L/\delta$. Since it is only the logarithm of v that enters into the result, the exact value of v is of minor importance. Using statistical methods to estimate the expectation value $\langle | \Sigma_{\max} | \rangle_{\mathrm{av}}$ [51], one obtains the following expression for H_c:

$$H_c = [(\ln v)^{1/2}/\sqrt{2}M_sF \cos \varphi]\{[\sum_j \bar{m}_j M^2(p_j)]^{1/2} + [\sum_j \bar{m}_j D(p_j)]^{1/2}\} \qquad (27.2)$$

where the quantities $M(p_j)$ and $D(p_j)$ may be calculated from Eqs. (26.1) and (26.2) in cases when the p_j's are functions only of the distance between the lattice defect and the center of the Bloch wall. In the important case where the mean values $M(p_j)$ vanish, one gets because of Eq. (26.2) (see also Vicena [28] and Dijkstra and Wert [56]):

$$H_c = [(\ln v)^{1/2}/\sqrt{2}M_s \cos \varphi] [\sum_j \bar{m}_j M(p_j^2)]^{1/2} \qquad (27.3)$$

In these formulas, $\bar{m}_j$ is the mean number of interactions between the Bloch wall and type-j lattice defects. If the ranges of the interaction forces are of the same magnitude as the Bloch wall thickness δ, we have

$$\bar{m}_j = F\delta\varrho_j \qquad (27.4)$$

where ϱ_j is the volume density of the interaction processes of type j. It should be emphasized that for dislocations, p_j is not equal to the dislocation density ϱ; instead, the term ϱ_i is to the interpreted as the volume density of the elementary interaction processes. If one wants to calculate ϱ_j from the dislocation density, it is necessary to have either a precise knowledge of the details of the interaction processes or to make reasonable assumptions. Since dislocations are mostly not exactly straight, they will, in general, run parallel to the Bloch walls for only short distances. This fact is taken into account by introducing a mean interaction length, l, which is the length of a dislocation parallel to the Bloch wall that would have the same effect [in terms of $M(p_j)$ and $D(p_j)$] on the Bloch wall as a curved dislocation. The approximate relation is

$$\varrho_j = (\varrho/3)l, \tag{27.5}$$

where ϱ is the dislocation density.

It should be emphasized that the considerations of this and the following section are valid for crystals containing *arbitrary* kinds of defects, i.e., also for samples that contain magnetic or nonmagnetic inclusions besides the dislocations. This treatise has been confined to dislocations only because in this case our knowledge of the interaction mechanisms is more advanced and most of the experimental data refer to dislocations. For interactions between domain walls and spherical inclusions see Dijkstra and Wert [56] (see also Chapter X).

28. Magnetization Changes by Bloch Wall Displacement and Magnetization Rotation. When the magnetization changes are due to both Bloch wall movements and rotations, further considerations become necessary. Conditions are simplest in the case of magnetically uniaxial crystals with high magnetocrystalline anisotropy, e.g., cobalt at temperatures $T \leq 400°C$. For intermediately oriented cobalt crystals at room temperature, for instance, approximately half the change in magnetization is due to rotation [51, 53]. Because of the high magnetocrystalline anisotropy the effects due to lattice defects are almost exclusively those on the Bloch wall movements. In this case, the occurrence of rotations can fairly easily be taken into account. As is shown elsewhere [57], one finds approximately

$$H_c = H_c'/[(M_s^2 N_\perp/2K_1 \cos \varphi_0) + \cos \varphi_0], \tag{28.1}$$

where φ_0 is the angle between the preferred axis and the field, and $N_\perp$ is the demagnetization factor perpendicular to the rod axis. H_c' is given by Eq. (27.2) with $\cos \varphi = 1$. In Sec. 36 ff this result will be compared with measurements on cobalt single crystals.

29. Summary. In conclusion, we shall compile those predictions from Eqs. (27.2) and (27.5) that may be considered as criteria for the validity of our model of the interactions between Bloch walls and lattice defects.

The Dependence of the Coercive Field on the Volume Density of Lattice Defects. From Eqs. (27.2) and (27.4) follows the general result

$$H_c \propto \varrho^{1/2}. \tag{29.1}$$

With Eqs. (5.3) and (6.2), the above yields the following relations for the increase in coercive field with shear stress, τ, during plastic deformation:
In Region I of the work hardening curve $\varrho \propto (\tau - \tau_0)$, and hence

$$H_c \propto (\tau - \tau_0)^{1/2}; \tag{29.2}$$

in Regions II and III of the work hardening curve $\varrho \propto \tau^2$ and hence (see Table III)

$$H_c \propto \tau \quad \text{or} \quad H_c \propto (\tau - \tau_0), \tag{29.3}$$

which is essentially the same.

The Superposition of Various Kinds of Lattice Defects. In general Eq. (27.2) must be evaluated. However an experimental test can easily be made in the special case of a superposition of (1) the grown-in dislocations (the crystal as grown) with the volume dislocation density ϱ_0 and (2) those dislocations that originate during plastic deformation (volume density $= \varrho_v$). According to Eq. (27.2) one may expect a superposition law of the form

$$H_c \propto (\varrho_0 + \varrho_v)^{1/2} \tag{29.4}$$

Relevant experimental results will be presented in Secs. 36–40.

Temperature Dependence of Coercive Field. Here, no generally valid principles can be given because the variation of H_c with temperature depends on the interaction functions p_j and hence on the nature and the geometrical arrangement of the lattice defects. For example, in the case of internal stresses, the $H_c(T)$ curve is determined by a combination (which must be calculated for each individual case) of the magnetostriction constants and elastic constants, the spontaneous magnetization M_s, and because of Eq. 27.4 the Bloch wall thickness δ. In Secs. 36–40 some examples will be discussed.

Initial Susceptibility, χ_i

30. Magnetization Changes by Bloch Wall Displacements. In the Bloch wall movement model (Sec. 25) which is a good approximation to magnetically multiaxial crystals, the initial susceptibility is closely related to the zero-point slopes α of the $\Sigma(z)$ curve (see Fig. 17). If we refer to displace-

TABLE III. THEORETICAL PREDICTIONS FOR THE DEPENDENCE ON TEMPERATURE AND DEFORMATION OF COERCIVE FIELD AND INITIAL SUSCEPTIBILITY OF MAGNETICALLY MULTIAXIAL SINGLE CRYSTALS WITH INTERNAL STRESSES[a]

	Coercive field, H_c	Initial susceptibility, χ_i
	A. *Magnetization changes by displacements of 180° Bloch walls in the presence of dislocations[b]*	
Variation with deformation	Region I: $H_c \propto (\tau - \tau_0)^{1/2}$ Region II: $H_c \propto (\tau - \tau_0)$	$\chi_i \propto (\tau - \tau_0)^{-1/2}$ $\chi_i \propto (\tau - \tau_0)^{-1}$
Variation with temperature	Model: interaction between Bloch walls and dislocation segments of length l parallel to the planes of the Bloch walls (example, nickel) $H_c \propto \dfrac{\lambda G(\delta l)^{1/2}}{M_s} \begin{cases} H_c \propto \lambda G/K^{1/4}M_s & \text{for constant } l(T) \\ H_c \propto \lambda G/K^{1/2}M_s & \text{for } l \propto \delta^{1/2} \end{cases}$ Model: interactions between (100) 180° Bloch walls and screw dislocations parallel to the $\langle 111 \rangle$ directions (example, iron) $H_c \propto \lambda_{100}\,\delta(c_{11}-c_{12})/M_s \propto \lambda_{100}(c_{11}-c_{12})/K^{1/2}M_s$	$\chi_i \propto \dfrac{M_s^2 \delta^{1/2}}{\lambda G l^{1/2}} \begin{cases} \chi_i \propto M_s^2/K_1^{1/4}G\lambda & \text{for constant } l(T) \\ \chi_i \propto M_s^2/K_1^{1/8}G\lambda & \text{for } l \propto \delta^{1/2} \end{cases}$
	B. *Magnetization changes by rotational processes at predominant stress anisotropy[c]*	
General relations	$H_c \simeq \lambda|\sigma_i|/M_s$	$\chi_i \simeq M_s^2/\lambda|\sigma_i|$
Temperature variation assuming $\sigma_i(T) \propto G(T)$	$H_c \simeq G\lambda/M_s$	$\chi_i \simeq M_s^2/\lambda G$
Variation with deformation assuming $\sigma_i \propto \varrho^{1/2}$	Region I: $H_c \propto (\tau - \tau_0)^{1/2}$ Region II: $H_c \propto (\tau - \tau_0)$	See text

[a] For results for magnetically uniaxial crystals, see the text (Secs. 24–33).
[b] (A): Case of predominant magnetocrystalline anisotropy (Bloch wall movements); the given relations follow from Eqs. (27.2) and (27.3).
[c] (B): Case of predominant stress anisotropy (the magnetization changes by nonhomogeneous rotational processes); $|\sigma_i|$ is the absolute value of the mean stress amplitude; λ stands for an expression composed of the magnetostriction constants λ_{ijk}; G is the shear modulus. The other parameters are defined in the text.

ments of 180° Bloch walls, according to the results of Sec. 19, we obtain the following definition:

$$\chi_a = (4\cos^2\varphi M_s{}^2 F/L)\langle|\,\alpha\,|\rangle_{\mathrm{av}}^{-1}. \tag{30.1}$$

The parameters in this equation have already been discussed above. The average $\langle|\,\alpha\,|\rangle_{\mathrm{av}}^{-1}$ in Eq. (30.1) is to be taken over all of the zero points in the $\Sigma(z)$ curve that are occupied by Bloch walls. It is more difficult to calculate $\langle|\,\alpha\,|\rangle_{\mathrm{av}}^{-1}$ than $\langle|\,\Sigma_{\max}\,|\rangle_{\mathrm{av}}$ because the distribution of the Bloch walls over the $\Sigma(z)$ curve must be taken into account. For a crystal that was demagnetized by an alternating field the statistical evaluation [51] yields

$$\chi_i = \frac{F}{L} \frac{(1.7\delta)M_s{}^2 \cos^2\phi}{[\sum_j \bar{m}_j M^2(p_j)]^{1/2} + [\sum_j \bar{m}_j D(p_j)]^{1/2}}. \tag{30.2}$$

The quantities entering into this equation are very much the same as those for the coercive field. With respect to Eq. (27.2), we find that in the 180° Bloch wall displacement model the product $H_c \cdot \chi_i$ is independent of the volume density of the lattice defects. If we further assume the Bloch wall surface F and the mean Bloch wall spacing L to be independent of temperature, the product $H_i \cdot \chi_i$, varies with temperature essentially as $M_s \times K^{-1/2}$. With respect to the dependence of χ_i^{-1} on deformation, the same statements are valid as for the coercive field (see Table III).

The parameters R_i introduced in Sec. 24 can be related similarly to the zero-point slopes α of the $\Sigma(z)$ curve. For example, for a two-phase domain structure consisting of 180° Bloch walls, one obtains the relationship

$$R^{-1} = \tfrac{1}{2}(F/L)\langle|\,a\,|\rangle_{\mathrm{av}}^{-1} \tag{30.3}$$

where $\langle|\,\alpha\,|\rangle_{\mathrm{av}}^{-1}$ follows from Eqs. (30.1) and (30.2).

31. *Magnetization Changes by Bloch Wall Displacement and Rotation.* Once the constant R is known, we can calculate the initial susceptibility and the reversible susceptibility for the general case that Bloch wall displacements and rotations occur simultaneously following the variational procedure outlined in Sec. 24.

This is most easily done for a magnetically uniaxial crystal having a high magnetocrystalline anisotropy. In this case the influence of the lattice defects (magnetoelastic coupling energy) on rotations may be neglected. The initial susceptibility is then determined by three factors: (1) by the influence of the magnetocrystalline anisotropy on the rotation, (2) the increase in stray field energy during magnetization, and (3) the interactions

XIII. CRYSTAL DEFECTS IN FERROMAGNETIC SINGLE CRYSTALS

between the Bloch walls and the lattice defects; these are described by the constant R_i. In the case of infinitely long crystals (the demagnetization factor $N_\parallel$ in the direction parallel to the rod axis being equal to zero) the calculations [53] give

$$\chi_i = \frac{N_\perp M_s^2 + 2K_1 \cos^2 \phi_0 + R \sin^2 \phi_0}{2K_1 N_\perp M_s^2 \sin^2 \phi_0 + R(2K_1 + N_\perp M_s^2 \cos^2 \phi_0)} M_s^2, \quad (31.1)$$

where ϕ_0 is the angle between the crystallographic c axis and the rod axis, and $N_\perp$ is the demagnetization factor perpendicular to the rod axis. For a crystal without lattice defects one would have instead (see Brown [42] and Kronmüller et al. [53])

$$\chi_i = \chi_i^\infty/(1 + N_\parallel \chi_i^\infty), \quad (31.2)$$

with χ_i^∞ given by

$$\chi_i^\infty = M_s^2/(2K_1 \sin^2 \phi_0) + \cot^2 \phi_0/N_\perp. \quad (31.3)$$

The relations (31.1) and (31.2) become identical when $R = 0$, and $N_\parallel = 0$. In a similar way one may calculate the initial susceptibility of magnetically multiaxial crystals taking into account the influence of dislocations not only on Bloch wall movements, but also on rotations [48]. For an infinitely long, rod-shaped nickel single crystal, with the rod axis parallel to a $\langle 100 \rangle$ direction, one obtains

$$\chi_i^\infty = \tfrac{1}{3} M_s^2 (R^{-1} + (K_{1,\text{eff}})^{-1}), \quad (31.4)$$

where R is defined by Eq. (30.3), and $K_{1,\text{eff}}$ by Eq. (18.3).

The Influence of the Dislocation Arrangement on Coercive Field and Initial Susceptibility

32. General Aspects. In the above, we have repeatedly considered the influence of the dislocation arrangement on the magnetic parameters. Thus, in Sec. 27 by introducing the interaction length l, we took into account the fact that the dislocations may be curved and hence may be parallel to a Bloch wall only over some mean distance l. Let us now consider these questions in somewhat greater detail.

It was shown in Secs. 13–15 how Bloch walls react to certain dislocation configurations such as dislocation dipoles and dislocation groups, when the effective dislocation spacing perpendicular to the Bloch wall is smaller than the Bloch wall thickness δ. Since δ is highly temperature dependent, a convenient experimental variable is available to study the influence of the dislocation arrangement on the magnetic properties.

We shall consider some special, not always realistic cases to illustrate how the dislocation arrangement can affect the magnetic parameters and conversely, how measurements of coercive field and initial susceptibility may be used to obtain information about the dislocation arrangement.

Procedure. The presence of certain dislocation arrangements in a crystal can be introduced into the theory of Secs. 25 and 26 in the following way. We shall regard an individual dislocation configuration (e.g., a dislocation dipole) as a unit, i.e., as a *single* defect. Then we shall calculate the corresponding interaction function p_j, its mean values $M(p_j)$ and $M(p_j^2)$ and introduce these values into Eqs. (27.2) and (30.2).

33. Examples. (1) As the simplest case let us consider a Bloch wall whose plane is parallel to the principal slip plane of plastic deformation, so that all of the dislocations run parallel to the plane of the Bloch wall (for example, a (110) Bloch wall in iron, and dislocation groups in (110) planes). A certain proportion of the dislocations, $\nu \varrho_{tot}$ (ϱ_{tot} is the total dislocation density), will be distributed statistically. These dislocations are assumed to interact individually with the Bloch walls. The remaining dislocations will be divided into groups containing n dislocations each (group density $= \varrho_{gr} = \varrho_{tot}(1 - \nu)/n$). The term *group* means that a Bloch wall moving in the direction of its normal will encounter n dislocations with the same interaction function p_j simultaneously. Let p_ν represent the interaction function of a single dislocation; then we have $p_{gr} = np_\nu$, $M(p_{gr}) = nM(p_\nu)$ and $M^2(p_{gr}) = n^2 M^2(p_\nu)$. For a crystal of this kind the coercive field can readily be calculated from Eq. (27.2) and the initial susceptibility from Eq. (30.2). For the coercive field one finds

$$H_c = [\nu + (1 - \nu)n]^{1/2} H_{c,\text{stat}} = W H_{c,\text{stat}} \tag{33.1}$$

where $H_{c,\text{stat}}$ represents the coercive field for the case of statistically distributed dislocations. In Eq. (33.1), n is the number of dislocations per group (group strength), ν is the fraction of individual dislocations, and $(1 - \nu)$ the fraction of dislocations in groups. Figure 18 shows the variation in the factor W in Eq. (33.1) as a function of ν and n. From this, the dependence of coercive field H_c on the group density and the group strength can be derived for constant total dislocation density.

When $\nu = 0$, all of the dislocations are in groups. The coercive field increases as $n^{1/2}$, i.e., a large number of small groups causes a smaller coercive field than a small number of correspondingly larger groups.

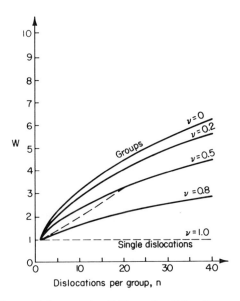

Fig. 18. Dependence of the coercive field on the dislocation configuration (group formation). The factor W $[W^2 = \nu + (1 - \nu)n]$ is defined by Eq. (33.1); ν is the fraction of individual dislocations; $(1 - \nu)$ is the fraction of dislocations in groups; n is the group strength. Model: 180° Bloch walls are shifted parallel to the principal slip plane; the dislocations are parallel to the plane of the Bloch wall. The dashed curve characterizes the change in H_c resulting from a disintegration of the groups at constant dislocation density (see text).

For $\nu = \frac{1}{2}$, only half of the dislocations are arranged in groups; the dependence on group size is correspondingly weaker. Thus, formation of dislocation groups raises the absolute value of the coercive field.

The dependence of the coercive field on deformation is also modified. For example, when $\nu = 0$, $\partial H_c/\partial \tau$ will be $n^{1/2}$ times as great as when the dislocations are distributed statistically. In cases where the factor W in Eq. (33.1) varies during the plastic deformation (variation of n or ν with deformation), deviations from the previously discussed $H_c \propto \varrho^{1/2}$ dependence [see Eq. (29.1)] are to be expected.

If all the dislocations in a group are in the same slip plane, the temperature dependence of the coercive field will not be affected by the occurrence of groups. The situation is quite different if the dislocations in a group do not lie in exactly the same plane, such that a mean vertical extension, a (normal to the principal slip plane), may be assigned to the dislocation group.

Let us consider two extreme cases. When the Bloch walls are sufficiently thick as compared to a, then the vertical range of the group is not significant.

If, however, the thickness of the Bloch walls is very small compared to a, the dislocations will act as individuals. If the dependence of the wall thickness δ on the temperature is sufficiently strong, both cases can be realized by varying the temperature. It is evident that the temperature dependence of H_c is then altered by the presence of groups.

The presence of dislocation groups can have an effect also on the recovery behavior of the coercive field and the initial susceptibility. Once again we shall consider two extreme cases: (a) When during recovery the groups (at constant total dislocation density) are dispersed within the slip planes, H_c will not be affected. (b) If, however, the dislocations climb vertically out of the slip planes to an extent that they must be considered as single dislocations with respect to their interactions with Bloch walls, then this corresponds to a decrease in group strength and an increase in the volume density of the individual dislocations. Here, the ratio $\varrho_{gr}/\varrho_{tot} = (1 - \nu)/n$ is invariant. The dashed curve in Fig. 18 gives the decrease of the coercive field for this case, assuming $n = 20$ and $(1 - \nu) = \frac{1}{2}$. Initially it is evident that the magnitude of this effect depends on the Bloch wall thickness.

(2) Second, we consider a crystal with dislocation groups, each of which consists of n equally spaced dislocations, whose Burgers vectors are parallel to the plane of the Bloch wall. The plane of the dislocation groups is assumed to be normal to the plane of the Bloch wall. The interaction functions p_j of the individual dislocations are given by Eq. (12.2) (see also Figs. 7 and 10). This corresponds roughly to the conditions in deformed cobalt single crystals.

The preceding example has shown how the simultaneous effects of individual dislocations and dislocation groups can be calculated. We shall now consider only the special case in which all of the dislocations are arranged into groups. The procedure for calculating the coercive field is the same as that used before. Since the mean value $M(p_{gr})$ vanishes, one obtains from Eq. (27.2)

$$H_c = \frac{(\ln \nu)^{1/2}}{\sqrt{2} M_s F \cos \phi} [\bar{m}_{gr} M(p_{gr})]^{1/2}. \tag{33.2}$$

Using the mean value for $M(p_{gr}^2)$ which was calculated in Sec. 15, one finds

$$H_c = g \varrho_{gr}^{1/2} [f(n, d)]^{1/2} = g \varrho_{tot}^{1/2} [f(n, d)/n]^{1/2}. \tag{33.3}$$

The factor g is a combination of the elastic constants, the magnetostriction constants, the spontaneous magnetization, and the magnetocrystalline anisotropy constants; its temperature dependence corresponds to that of the coercive field in the case where the dislocations are randomly distributed. According to Fig. 10b, the function $[f(n, d)]^{1/2}$ changes as the group size n and the dislocation spacing d change.

It is of interest to know the dependence of the coercive field on group size n and dislocation spacing d; these are the two parameters that one wants to determine from the measurements. In order to illustrate the specific group effect, we shall consider two cases, which may also be realized experimentally.

First, we consider the Bloch wall thickness δ as a variable (d is a contant) this case may be realized by changing the temperature.

Second, we consider the dislocation spacing within the groups as a variable; this may be realized by dispersion of dislocation groups by annealing.

Case I. As the Bloch wall thickness increases, the curves in Fig. 10b are followed from right to left. While $d > \delta$, nothing unusual can be expected. The dislocation groups become effective as soon as $d \leq \delta$. The coercive field first decreases as shown in Fig. 10b, reaches a minimum at $\delta = 2d$, and then begins to increase again. The parameters d and n may be calculated from the corresponding points on the $H_c(T)$ curve. Furthermore, from measurements of $H_c(T)$ as a function of plastic deformation, conclusions can be drawn concerning the behavior of the parameters d and n at plastic deformation, provided the coercive field is determined by the interaction between Bloch walls and dislocations.

Case II. As the dislocation spacing increases, the curves in Fig. 10b are followed from left to right. If the measurements are made at temperatures at which $\delta < d$, the dispersing of the groups has no effect on the coercive field. If, however, the temperature of measurement is such that $\delta \gg d$, then, as the dislocation spacing d increases, H_c will first decrease down to a minimum at $d = \delta/2$, followed by an increase up to a constant value.

Similarly, the effect of dislocation dipoles on the magnetic parameters may be discussed using Figs. 7 and 9.

The only experimental investigations which have been interpreted thus far using the concept of dislocation groups and their disintegration during annealing are due to De Wit [58] (recovery of the coercive field of polycrystalline nickel below room temperature). No suitable measurements of the coercive field or initial susceptibility have been made, with the aim of obtaining information about the dislocation configuration. The procedure would be to measure the coercive field H_c at two temperatures, T_1 and T_2, chosen such that $\delta > d$ and $\delta < d$.

Let us consider briefly whether there is any chance at all of observing configuration effects. As an example we take a deformed cobalt single crystal. According to electron microscopy studies by Thieringer [18, 59], the dislocation spacings in the basal plane are between 500 and 1000 Å. On the other hand, $\delta \simeq 200$ Å at room temperature. However, as the temperature increases, the Bloch wall thickness increases rapidly. For example, $\delta \simeq 500$ Å at 250°C, and theoretically, δ becomes infinite at about 300°C. It is therefore definitely possible, on the basis of measurements of coercive field (temperature dependence, dependence on deformation at two suitably chosen temperatures, measurements of annealing behavior at various temperatures) to obtain information about the dislocation arrangement. The conditions are favorable also in the case of iron single crystals. In the case of nickel, on the other hand, the temperature range in which Bloch wall displacements predominate is relatively small.

Approach to Saturation

34. The range of approach to saturation is the portion of the magnetization curve at very high fields where the spontaneous magnetization is nearly parallel to the external field, and where changes in magnetization occur by magnetization rotation only.

For field intensities $H \gg 4\pi M_s$ (where M_s is the saturation magnetization), the susceptibility $\chi_S = dM/dH$ of single crystals as well as polycrystals may be approximated by the series,

$$\chi_S = a/H^2 + b/H^3 + \alpha/H^{1/2}. \tag{34.1}$$

At relatively low fields, the term varying as H^{-3} predominantes. At moderate fields, the H^{-2} term determines the shape of the magnetization curve. Finally, at very high fields, χ_S varies as $H^{-1/2}$; this last term describes magnetization changes not due to classical rotation but rather a paramagnetic change of M beyond $M_s(T)$ (paraprocess [60]).

In the field range where the H^{-3} term is the dominant one, the susceptibility χ_S is very sensitive to internal stresses (see Sec. 17). The magnetization inhomogeneities caused by magnetoelastic coupling act such as to increase the susceptibility χ_S. In this region, the dislocation configuration plays an essential role. This, in turn, makes it possible to obtain information on the dislocation configuration from measurements of χ_S.

As has already been mentioned, Seeger and Kronmüller [44, 45] have extended an largely generalized the theory of the approach to saturation,

XIII. CRYSTAL DEFECTS IN FERROMAGNETIC SINGLE CRYSTALS

first developed by Brown [40, 41], and have applied this theory to calculate the susceptibility χ_S of plastically deformed single crystals. One of the models discussed by these authors is the following. Dislocation groups consisting of n parallel dislocations with a mean length L lie in parallel slip planes. Let R_0 be the half mean spacing between dislocations, and R_{ij} be the spacing between the dislocations within a group. The most important term in the approach to the saturation susceptibility equation for nickel at room temperature is then given by

$$\chi_S = 4.4 \times 10^{-9} \frac{G^2 b^2 NL}{2\pi M_s H^3} \left\{ n \ln \frac{R_0}{l_H} - 2 \sum_{i>j} \ln \frac{R_{ij}}{R_0} - 2 \sum_{i>j} K_0 \frac{R_{ij}}{l_H} \right\}, \quad (34.2)$$

where G is the shear modulus, b is the Burgers vector, N the number of dislocation groups per cubic centimeter, K_0 is the zero-order modified Hankel function, and l_H is the exchange length of the field, as discussed in Sec. 16.

The effect of dislocation configuration on χ_E can easily be derived quantitatively by considering two extreme cases: (1) $R_{ij} \ll l_H$ (the superdislocation case), and (2) $R_{ij} \gg l_H$ (individual dislocation case). For $R_{ij} \ll l_H$, Eq. (34.2) reduces to

$$\chi_S = 4.4 \times 10^{-9} (Gb^2 NLn^2 / 2\pi M_s H^3) \ln(R_0/l_H), \quad (34.3)$$

while for $R_{ij} \gg l_H$ (individual dislocation cases), one obtains, with $R_{ij} \simeq R_0$,

$$\chi_S = 4.4 \times 10^{-9} (Gb^2 NLn / 2\pi M_s H^3) \ln(R_0/l_H). \quad (34.4)$$

Provided the total dislocation density is equal, the approach to saturation susceptibility, χ_S, differs in these two cases by the factor n. Since n is the number of dislocations per group and is of the order of 15–30, and since the exchange length l_H can be varied by varying the field, measurements of the approach to saturation susceptibility will yield valuable information about the arrangement of the dislocations. Some characteristic experimental results will be discussed in Sec. 40.

Experimental Results

General

35. The above theoretical model was tested using measurements on single crystals of cobalt, iron, and nickel. The investigation was primarily concerned with the effect of plastic deformation on the parameters H_c, χ_i, χ_S, as well as the field dependence of the reversible susceptibility and of the magnetic domain structure (Secs. 19–23). Table III summarizes the theoretically expected dependence of the coercive field and initial susceptibility on temperature and deformation. The formulas given under Sec. A in Table III result from Eqs. (27.2) and (30.2). The differences in the functional dependence on temperature are due to different assumptions about the elementary processes of the interactions between Bloch walls and dislocation structure. Since the theory of magnetization changes due to nonhomogeneous magnetization rotations with predominating stress anisotropy is not very advanced as yet, the formulas presented under Sec. B in Table III [5] may at best be considered as rule of thumb estimates.

Coercive Field, H_c

The coercive field of cobalt, iron, and nickel single crystals are presented as functions of temperature in Figs. 19–22, and as functions of deformation in Figs. 23–26. As to the behavior of the coercive field of nickel single crystals, similar results were obtained by numerous other authors [63–66].

36. Temperature Dependence. The temperature dependence of the coercive field is similar in many respects for deformed iron and nickel single crystals. In the case of deformed iron (Fig. 19), the magnetocrystalline anisotropy below 600°C (maximum of the $H_c(T)$ curve) is considerably larger than the stress anisotropy caused by inhomogeneous internal stresses. In this temperature range, there exists a well-defined magnetic domain structure with $\langle 100 \rangle$ preferred magnetization directions. If the decisive process that determines the coercive field is assumed to be the inhibition of (100) 180° Bloch wall displacements by screw dislocations [61],* Eq. (27.2) for the coercive field as a function of temperature leads to the expression

$$H_c \propto \delta\lambda_{100}(c_{11} - c_{12})/M_s. \tag{36.1}$$

This curve is the dashed one in Fig. 19 (temperature $T \leq 600°C$). In this the temperature dependence of the magnetostriction constant λ_{100} and the spontaneous magnetization M_s, as given by Tatsumoto and Okamoto [71] and Patter [72] were included, while the temperature dependence of the

---

* At least at room temperature [64, 67–69], the (100) 180° walls have the smallest specific surface energy.
Transmission electron microscopy investigations [24, 70] show that screw dislocations predominate in the dislocation structure of deformed iron single crystals.

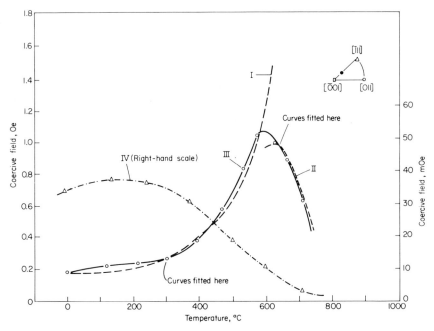

FIG. 19. Temperature dependence of the coercive field of a rod-shaped iron single crystal in the undeformed state as well as after 24.6% deformation at room temperature (tensile test) [61]. According to Allen et al. [62], single crystals with this orientation deform in the slip system (101), [111]. The theoretical curves in the region $T \leq 600°C$ were calculated from Eqs. (27.3) and (36.1), and those for $T \geq 600°C$ from Eq. (36.2). Theoretical curve (I): interaction between moving (001) 180° walls and screw dislocations in the (011) plane. Theoretical curve (II): rotation processes against internal stresses. Experimental curve (III): deformed crystal. Experimental curve (IV): undeformed crystal (right-hand scale).

elastic constants c_{11} and c_{12} was approximated by the temperature dependence of the Young's modulus [73]. If the temperature dependence of the Bloch wall thickness is calculated taking into account the magnetoelastic coupling [48, 61], the agreement with experimental results is not as good. The maximum of the coercive field moves toward lower temperatures as deformation is increased [61]. At temperatures above that of the maximum, the nonhomogeneous stress anisotropy is larger than the magnetocrystalline anisotropy. Here the magnetization changes occur predominantly by nonhomogeneous rotation. The coercive field as a function of temperature is then approximately given by the equation

$$H_c \propto \lambda_{100} G / M_s, \qquad (36.2)$$

(see also Table III and Becker and Döring [6] and Bilger and Träuble [61]). G is the shear modulus. In fact, this relation agrees quite well with experimental results (see Fig. 19).

In the case of undeformed, well-annealed iron single crystals with very low dislocation densities, the $H_c(T)$ curve has a completely different shape, as shown in Fig. 19. This temperature dependence can be explained by assuming that the impeding of Bloch wall displacements by the surface domain structure determines the coercive field [61].

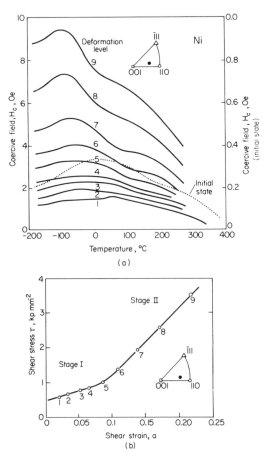

FIG. 20. (a) Temperature dependence of the coercive field of an intermediately oriented, rod-shaped nickel single crystal as a function of plastic deformation: measurements were made up to about 300°C only, in order to eliminate the possibility of recovery. (b) Corresponding work hardening curve: the crystal was deformed at room temperature by tensile stress. The curves shown in Fig. 20a were measured at the breaking points indicated.

For deformed *nickel* single crystals, similar considerations apply as for deformed iron. The temperature range in which the Bloch wall motions predominate (the temperature range below the maximum of $H_c(T)$ in Fig. 20a; the preferred directions are $\langle 111 \rangle$) is certainly much narrower than in the case of iron, due to the lower magnetocrystalline anisotropy and the larger magnetostriction of nickel. If we consider the dominant process to be the impeding of 180° Bloch wall movements by dislocation segments running parallel to the Bloch walls along a mean length l, then Eq. (27.2) for H_c as a function of temperature essentially reduces to

$$H_c \propto G |\lambda| (\delta l)^{1/2}/M_s, \tag{36.3}$$

where δ is the Bloch wall thickness, G is the shear modulus, and λ is a constant representing the magnetostriction, which is assumed to be isotropic. The interaction length l is dependent on the thickness of the Bloch wall, δ, and hence on temperature in a way that depends somewhat on the dislocation configuration [51]. For the extreme cases characterized by $l = $ constant (T), and $l \propto \delta^{1/2}$, we obtain the following relations:

$$H_c \propto G\lambda/M_s K_1^{1/4}, \qquad H_c \propto G\lambda/M_s K_1^{1/2}. \tag{36.4}$$

Below the maximum of $H_c(T)$, the measured curves always lie between these two limiting cases. A detailed discussion of this subject is presented elsewhere [51].

At temperatures sufficiently far above the maximum of $H_c(T)$ (more precisely, above the hump in the $H_c(T)$ curve), the shape of the $H_c(T)$ curve can be described, as in the case of iron, by a relation of the form of $H_c \propto \lambda G/M_s$ (magnetization changes due to rotations against stress anisotropy). The shift at increasing deformation of the maximum of the $H_c(T)$ curve towards lower temperatures, as revealed by Fig. 20a, may be deduced from the fact that at this maximum the strain anisotropy and the crystalline anisotropy become equal [51, 66, 74].

For cobalt single crystals (Fig. 21) the $H_c(T)$ curve has a completely different character. This is connected with the strong temperature dependence at higher temperatures of the magnetocrystalline anisotropy [36, 75]. For $T \leq 245°C$, cobalt is magnetically uniaxial (the preferred axis is the crystallographic c axis). For the intermediately oriented single crystals used in this investigation, approximately half the magnetization changes are due to rotation processes [51, 53]. The temperature dependence of the coercive field is therefore determined to a large extent by the temperature depend-

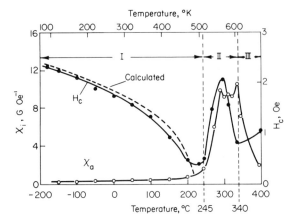

FIG. 21. Temperature dependence of the coercive field and initial susceptibility of an undeformed, rod-shaped cobalt single crystal [36]. The angle ϕ_0 between the c axis and the rod axis was 50°. The three intervals I, II, and III are distinguished by differences in the magnetic anisotropy (see text).

ence of the magnetocrystalline anisotropy. The dashed curve in Fig. 21, which was calculated from Eq. (28.1) with Eq. (12.2) agrees well with the measurements. For 245°C $\leq T \leq$ 340°C, the magnetically preferred axes form a cone whose opening angle with respect to the c axis, ϕ_0, increases continuously with rising temperature, and which reaches the value $\phi_0 = 90$ at approximately 340°C. The coercive field maximum at about 300°C occurs because at that temperature one of the preferred directions becomes parallel to the sample rod axis. Here changes in magnetization occur exclusively by Bloch wall displacements. In a third temperature interval, $T > 340$°C, any direction in the basal plane is a magnetically preferred direction. Among these the one is distinguished which forms the smallest angle with the rod axis because in that case the stray field energy is at a minimum. Thus, it follows that considerations similar to those employed for temperatures $T \leq 245$°C (magnetization changes by a combination of wall motions and rotations) lead to an understanding of the experimental curve [57].

In magnetically uniaxial crystals the relative contributions to magnetization changes of rotational processes and of Bloch wall movements are very sensitive to the orientation of the crystal. This would lead one to expect a strong dependence of the form of the $H_c(T)$ curve on sample orientation. The measurements shown in Fig. 22 confirm this prediction. When Bloch wall displacements predominate, the $H_c(T)$ curve is essentially determined by the temperature dependence of the magnetostriction constants; since

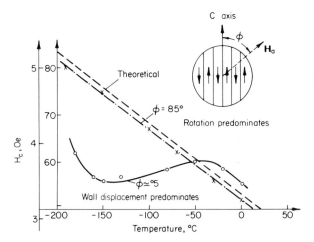

FIG. 22. The influence of crystal orientation on the temperature dependence of the coercive field in magnetically uniaxial cobalt single crystals. The measurements were taken on a short cylindrical crystal with the c axis perpendicular to the axis of the cylinder, for various directions between the c axis and the field.

these constants have not yet been measured for cobalt as a function of temperature, we cannot at present calculate $H_c(T)$ for a crystal with $\phi_0 \simeq 5°$.

37. Dependence on Deformation. The room temperature measurements shown in Figs. 23–26 fully confirm the theoretical expectations (see Table III).

For nickel (Figs. 23 and 24) the coercive field in region I of the work hardening curve is a linearly increasing function of $(\tau - \tau_0)^{1/2}$ (see Fig. 24a), and in regions II and III it is a linear function of $\tau - \tau_0$. The results are in satisfactory quantitative agreement with the theory. As an example we shall calculate the dependence of $H_c(\tau)$ on deformation in region II of the work hardening curve. The discussion is based on the same model that was described previously in connection with the temperature dependence of H_c. Assume that the 180° Bloch wall motion is impeded by dislocation segments of length l, and in each of the interaction processes experiences a mean force of magnitude $\pm 3Gb\lambda l$, where the sign of the force is determined by the sign of the dislocation. Here, G is the shear modulus, b (2.5×10^{-8} cm) is the magnitude of the Burgers vector, and λ (34×10^{-6}) is the magnetostriction constant, assumed to be isotropic. If ϱ is the total dislocation density, then the volume density of the obstacles is approximately equal to $\varrho/3l$ [see Eq. (27.5)]. The relation between ϱ and the shear stress τ in region II of the work hardening curve is given by Eq. (6.1). According to Kron-

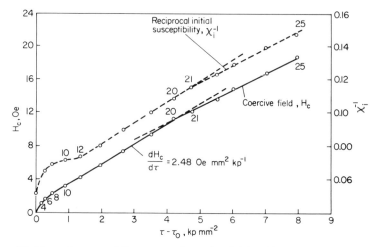

Fig. 23. Dependence on deformation of coercive field and initial susceptibility of an intermediately oriented nickel single crystal (room temperature). The corresponding work hardening curve is shown in Fig. 1.

müller [76] the number of dislocations per group, n, is about 20 in this case. From observations of Bloch walls [51], the mean Bloch wall area is $F \simeq 25 \times 10^{-6}$ cm², and the mean Bloch wall spacing $L \simeq 10^{-3}$ cm (both of these figures are somewhat uncertain). If further $l = 10^{-4}$ cm, which is a reasonable value according to the results of transmission electron microscopic studies, then Eq. (27.2) yields

$$(\partial H_c/\partial \tau)_{\text{II}} \simeq 0.5 \text{ Oe mm}^2 \text{ kp}^{-1}.$$

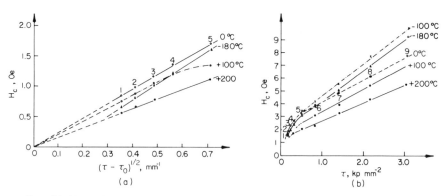

Fig. 24. Dependence on deformation of the coercive field of an intermediately oriented nickel single crystal at various temperatures corresponding to the measurements reproduced in Fig. 20. The corresponding work hardening curve is shown in Fig. 20b: (a) region I of the work hardening curve; (b) region II of the work hardening curve.

The experimental values range between 1.0 and 3.0 Oe mm² kp⁻¹. The dependence on deformation of the coercive field in region I of the work hardening curve can be calculated in the same way [51].

In the high-temperature region (e.g., at 200°C) magnetization changes occur mostly by inhomogeneous magnetization rotations against the stress anisotropy. As may be seen from Fig. 24, one finds $H_c \propto (\tau - \tau_0)^{1/2}$ in region I of the work hardening curve, and $H_c \propto (\tau - \tau_0)$ in region II of the work hardening curve. These results may be brought into agreement

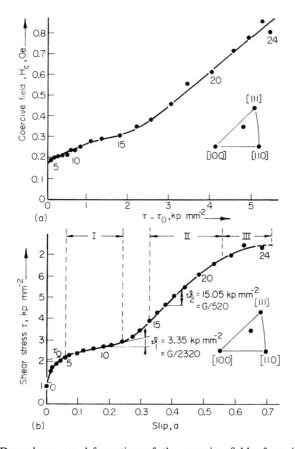

FIG. 25. Dependence on deformation of the coercive field of an intermediately oriented, rod-shaped iron single crystal at room temperature [61]: (a) the $H_c(\tau)$ function; (b) the corresponding work hardening curve measured at room temperature (tensile test). For the calculation of the work hardening curve, the slip system $\{(\bar{3}12), [111]\}$ has been used, according to the orientation change of the crystal during deformation. τ_0 is the critical shear stress; ϑ_1, ϑ_2 are work hardening coefficients in the regions I and II, respectively, of the work hardening curve; G is the shear modulus.

with the relation $H_c \propto |\sigma_i|$ (see Table III) if one assumes that the absolute value of the mean amplitude of the internal stresses $|\sigma_i|$ is proportional to $\varrho^{1/2}$ (ϱ is the dislocation density) [66]. However, at present there is no rigorous theory for this case.

In Fig. 25a, $H_c(\tau)$ is shown for an intermediately oriented iron single crystal. The corresponding work hardening curve is shown in Fig. 25b. It is similar to the work hardening curve of intermediately oriented fcc single crystals (see Fig. 1). $H_c(\tau)$ clearly reflects the work hardening regions I and II. It is seen, however, that the transition region between region I and region II is wider in the case of $H_c(\tau)$ than in the case of the work hardening curve. This is connected with the fact that in bcc crystals several glide systems are being activated. In region I, one has

$$H_c \propto (\tau - \tau_0)^{1/2},$$

and in region II,

$$H_c \propto \tau - \tau_0,$$

as for fcc crystals.

These results indicate that at least for intermediately oriented iron single crystals, the total dislocation density depends on shear stress in the same characteristic manner as in fcc metals. Hence, the theoretical interpretation will be analogous to that given in the case of nickel. A quantitative analysis between theory and the experiment in Fig. 25b is not possible at present, because as yet the relation between shear stress τ and dislocation density ϱ has not been investigated quantitatively. Conversely, however, the measurements of $H_c(\tau)$ yield information about the dependence of τ on ϱ [61].

Finally, Fig. 26 shows the coercive field as a function of deformation for a *cobalt* single crystal. As has already been mentioned in Sec. 5, cobalt is characterized by a high density of grown-in dislocations. This makes it possible to test the relations concerning the interplay of various types of lattice imperfections, as was discussed in Secs. 27–29. In deformed crystals, the first type of obstacle is the grown-in dislocations with a density ϱ_0, which are in effect independent of deformation. The second type of obstacle consists of those dislocations which are formed during deformation. According to Eq. 5.1, their density increases with increasing deformation. The relation between H_c and $(\tau - \tau_0)^{1/2}$ calculated from these considerations [57] is in such good agreement with experiment (see Fig. 26) that there can be hardly any doubt about the correctness of the initial Eq. (27.2) or Eq. (29.4). The results with respect to the dependence on deformation of the coercive field (and the initial susceptibility, as well: see Sec. 38)

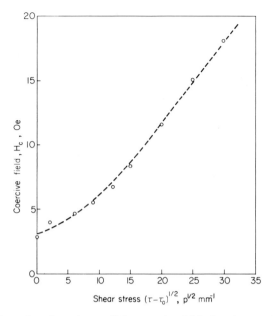

FIG. 26. Deformation dependence of the coercive field of an intermediately oriented, rod-shaped cobalt single crystal at room temperature, where H_c is plotted against $(\tau - \tau_0)^{1/2}$; (○) experimental; (----) theory, $H_c = \text{const} \, (\varrho_0 + \varrho_v)^{1/2}$. The angle between the rod axis and the c axis is 50°. The corresponding work hardening curve is linear, with a hardening coefficient $\theta = d\tau/da = 1$ kp mm^{-2} [36].

provide strong support for the applicability of our statistical method of treating the Bloch wall displacements. If the Bloch walls would bulge out during their motion at individual lattice defects [55], the expected relation would have the form $H_c \propto \varrho$, in contrast to the experimental results.

Initial Susceptibility, χ_i

38. The dependence on temperature and deformation of the initial susceptibility was measured on nickel and cobalt single crystals. Figure 27 shows the temperature dependence of the initial susceptibility of an intermediately oriented nickel single crystal after various degrees of plastic deformation. In Fig. 28 an interpretation of this curve is proposed, starting with the simplifying assumption that the contributions to magnetization changes of Bloch wall movements, χ_i^ω, and rotational processes, χ_i^R, is additive. The intermediate maximum of χ_i (at 210°C) in an undeformed crystal, occurs at the zero point of the anisotropy constant K_1 [78, 79]. Below that temperature (low-temperature range) the magnetocrystalline

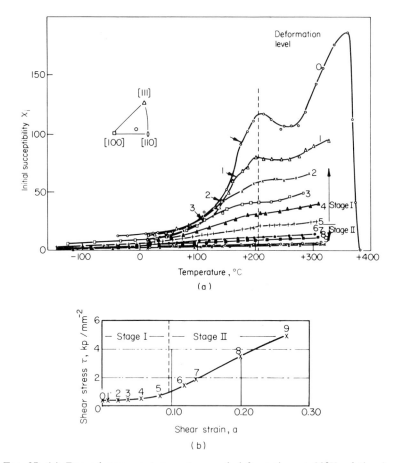

FIG. 27. (a) Dependences on temperature and deformation at 90°K of the (true) initial susceptibility of an intermediately oriented, rod-shaped nickel single crystal. Ballistic measurements by Rieger [77]. The arrows indicate the shift of a kink in the $\chi_i(T)$ curves as the deformation increases (see Träuble [51]). (b) Corresponding work hardening curve for $T = 90°K$.

anisotropy is considerably larger than the stress anisotropy. The magnetization changes occur essentially by Bloch wall displacements. At temperatures sufficiently far above that at the maximum, the magnetization changes by rotational processes against the stress anisotropy. This concept, which is to a large extent analogous to that employed in the interpretation of the $H_c(T)$ curve, is supported by the following results:

(1) In the low-temperature range, the $\chi_i(T)$ curve follows Eq. (30.2). With the model described in Sec. 37 for the interaction between Bloch

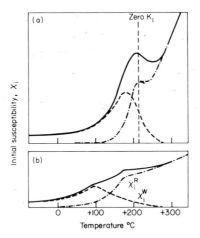

FIG. 28. Interpretation of the temperature dependence of the initial susceptibility of nickel single crystals. Superposition of the contributions of magnetization changes of rotational processes and Bloch wall movements. (a) Undeformed crystal: χ_i^R = contribution of rotation processes; χ_i^W = contribution of wall motion. (b) Deformed crystal. [$\chi_i(T)$ curve taken from Fig. 27a in (a) and (b).]

walls and dislocation structures, Eq. (30.2) yields the following relation for the temperature dependence of the initial susceptibility:

$$\chi_i \propto M_s^2 \delta^{1/2}/|\lambda| G l^{1/2}. \tag{38.1}$$

Equation (38.1) gives a satisfactory description of the experimental curve (see Table III).

(2) The deformation dependence of χ_i in the low-temperature range obeys the law $\chi_i^{-1} \propto \varrho^{1/2}$ as expected from Eq. (27.2). In particular, the quantity χ^{-1} increases linearly with $(\tau - \tau_0)^{1/2}$ in region I of the work hardening curve, and linearly with $(\tau - \tau_0)$ in regions II and III (see Fig. 23). The calculated slopes [50] are in good agreement with experiment.

In the high-temperature range (above ~220°C) the temperature dependence of χ_a can be described by the relation,

$$\chi_a \propto M_s^2/|\lambda| G, \tag{38.2}$$

which is expected when magnetization changes occur by rotations against the stress anisotropy energy [5]. The rapid decrease in χ_i shortly below the Curie temperature point is connected with the rapid decrease of the spontaneous magnetization in that region. Let us now briefly consider the deformation dependence in the high-temperature range. According to

Fig. 29, χ_i^{-1} at about 220°C increases linearly with $(\tau - \tau_0)$ along the entire work hardening curve; only the rates of increase are different in regions I and II. This means [see Eqs. (5.3) and (6.2)] that the initial susceptibility in work hardening regions I and II varies in different ways with the dislocation density. A theoretical interpretation of this behavior thus far could be given for the work hardening region I only (see Eq. (5.6) as well as Seeger et al. [37] and Kronmüller [48]).

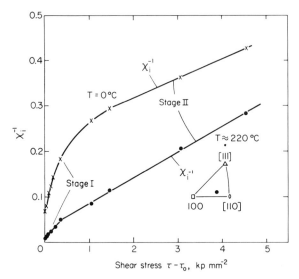

FIG. 29. Dependence on deformation of the initial susceptibility of an intermediately oriented nickel single crystal at room temperature and at 220°C. (After Rieger [77].)

In Fig. 28b an attempt is made to explain the behavior of the $\chi_i(T)$ curve as the deformation, i.e., the stress anisotropy increases (Fig. 27). According to Fig. 27, the maximum of $\chi_i(T)$ first transforms into a plateau as the deformation is increased. At further deformation the plateau spreads toward lower temperatures, and eventually vanishes entirely. The explanation sketched in Fig. 28 starts with the concept that the magnetocrystalline anisotropy decreases rapidly as the temperature increases [78, 79], whereas the stress anisotropy remains nearly constant. Hence, with increasing deformation, the upper bound of the temperature region where Bloch wall movements are dominant ($\Phi_K > \Phi_0$) is shifted towards lower temperatures, while the relative contribution of the rotational processes increases.

In Fig. 30 we present the temperature dependence of the initial susceptibility of an intermediately oriented, undeformed cobalt single crystal with a relatively high dislocation density ($\varrho_0 \simeq 10^8$ cm^{-2}). In the attempt to explain this curve, the reader is reminded to the previously discussed dependence of the preferred magnetic directions on temperature. In the temperature ranges I and III, magnetization changes occur by Bloch wall movements and rotations simultaneously. Because of the large magnetocrystalline anisotropy of cobalt, the effect of lattice imperfections on the initial susceptibility is small in this region. Except for the temperature

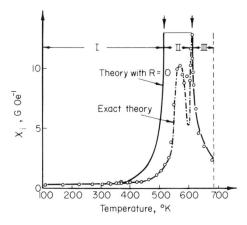

FIG. 30. Temperature dependence of the initial susceptibility of an undeformed, intermediately oriented, rod-shaped cobalt single crystal [36]: (○) experimental; (———) theoretical ideal crystals; (– – – –) rigorous theory in which the magnetoelastic coupling is taken into account [53].

interval between 400 and 500°K (see below), the simple theory, neglecting magnetoelastic coupling [Eq. (31.2) temperature region I], describes the measured curves (solid curves in Fig. 30). In the temperature region II, on the other hand, there exists a multiphase domain structure. Here magnetization changes occur predominantly by Bloch wall movement, and the initial susceptibility is very sensitive to crystal imperfections. Therefore the experimental curves within this region deviate considerably from the ideal curve. For example, the first maximum does not occur at $K_1 = 0$, but at 570°K instead. Furthermore, the initial susceptibility is strongly temperature dependent between 500 and 600°K. These results can be related quantitatively to the effect of the magnetoelastic coupling and to the gradual change in the preferred magnetic directions as the temperature rises [51, 53].

If one introduces a force constant R which takes into account the effect of dislocations on the Bloch wall mobility, then the calculations based on Eq. (31.1) will correctly reproduce the experimental $\chi_i(T)$ curve up to the beginning of temperature region II. The first maximum of the initial susceptibility at 570°K is due to the fact that at this temperature one of the preferred directions becomes parallel to the axis of the sample rod (see also the $H_c(T)$ curve in Fig. 21). The presence of a second maximum at $K_1 = -2K_2$ is connected with the fact that the Bloch wall thickness becomes theoretically infinite at this temperature, i.e., the force constant R approaches zero, and the initial susceptibility assumes its limiting value $1/N_\parallel$ which is determined by the demagnetizing energy ($N_\parallel$ is the demagnetization factor parallel to the axis of the sample rod). A detailed discussion of the dashed theoretical curve in Fig. 30 will be found in the original paper [53].

The dependence on deformation of the initial susceptibility of cobalt is, at least at room temperature (and presumably in the entire temperature range I) considerably smaller than that of nickel. This is due to the high magnetocrystalline anisotropy which must be overcome by the rotational processes. The experimental $\chi_i(\tau)$ curve [36] can be analyzed quantitatively [51, 53] as the consequence of a change in crystal orientation (change of the contribution of rotational processes to magnetization changes), as well as that of an increase of the dislocation density (reduction of the Bloch wall mobility) due to deformation.

Reversible Susceptibility, χ_{rev}

39. Figure 31 shows the behavior of the reversible susceptibility, χ_{rev}, of an intermediately oriented nickel single crystal as a function of field strength and plastic deformation. In the stepwise reduction of the $\chi_{\text{rev}}(H)$ curve, which is most clearly seen for undeformed crystals, those rules become evident that are expected from the phase theory (Sec. 24) for the sequence of the magnetization processes in an ideal crystal. Following a first section (plateau) where the susceptibility is high (Bloch wall movements being dominant), there is a field range in which Bloch wall movements and rotational processes play considerable roles. In this field range, the number of magnetic phases decreases stepwise as the field increases. In strong fields, the changes in magnetization occur predominantly by rotations (range of approach to saturation).

Following plastic deformation, the mobility of the Bloch walls is considerably lower (lowering of the plateau: see also Fig. 23) whereas the rotational processes are only slightly affected. The plateau, which characterizes the field range of Bloch wall movements, increases with increasing deformation. The location of the kink at the end of the plateau in many

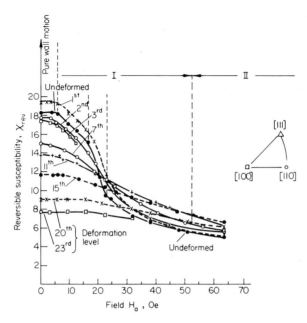

FIG. 31. Dependence on field and deformation of the reversible susceptibility, χ_{rev}, of an intermediately oriented, rod-shaped nickel single crystal at room temperature: (I) wall motion and rotation processes (number of phases decreasing in steps); (II) predominant rotation processes (at higher fields exclusively). The measurements were taken along the virgin magnetization curve. The corresponding work hardening curve is shown in Fig. 1. The deformation dependence of χ_i is presented in Fig. 23.

of the $\chi_{\text{rev}}(H)$ curves depends on deformation in the same way as the coercive field (compare Figs. 1, 23 and 31).

It is interesting that the reversible susceptibility in the region of pure rotational processes is larger *after* plastic deformation than *before*. This is due to the fact that the magnetization distribution is not homogeneous in the vicinity of dislocations, whereby the susceptibility increases. Further measurements of the dependence of the reversible susceptibility of nickel single crystals on orientation and deformation have been published by Rieger [80] and by Mehrer et al. [81].

Approach to Saturation

40. According to Sec. 34, susceptibility measurements in the range of the approach to saturation may be very useful for (1) studies of the effects of internal stresses on reversible rotation processes, and (2) for investigations of the dislocation configurations in plastically deformed crystals. At this point recent measurements of the field dependence of χ_S of nickel single crystals [45, 76, 82] should be cited. These measurements make it possible to verify the models of dislocation configurations in plastically deformed crystals, as discussed in Secs. 4–8. In region II of the work hardening curve, for example, the number of dislocations delivered from a given dislocation source should be independent of shear strain, and the total dislocation density ϱ should increase as the square of the flow stress, τ^2. Hence, Eq. (34.3) yields the following relation for χ_S as a function of deformation in region II of the work hardening curve:

$$\chi_S \propto \tau^2. \tag{40.1}$$

Corresponding measurements taken on a nickel and a 80–20% Ni–Co single crystal in region II of the work hardening curve are reproduced in Fig. 32. The theoretical relation of Eq. (40.1) is well confirmed by these experiments.

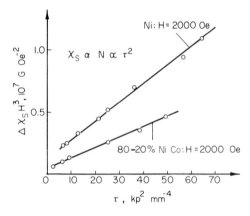

FIG. 32. Susceptibility at approach to magnetic saturation, χ_S, as a function of the shear stress τ in region II of the work hardening curve for deformed single crystals of nickel and 80–20% Ni–Co at room temperature [45, 76].

Moreover, a comparison of the measured with the calculated proportionality factor in Eq. (40.1) leads to the conclusion that the dislocations are arranged spatially discontinuous and thus create strong, long-range elastic stress fields. The calculated number n of dislocations with the same sign in a group agrees well with electron microscopy observations [45, 76].

Further, measurements of the variation of χ_S with temperature permit a decision on the question of whether the interactions between the dislocations and the magnetization are due to magnetoelastic coupling as was assumed earlier, or whether stray fields that may be present in the dislocation core (Néel [83]) may perhaps play an essential role. In fact, the experiments unequivocally support the first concept.

Finally, it should be mentioned, that from the annealing behavior of the susceptibility χ_c and the parameters χ_i and H_S, important conclusions may be drawn concerning dislocation reactions during annealing (see Sec. 8). As to these problems, the reader may refer to the original literature [84–86].

REFERENCES

1. J. A. Brinkman, *J. Appl. Phys.* **25**, 961 (1954).
2. J. A. Brinkman, *Am. J. Phys.* **24**, 246 (1956).
3. A. Seeger, *Proc. Intern. Conf. Peaceful Uses Atomic Energy, 2nd, Geneva* **6**, 250, United Nations, New York (1958).
4. A. Seeger, "Radiation Damage in Solids," Vol. 1, p. 101. Intern. Atomic Energy Agency, Wien, 1962.
5. R. Becker and W. Döring, "Ferromagnetismus." Springer, Berlin, 1938.
6. A. Seeger, H. Kromüller, and H. Rieger, *Z. Angew. Phys.* **18**, 377 (1965).
7. A. Seeger, Kristallplastizität, *in* "Handbuch der Physik" (S. Flügge, ed.), Vol. VII, p. 2. Springer, Berlin, 1958.
8. S. Mader, "Electron Microscopy and Strength of Crystals," p. 183. Wiley (Interscience), New York, 1963.
9. A. Seeger, S. Mader, and H. Kronmüller, "Electron Microscopy and Strength of Crystals," p. 665. Wiley, New York, 1963.
10. S. Mader, *in* "Moderne Probleme der Metallphysik" (A. Seeger, ed.), Vol. I, p. 192. Springer, Berlin, 1965.
11. R. Berner and H. Kronmüller, *in* "Moderne Probleme der Metallphysik" (A. Seeger, ed.), Vol. I, p. 35. Springer, Berlin, 1965.
12. U. Essmann, *Acta Met.* **12**, 1468 (1964).
13. J. W. Steeds and P. M. Hazzledine, *Discussions Faraday Soc.* **38**, 103 (1964).
14. Z. S. Basinski, *Discussions Faraday Soc.* **38**, 93 (1964).
15. U. Essmann, *Phys. Status Solidi* **12**, 707, 723 (1965); **17**, 725 (1966).
16. A. Seeger, H. Kronmüller, S. Mader, and H. Träuble, *Phil. Mag.* **6**, 639 (1961).
17. A. Seeger, H. Kronmüller, O. Boser, and M. Rapp, *Phys. Status Solidi* **3**, 1107 (1963).
18. H.-M. Thieringer, *European Regional Conf. Electron Microscopy, 3rd, Prague, 1964*, p. 229.
19. J. W. Steeds, cited by P. B. Hirsch, *in* "The Relation between the Structure and Mechanical Properties of Metals." H. M. Stationary Office, London, 1963; See also Steeds and Hazzledine [13].
20. S. Mader and H.-M. Thieringer, *Conf. Electron Microscopy, 5th, Philadelphia*, J3. Academic Press, New York (1962).
21. Z. S. Basinski, and S. J. Basinski, *Phil. Mag.* **9**, 51 (1964); see Basinski [14].
22. H. Kronmüller, *in* "Moderne Probleme der Metallphysik" Vol. I, (A. Seeger, ed.), Vol. I, p. 126. Springer, Berlin, New York, 1965.
23. H. G. van Bueren, "Imperfections in Crystals." North-Holland Publ., Amsterdam, 1960.

24. A. S. Keh, *Phil. Mag.* **12**, 9 (1965).
25. L. Néel, *J. Phys. Radium* **5**, 241 (1944).
26. H. Rieger, H. Kronmüller, and A. Seeger, *Z. Metallk.* **54**, 553 (1963).
27. A. Seeger and H. Kronmüller, *Phil. Mag.* **7**, 897 (1962).
28. F. Vicena, *Czech. J. Phys.* **5**, 480 (1955); **4**, 419 (1954); **5**, 11 (1954).
29. M. Peach and J. S. Koehler, *Phys. Rev.* **80**, 436 (1956).
30. R. De Wit, *Acta Met.* **13**, 1210 (1965).
31. G. Rieder, *Abhandl. Braunschweig. Wiss. Ges.* **11**, 20 (1959).
32. H. Träuble, *Z. Metallk.* **53**, 211 (1962).
33. C. Radeloff, *Z. Angew Phys.* **17**, 247 (1964).
34. E. W. Lee, *Rept. Progr. Phys.* **18**, 184 (1955).
35. B. A. Lilley, *Phil. Mag.* **41**, 792 (1950).
36. H. Träuble, O. Boser, H. Kronmüller, and A. Seeger, *Phys. Status Solidi* **10**, 283 (1965).
37. A. Seeger, H. Kronmüller, H. Rieger, and H. Träuble, *J. Appl. Phys.* **35**, 740 (1964).
38. J. D. Eshelby, F. C. Frank, and F. R. N. Nabarro, *Phil. Mag.* **42**, 351 (1951).
39. G. Leibfried, *Z. Phys.* **130**, 214 (1951).
40. W. F. Brown, Jr., "Magnetostatic Principles in Ferromagnetism." North-Holland Publ., Amsterdam, 1962.
41. W. F. Brown, Jr., "Micromagnetics." Wiley (Interscience), New York, 1963.
42. W. F. Brown, Jr., *Phys. Rev.* **58**, 736 (1940).
43. W. F. Brown, Jr., *Phys. Rev.* **60**, 132, (1941).
44. A. Seeger and H. Kronmüller, *J. Phys. Chem. Solids* **12**, 298 (1960).
45. H. Kronmüller and A. Seeger, *J. Phys. Chem. Solids* **18**, 93 (1961).
46. E. Kröner, "Kontinuumstheorie der Versetzungen und Eigenspannungen." Springer, Berlin, 1958.
47. H. Kronmüller, *Z. Angew. Phys.* **15**, 197 (1963).
48. H. Kronmüller, in "Moderne Probleme der Metallphysik" (A. Seeger, ed.), Vol. II. Springer, Berlin, 1966.
49. F. Bitter, *Phys. Rev.* **38**, 1903 (1931); **41**, 507 (1932).
50. H. J. Williams, R. M. Bozorth, and W. Shockley, *Phys. Rev.* **75**, 155 (1949).
51. H. Träuble, in "Moderne Probleme der Metallphysik" (A. Seeger, ed.), Vol. II, p. 157. Springer, Berlin, 1966.
52. H. Lawton and K. H. Stewart, *Proc. Roy. Soc. (London)* **A193**, 72 (1948).
53. H. Kronmüller, H. Träuble, A. Seeger, and O. Boser, *J. Mat. Sci. Eng.* **1**, 91 (1966).
54. Y. Barnier, R. Pauthenet, and G. Rimet, *Compt. Rend.* **252**, 3024 (1961).
55. M. Kersten, *Z. Angew. Phys.* **8**, 313, 382 (1956).
56. L. J. Dijkstra and C. Wert, *Phys. Rev.* **79**, 979 (1950).
57. H. Träuble, H. Kronmüller, A. Seeger, and O. Boser, *J. Mat. Sci. Eng.* **1**, 167 (1966).
58. H. J. De Wit, dissertation, Univ. Amsterdam, 1965.
59. H.-M. Thieringer, *Z. Metallk.* **59**, 400 (1968); **59**, 476 (1968).
60. T. Holstein and H. Primakoff, *Phys. Rev.* **58**, 1098 (1940).
61. H. Bilger and H. Träuble, *Phys. Status Solidi* **10**, 755 (1965).
62. M. P. Allen, B. E. Hopkins, and J. E. McLennan, *Proc. Roy. Soc. (London)* **A234**, 221 (1956).
63. T. Okamura and T. Hirone, *Phys. Rev.* **55**, 102 (1939).
64. H. Dietrich and E. Kneller, *Z. Metallk.* **47**, 716 (1956).

65. E. Kneller and G. Schmelzer, *Z. Metallk.* **51**, 342 (1960).
66. D. Krause, *Z. Physik* **168**, 239 (1962).
67. C. D. Graham, Jr., and P. W. Neurath, *J. Appl. Phys.* **28**, 888 (1957).
68. C. D. Graham, Jr., *J. Appl. Phys.* **29**, 1451 (1958).
69. L. Špaček, *Ann. Phys.* **5**, 217 (1960).
70. J. R. Low and A. M. Turkalo, *Acta Met.* **10**, 215 (1962).
71. E. Tatsumoto and T. Okamoto, *J. Phys. Soc. Japan* **14**, 1588 (1959).
72. H. H. Patter, *Proc. Roy. Soc. (London)* **A146**, 362 (1934).
73. W. Köster, *Z. Metallk.* **39**, 1 (1948).
74. E. Kneller, *in* "Berichte der Arbeitsgemeinschaft Ferromagnetismus," 1958, pp. 35ff. Riederer, Stuttgart, 1959; see also "Ferromagnetismus." Springer, Berlin, 1962.
75. Y. Barnier, R. Pauthenet, and G. Rimet, *Compt. Rend.* **252**, 2839 (1961).
76. H. Kronmüller, *Z. Physik* **154**, 574 (1959).
77. H. Rieger, *Z. Angew. Phys.* **17**, 166 (1964).
78. U. Hofmann, *Phys. Status Solidi* **7**, K145 (1964).
79. I. N. Pussei, *Isv. Akad. Nauk SSSR Ser. Phys.* **21**, 1088 (1957).
80. H. Rieger, *Z. Angew. Phys.* **15**, 232 (1963).
81. H. Mehrer, H. Kronmüller, and A. Seeger, *Phys. Status Solidi* **11**, 313, 323 (1965).
82. H. Dietrich and E. Kneller, *Z. Metallk.* **47**, 672 (1956).
83. L. Néel, *J. Phys. Radium* **9**, 185, 193 (1948).
84. H. Rieger, H. Kronmüller, and A. Seeger, *Z. Metallk.* **54**, 553 (1963).
85. A. Seeger and H. Kronmüller, *Phil. Mag.* **7**, 897 (1962).
86. H. Mehrer, H. Kronmüller, and A. Seeger, *Phys. Status Solidi* **10**, 725 (1965).
87. H. Kronmüller, "Nachwirkung in Ferromagnetika" (Springer Tracts in Natural Philosophy, Vol. 12). Springer, Berlin, 1968.
88. H. Mughrabi, *Phil. Mag.* **18**, 1211 (1968).
89. K.-H. Pfeffer, *Phys. Status Solidi* **20**, 395 (1967); **21**, 837 (1967).
90. A. Hubert, *Phys. Status Solidi* **22**, 709 (1967); *J. Appl. Phys.* **39**, 444 (1968).

CHAPTER XIV

Recovery and Recrystallization

V. V. DAMIANO
The Franklin Institute Research Laboratory
Philadelphia, Pennsylvania

C. DOMENICALI
Temple University
Philadelphia, Pennsylvania

E. W. COLLINGS*
The Franklin Institute Research Laboratory
Philadelphia, Pennsylvania

SURVEY OF MAGNETIC AND STRUCTURAL CONCEPTS	689
FERROMAGNETIC METALS AND ALLOYS	697
NONFERROMAGNETIC METALS AND ALLOYS	712
SIGNIFICANCE AND USES OF MAGNETIC ANALYSIS FOR METALLURGY	718
References	720

Survey of Magnetic and Structural Concepts

1. The magnetic properties of metals and alloys can be divided into two categories according to their dependence upon the detailed metallurgical structure or physical state. In the first category are those few properties such as the saturation magnetization and Curie temperature of ferromagnetic metals that are almost entirely independent of the structure and which instead depend upon the chemical or compositional nature of the metal or alloy. In the second and larger category are those properties which are structure sensitive and depend upon the mechanical history of the metal. These properties, because of this structure sensitivity, often also show a time dependence which is a measure of the speed with which changes in the physical state of the material occur.

* Present Address: Battelle Memorial Institute, Columbus, Ohio.

All the magnetic properties of a metal depend ultimately upon the manner in which the atomic magnetic dipoles become aligned with an externally applied magnetic field. If the metal or alloy is ferromagnetic, the material consists of ferromagnetic domains whose orientations depend upon numerous factors of both a magnetic as well as a mechanical nature. As has been discussed in Chapter XIII, the resultant magnetic structure and magnetic behavior of a given piece of metal depend on an interplay between these various magnetic and mechanical factors. Such an interplay often gives rise to an enormously complicated overall magnetic behavior. If the metal or alloy is nonferromagnetic, magnetic effects are often present on a considerably smaller scale, and while a certain structure sensitivity may very well occur, it is very difficult to properly separate the observed effects into a part due to the intrinsic properties of the nonferromagnetic metal, and a part due to any minute ferromagnetic inclusions contained in the material.

2. *Saturation.* The magnetic behavior of a ferromagnetic metal within a particular microscopic region involves an interplay between the "local" magnetic and mechanical parameters within this small region; it then is clear that if the externally applied magnetic field is large enough, the magnetic behavior ideally becomes relatively simple. As a result, we are concerned with the *"saturation"* properties of the material. In such a case, if a ferromagnet is magnetized to saturation, the mechanical and magnetomechanical parameters (such as internal stresses and magnetostriction effects, etc.) are overwhelmed, so to speak, and the saturation magnetization is determined by the temperature and the chemical composition of the material. Thus, the saturation magnetization itself would be unaffected by the physical or mechanical state of the ferromagnetic material; conversely, little if anything could be learned about the physical or mechanical state of a ferromagnet from a study of its saturation magnetization [1].

3. *Approach to Saturation.* On the other hand, one can appreciate the fact that experimentally we cannot actually completely magnetize a ferromagnet to saturation; we can only get arbitrarily close to the state of saturation, and it is of importance to know how the magnetization depends upon the applied magnetic field at high fields. This *approach to saturation* with increasing magnetic field depends upon the detailed mechanical and physical state of the ferromagnet. For this reason, it is of interest and importance to study the effects of imperfections upon the magnetic state of ferromagnetic metals and alloys. In particular, we shall discuss in Section 15 the work of Brown [2] and Néel [3], who discuss the influence of dislocations and nonmagnetic inclusions, respectively, upon the dependence of magnetization on magnetic field at very high fields (the so-called law of approach to saturation; see also Chapter XII).

4. Initial Susceptibility and Permeability. We return now to a consideration of a ferromagnetic specimen, single crystalline or polycrystalline, whose net magnetization is zero. Such a specimen is made up in general of many ferromagnetic domains which are individually magnetized to saturation (for the given temperature) but whose total vector magnetic moment per unit volume vanishes. Upon the application of a small magnetic field, the domain walls shift about in such a manner as to give a finite component of magnetic moment in the direction of the applied field. The motion of these domain walls is usually extremely complicated and has been shown to depend very strongly upon the precise state of stress throughout the specimen, the magnetostriction properties of the substance, the density and distribution of dislocations and of other imperfections, and other characteristics such as crystallite size, etc. [4]. Many of these variables in turn depend upon the previous history of mechanical deformation and heat treatment. Furthermore, these variables, as well as their related magnetic properties, may in general be time dependent. It is clear that metallurgical processes such as recovery and recrystallization will generally be accompanied by corresponding changes in certain magnetic properties, and conversely, that these magnetic properties may be used to study the metallurgical processes themselves. If a long, thin, needle-shaped ferromagnetic specimen has vanishing magnetization in the absence of any applied magnetic field, the material of the specimen is said to be in the virgin state. The application of a very small field ΔH will give rise to a very small magnetization ΔM and very small induction ΔB within the material. The *initial susceptibility* is then defined as

$$\chi_0 = \lim_{\Delta H \to 0} \Delta M / \Delta H,$$

and the *initial permeability* as

$$\mu_0 = \lim_{\Delta H \to 0} \Delta B / \Delta H.$$

We see then that χ_0 and μ_0 are highly structure sensitive, so that these parameters not only can be altered drastically by metallurgical treatment, but may also be used to investigate the changes accompanying metallurgical treatment.

5. Incremental Susceptibility. If a specimen in the virgin state is subjected to a monotonically increasing magnetic field H, it reaches states of magnetization $M(H)$ which are no longer reversibly connected, in contrast to the states in the near vicinity of $H = 0$, $M = 0$. If at a given point (M, H) on the virgin curve the field H is now decreased slightly, the magnetization does not follow the original virgin curve, but rather, exhibits a small M–H "loop" as H is cycled through values near the initial point of departure from the original virgin curve. The inclination of this loop relative to the H axis gives a measure of the so-called *"differential susceptibility"* or *"incremental susceptibility,"* χ_{diff}, corresponding to the value H at which the field was originally diminished at the start of the loop. In terms of a virgin curve giving B vs. H these quantities are called the *differential* or *incremental permeability* μ_{diff}. Here, too, the dependence of χ_{diff} or μ_{diff} upon H is determined by the metallurgical state of the ferromagnetic material, and the differential permeability has been used to study the metallurgical processes of recovery and recrystallization.

6. Hysteresis Loop. It is well known that if the magnetic field H is slowly cycled between positive and negative values sufficiently large to achieve near saturation, one obtains a closed curve usually called a *"B–H curve"* or *"M–H curve"* on which the magnitude of the intercept on the B or M axis is called the *remanence* B_R or M_R, and the field at the intercept of the H axis is called the *coercive force* H_c. The shape of this B–H curve is highly structure sensitive, as would be expected, and consequently the parameters H_c and B_R (which are of great technological importance) are clearly also highly structure sensitive.

7. Hysteresis Loss. In many technological applications the ferromagnetic materials are driven rapidly through a B–H loop many times per second (for example, in transformers, chokes). If the nature and even the very existence of an "open" B–H loop originates in irreversible interactions between moving domain walls and the crystal lattice, it is to be expected that there would result a heating effect which increases as the size of the loop (extent of the irreversibility) increases. It can be shown that the area enclosed by such a loop is, in fact, a direct measure of the heat given off to the lattice during the passage of H through a complete cycle (Warburg's law), and this heat loss is called the *hysteresis loss* per cycle. Thus, hysteresis loss is also a structure-sensitive property which has been used to follow metallurgical changes of ferromagnetic materials during recovery and recrystallization. It should be noted that hysteresis losses, strictly speaking, refer to losses arising from the irreversibility of magnetization changes. These in turn give rise to B–H loops traversed very slowly and if the B–H cycle is carried out more rapidly (60 Hz transformers, for example), there will be additional (eddy current) losses involved.

8. Aftereffect. It is a well established fact that the magnetization of ferromagnets proceeds mainly by domain wall motion at low fields and by magnetization rotation against anisotropy forces at high fields. The fact that these motions do not *instantaneously* follow an applied field is due to the phenomenon called *magnetic aftereffect*, first observed experimentally by Ewing [5] and explained by Snoek [6]. Actually, there are various time effects related to the motions of domain walls and their delay in following an applied field. For example, the ac permeability may change with time after the application of a magnetic field, an effect which was named "disaccomodation" by Snoek. Magnetic aftereffects are considered to arise from the interactions between domain walls and various lattice imperfections.

9. *Nuclear Magnetic Resonance.* All the magnetic properties discussed above represent macroscopic or large-scale consequences of the interaction between lattice imperfections and magnetization (except for the saturation magnetization itself, which is essentially independent of lattice imperfections). In other words, the parameters which are measured (for example, the initial permeability or the hysteresis loss) represent a kind of "average" value over the entire crystal and do not permit us to "probe" into the lattice structure to observe the "local" field on an *atomic scale*. It would be most helpful if one had some means for inserting an atomic probe near a dislocation, for example, to investigate the local magnetic field in the vicinity of such an imperfection. The phenomenon of *nuclear magnetic resonance*, [7] first observed in 1946, furnishes such a probe. A nucleus located at a particular point within a crystal finds itself subject to electric and magnetic fields. The magnetic field is usually the applied field H_0, but in ferromagnetic materials this field exists as a kind of "Weiss internal field" because of the spontaneous magnetization within a domain. In some situations, if a nucleus has a nonspherical "shape," and if this nucleus finds itself in an electric field which is of appropriately low symmetry, there will be a torque on the nucleus tending to line it up along certain directions. Both of these situations, the first consisting of a nuclear magnetic dipole moment being aligned by a magnetic field and the second consisting of a nuclear *electric quadrupole* moment being aligned by an appropriately asymmetric electric field, can give rise to a set of closely spaced, quantized energy levels for the nuclear magnetic moment. By use of an externally applied magnetic field of the proper frequency it is possible to observe absorption of energy from this weak field by the nuclear magnetic moment. Thus, the nucleus acts like an "atomic probe" whose magnetic resonance behavior is determined by the detailed structure and configuration of the magnetic and electric fields in its vicinity. Therefore, magnetic resonance techniques can be used to investigate the environment of vacancies, interstitials, dislocations, etc. and to attack problems of metallurgical interest.

10. We are concerned in this chapter with the twofold problem of the effects of recovery and recrystallization on the magnetic properties of metals and alloys, and conversely, how magnetic effects can be used to investigate recovery and recrystallization. Thus far, we have described briefly and qualitatively the way in which various important magnetic parameters may be affected by the physical or metallurgical state of metal. We now wish to describe briefly and again qualitatively the meaning and current interpretation of the two metallurgical processes of recovery and recrystallization.

Let us imagine that we begin with an annealed polycrystalline sample of some pure ferromagnetic metal or alloy, and that we then cold-work the specimen very severely. A small part of the work, which was done in deforming the specimen, is stored in the metal as the energy of lattice defects introduced by plastic deformation, elastic energy, magnetostrictive energy, energy associated with local disorder of atoms from their original lattice positions, magnetostatic energy, and others. The structure of the specimen is now rather complicated and there are large numbers of deformed grains. These grains contain large numbers of dislocations and regions of elastic strain apart from what is associated with the dislocations, as well as large numbers of vacancies and interstitial atoms distributed throughout the metal. The crystallographic orientations of certain adjacent regions within the metal may differ slightly so that there exists a *low-angle boundary* between them. If, however, two adjacent regions have drastically different (say, more than 10 or 15°) crystallographic orientations, then there is a *grain boundary* between them. The entire specimen is in a metastable state on both a macroscopic and atomic scale. Interstitial atoms could lower the overall energy of the specimen if they could get over a "potential barrier" which prevents them from taking their normal places in the regular crystal lattice; the filling of excess vacancies would lower the overall energy if neighboring atoms could move in to fill the vacancies as they diffuse to the surface; the redistribution and/or annihilation of dislocations would lower the energy if the corresponding potential barriers could be surmounted; ferromagnetic domain walls ("Bloch walls") are held in position by corresponding "barriers", (and so on for the various deviations from perfection within the severely cold worked specimen). Now, the various potential barriers can be overcome by raising the temperature of the specimen to the appropriate level. There are a variety of effects of enormous complexity which may take place in the "relaxation" of the cold-worked specimen when its temperature is raised.

11. It is very difficult to give universally suitable definitions of terms such as recovery and recrystallization. For our purposes the following definitions will be reasonable and for the most part satisfactory. We consider *recovery* to consist of all those processes which take place during heat treatment primarily within the grains of the material; the overall grain structure of the specimen produced by cold working is considered to be essentially unchanged during the recovery processes. We consider *recrystallization* to consist of all those processes which take place at higher temperatures than those required for recovery, and consisting of the nucleation and growth of undeformed grains at the expense of the deformed structure.

12. *Recovery.* Recovery processes are distinguished from recrystallization processes by the kinetics associated with changes of some particular macroscopic property. Recovery is generally found to follow first-order reaction kinetics. Therefore, if x represents the deviation of some macroscopic parameter from its value in the completely annealed state, then the time evolution of x is expressed by $dx/dt = -cx$, where t is time and c is a constant characteristic of the parameter chosen, as well as the temperature (heat treatment) and the material itself. The temperature dependence of c is ordinarily that characteristic of thermally activated processes and has the usual form $c(T) = c_0 \exp(-E/kT)$, where k is Boltzmann's constant and E represents the energy barrier height (the "activation energy") for the particular process in question. The experimental determination of the activation energy furnishes an insight into the mechanisms of recovery, and often several different activation energies are found to apply in the recovery range, suggesting that several different processes are taking place simultaneously. In some cases, for example, in aluminum studied by Masing and Raffelsieper [8] and discussed theoretically by Kuhlmann *et al.* [9], the activation energy itself is found to decrease linearly with progressive recovery.

The first stages of recovery are believed to be associated with the disappearance of excess vacancies. There are at least four mechanisms for the disappearance of vacancies [10]: (1) vacancies disappear at the free surface of the specimen and at grain boundaries; (2) vacancies disappear upon recombination with interstitials; (3) vacancies disappear by reacting with dislocations; and (4) vacancies combine to form vacancy disks, which collapse into dislocation loops. Because point defects give rise to considerable scattering of conduction electrons, the early stages of recovery can often be detected by electrical resistivity changes before other magnetic or mechanical changes become apparent. Mott [11] has suggested, as a basic process in recovery, the "climbing" of dislocations as a result of vacancies migrating to the dislocations. Kuhlmann [12] has discussed a mechanism of recovery in which dislocations glide over obstacles with the aid of thermal activation. She finds the activation energy for recovery by this process to be a decreasing function of stress in the material. Sometimes a grain consists of many subgrains which are only slightly misoriented crystallographically with respect to one another and thus separated by low-angle grain boundaries. In such cases, during recovery the low-angle boundaries move so as to allow one subgrain to grow at the expense of its neighbors, leading to the formation of fewer and larger subgrains which, in an electron microscope, appear as clearly defined cells. This process is called polygonization. During the growth of subgrains, the main process is migration of low-angle boundaries, and the kinetics of recovery of various macroscopic properties differ for subgrain growth from those observed at lower temperatures. Some investigators [13] treat subgrain growth as separate from "true" recovery. Direct observations of the recovery process in the electron microscope by Bailey and Hirsch [14] reveal that the density of dislocations within the subgrains decreases during the final stages of recovery, indicating actual annihilation of dislocations within the subgrains.

13. *Recrystallization.* The recrystallization process consists of the growth of undeformed grains or crystallites, and can usually be followed, for example, by various metallographic techniques, as well as hardness measurements. The kinetics of recrystallization are typically rather complicated, and the fraction of material recrystallized can often be expressed as some exponential function of the time. Although we have described subgrain growth via low-angle boundary migration as a recovery process, it is really open to question whether strain-free nuclei are formed in the deformed matrix with release of stored energy, or whether nucleation leading to crystallization is associated with the migration of low-angle boundaries as in subgrain growth. The work of Bailey and Hirsch [14] indicates that the value of the activation energy for nucleation agrees with the value of the activation energy for boundary migration. These results are supported by observations of recrystallization occurring by boundary migration in foils observed in the electron microscope. Finally, since recrystallization involves the movement of boundaries and since these boundaries can become "attached" to impurity atoms, the rate of recrystallization depends in general upon the concentration of such impurities. Blade *et al.* [15] have found that as little as 50 ppm of iron, silicon, or copper in pure aluminum drastically lowers the rate of recrystallization of the aluminum.

Ferromagnetic Metals and Alloys

14. *Saturation Magnetization.* We have already mentioned in Sec. 2 that the saturation magnetization under certain ideal conditions is not a structure-sensitive property. Strictly speaking, this is not quite accurate, since the saturation magnetization measures essentially the number of magnetic dipoles per unit volume, and the density of a ferromagnetic specimen quenched from high temperatures may be slightly decreased because of vacancies. Therefore, if the total magnetic moment of the sample is fixed but its volume is slightly increased because of vacancies, the magnetization will probably be slightly lower than for the same specimen without vacancies. Of course, this is a rather small effect, at most perhaps 0.001% under conditions of an extremely rapid quench. Similarly, a heavily cold-worked specimen contains relatively high concentrations of vacancies and interstitials, and should give a low value of saturation magnetization M_s, while the annealed specimen should give a high value of M_s. In practice this dependence of M_s upon the concentration of point defects does not usually arise, and it becomes more important to consider the approach to saturation as the field is increased. It is an interesting question whether or not a vacancy has the simple "volume" effect implied above upon the saturation magnetization. The contribution of a given atom (say, of nickel) to M_s depends upon the electronic nature of its surroundings, and an atom in the interior of a crystal does not have the same surroundings as an atom on the surface. Although an experimental investigation of this question would be quite difficult, it might nevertheless be quite illuminating. The experiment might best be done with single crystals of nickel quenched from various high temperatures in order to observe the effects of quenched-in vacancies and yet such that saturation could be nearly achieved in reasonable and attainable magnetic fields. Experiments on recovery of high-purity nickel wire plastically deformed by loading have recently been reported by Reits *et al.* [16]. The loading and all measurements were carried out at $-196°C$. Among other parameters (electrical and mechanical), they measured the magnetic induction B after various low-temperature anneals of the wire pulled at $-196°C$, and they found a linear change in B (for an applied field of 18,200 Oe) with annealing temperatures between $-196°C$ and room temperature. They give only the absolute value of the change in B, and presumably the B increases with increasing annealing temperature. However, their specimen was a polycrystalline wire so that they are actually observed the approach to saturation rather than the effect suggested above. We shall return later to this work.

15. *Approach to Saturation and High-Field Susceptibility.* For a single crystal of a ferromagnet without crystalline imperfections, neglecting shape effects, the approach to saturation would depend upon the relative directions of the applied magnetic field and the crystallographic axes. However, even for the so-called "hard" directions, the approach to saturation would depend entirely upon the magnetocrystalline anisotropy coefficients which measure the ease or difficulty of magnetizing the crystal in various crystalline directions. These coefficients are not structure sensitive, so that their measurement and their effect on the approach to saturation of ferromagnets is not, in itself, of interest to the metallurgist. However, any imperfection, in particular, a dislocation, can in principle affect the way in which the saturation state is approached in large magnetic fields. In very large magnetic fields, the various domain wall motions have presumably been exhausted, leaving only the final twisting or rotation of all local magnetic dipoles into alignment with the large applied field. In accord with the views of Brown [2] and Néel [3], we consider the effect of an imperfection on this final rotational alignment to be that of setting up a preferred direction for the "local" magnetization. If this local direction does not happen to coincide with that of the applied field, this local competition between applied field and local "effective" field will prevent perfect alignment. A point defect "occupies" a very small volume compared with a dislocation or a large-scale void, so that the law of approach to saturation in ferromagnetics may be attributed primarily to the presence of dislocations and voids.

Considering only the rotation of the magnetization against crystal anisotropy torques, Becker and Döring [4] and Chikazumi [1] derive the following formulas for the magnetization M vs. H:

$$M = M_s(1 - b/H^2 - c/H^3 - \cdots) + \chi_p H \qquad \text{Becker and Döring} \qquad (15.1)$$

$$M = M_s(1 - b/H^2 - d/H^4 - \cdots) + \chi_p H \qquad \text{Chikazumi.} \qquad (15.2)$$

Here, M is the magnetization at field H, M_s is the saturation magnetization (dependent upon the temperature), and the term $\chi_p H$ represents the paramagnetic contribution. The difference between the two expressions results from the different approximations made by these authors. Chikazumi assumes that the torque aligning the magnetization is independent of H, leading to an angular deviation $\theta \sim H^{-1}$ for the angle between the magnetization and the field. If instead, we take $\theta = A/H = B/H^2$, we find a formula for M vs. H which contains a H^{-3} term as well, in agreement with Becker and Döring [17]. In any case, this discrepancy may be academic for two reasons: first, at high fields the terms in H^{-3} or H^{-4}, etc. are negligible in comparison to the term b/H^2 which is common to both formulas and second, it was found experimentally [17] by Weiss, Weiss and Forrer, and others that M actually depends on H in the manner

$$M = M_s(1 - \alpha/H), \qquad (15.3)$$

in the case of direct high-field measurements of M vs. H; thus, neither of the formulas cited above, which are based on crystal anisotropy, includes the experimentally predominant term α/H. It is, in fact, this term in α/H, as well as a part of the b/H^2 term, which Brown attributes to dislocations and with Néel attributes to voids. (Néel has pointed out that a strict H^{-1} dependence as $H \to \alpha$ leads to an infinite amount of work required to magnetize a specimen to complete saturation, so that at very high fields the dependence of M on H must be dominated by a term decreasing at least as fast as H^{-2}.) On the other hand, medium high-field measurements of dM/dH by Czerlinsky [18] indicate an M vs. H^{-3} variation which is linear over a large range of field in the case of nickel and reasonably linear in the case of iron. Work of Polley [18, 18a] indicates a variation of the form $M = A/H + B/H^2$ for nickel in fields of 500–3000 Oe.

In Brown's theory, the deviation from saturation is assumed to be due to magnetostrictive forces localized in the stress field around the dislocation. Pairs of dislocations of opposite sign, separated by a short distance, contribute a term α/H to the deviation from saturation, while pairs separated by a long distance and surplus dislocations of one sign contribute a term b/H^2, where b is a logarithmically varying function of H. This b/H^2 term would be an additional contribution (due to dislocations) to the expression for M over and above the H^{-2} term due to crystalline anisotropy and discussed above (equations of Becker-Döring and Chikazumi).

More recently, Seeger and Kronmüller [19] (see also Chapter XIII) have extended the theoretical work of Brown and have derived expressions for the high-field differential magnetic susceptibility χ_d in the case of deformed fcc metals. In their treatment, the susceptibility is separable into contributions arising from individual dislocations and contributions arising from interactions between all the dislocations in a piled up group.

In particular, they emphasize that great care is required in simply treating the stress field of dislocation pileups as if the pileup were a "super dislocation" of greater strength than that of individual dislocations. By effectively separating the stress field of a pileup into a "near field" and a "distant" field, they are able to obtain expressions for stored energy as well as for high-field susceptibility in which the effects of *arrangements* of dislocations can be separated from the effects of *total number* of dislocations. They find that stored energy in a deformed fcc crystal, as with most physical properties, is not very sensitive to the arrangement of dislocations, whereas the differential susceptibility χ_d in the saturation region should be sensitive to the actual relative arrangement or spatial distribution of dislocations. Seeger and Kronmüller quote the (unpublished) work of Rieger (1962), who carried out annealing experiments between 300 and 1300°K on nickel single crystals deformed in tension (see Fig. 1). The heat treatment was reported to be

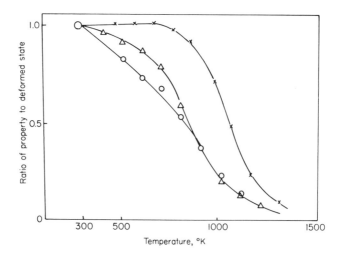

FIG. 1. Recovery of differential susceptibility in the saturation range and the coercive force for nickel single-crystal deformed to stage II: (×) coercive force; (△) differential suceptibility at 700 Oe; (○) differential susceptibility at 1900 Oe.

such that no recrystallization took place, so that the investigation was concerned with the processes involved in recovery. It was found, for example, that after the 1000°K anneal about three-quarters of the differential susceptibility (at 700 and 1900 Oe) has recovered, whereas only about one-quarter of the coercive force has recovered. Seeger and Kronmüller state that "the annealing behavior of χ_d below 900°K depends upon the measuring field in the sense that recovery is faster for larger values of H." The quoted work of Rieger clearly indicates that the *law of approach to saturation* depends markedly upon the processes that have taken place during recovery.

Kronmüller et al. [20] have also reported on electrical, magnetic, and mechanical studies of recovery in nickel. They have used the following three methods for generating lattice defects: (1) plastic deformation at room temperature—producing interstitial atoms, vacancies and dislocations; (2) quenching from high temperatures to room temperature—producing vacancies and vacancy clusters only; and (3) neutron irradiation at room temperature—producing point defects and depleted zones. Among other magnetic properties they measured the differential (reversible) susceptibility χ_d in the approach to saturation. They propose that fcc metals may exhibit five stages of annealing, three of which are found in nickel above room temperature. The annealing processes ascribed to these three stages are migration of interstitials (stage III), migration of vacancies (stage IV), and the rearrangement and annihilation of dislocations (stage V). Kronmüller et al. gave a table showing the direction of the change (i.e., increase or decrease) in various electrical and magnetic properties, as well as the measured activation energies, for the three stages of recovery and for the three methods for generating lattice defects listed above. The differential susceptibility only decreases during stages IV and V, and it changes only in originally deformed specimens. Quenched and deuteron-irradiated specimens showed no changes in χ_d at any stage, in accordance with the notion that the high-field susceptibility χ_d is sensitive primarily to the rearrangement and annihilation of dislocations and that point defects can be detected (by measurements of χ_d) only in an indirect manner if they cause recovery of long range stresses. (A more quantitative treatment of the approach to saturation is given in Chapter XIII.)

16. *Initial Susceptibility and Initial Permeability.* Ferromagnetic materials usually exhibit an appreciable change in the initial susceptibility when they are subjected to applied stresses. In the elastic range of stresses and strains, tensile strains produce an increase in the initial susceptibility χ_i of ferromagnets with positive magnetostriction and a decrease in χ_i of ferromagnets [5] with negative magnetostriction. When the elastic strains are relaxed, the susceptibility of the unstrained material is restored. It should be noted, however, that cold working of a metal involves plastic deformation and leaves residual elastic strains which do not disappear upon removal of the externally applied stresses during cold working. As previously mentioned, cold working introduces numerous imperfections into the lattice including vacancies, interstitials, dislocations, and others.

Annealing of the cold-worked ferromagnetic metal results in the gradual restoration of the magnetic properties to the original stage prior to cold working unless, of course, the situation is even further complicated by questions of solid solubility of the constituents. Bumm and Müller [21] were able to distinguish three distinct stages in the restoration process of iron–nickel which were associated with phenomena which they called

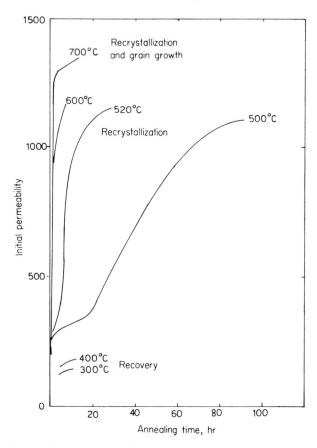

FIG. 2. Change of initial permeability on annealing and the relationship to recovery, recrystallization, and grain growth for iron–nickel alloy cold-rolled 97%.

recovery, recrystallization, and grain growth. The results of their work are shown in Fig. 2. In the "recovery" range 300–400°C, only small changes in the initial permeability μ_0 were observed. In the range 500–520°C, which they took to be the range of recrystallization, the largest changes took place in the initial permeability. Further increases in μ_0 observed after annealing in the temperature range 600–700°C were attributed to recrystallization and grain growth.

Effects of dislocations and their associated stresses upon the initial susceptibility, and the effects of annealing, have been treated theoretically in recent years by Kersten [22] and have been investigated experimentally by Asanuma [23].

Asanuma investigated the recovery of initial susceptibility in pure nickel wires which had been plastically deformed in tension. His results for the behavior of initial susceptibility with time of anneal are shown in Fig. 3.

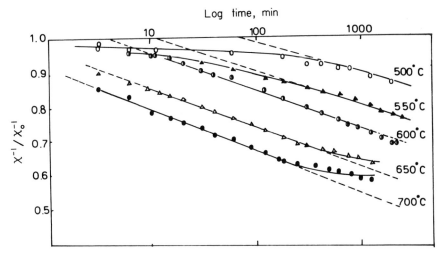

FIG. 3. Isothermal recovery of the initial susceptibility of pure nickel.

It was found that no change in χ_0 took place at temperatures below 450°C for annealing times up to about 20 or 30 hr (judged by Fig. 3). On the other hand, significant changes in electrical resistivity were observed even at 350°C, indicating that the initial recovery processes involve motions of point defects but not of dislocations. The activation energy determined from the resistivity data was found to be 0.9 eV, which is in agreement with the activation energy for vacancy migration in nickel. It was concluded that recovery at the lower temperatures proceeds by the disappearance of excess lattice vacancies. The recovery of the initial susceptibility, which began to appear above 500°C, followed a logarithmic time dependence which was in accord with the analysis of Kuhlmann [24]. The mechanism of recovery of the initial susceptibility was considered to be the disappearance of dislocations followed by climbing of dislocations through self-diffusion.

Support for this conclusion is given by the fact that Asanuma's analysis of his initial susceptibility, based on Kuhlmann's formulation, gives an activation energy of 3.0 eV in agreement with the same value found by others for the self-diffusion process in nickel. These results are similar to those of Boas [25], who investigated the annealing kinetics of nickel by measurements of internal energy, electrical resistivity, density, and hardness.

Träuble and Seeger [26] have shown that in nickel, the governing processes for the initial susceptibility χ_0, as well as those for the coercive force H_c, are domain wall movements and that these wall movements are strongly impeded by long-range stresses. Their conclusions are thus in accord with the experimental findings of Bumm and Müller and of Asanuma cited above. (A quantitative treatment of the initial susceptibility problem and related measurements are given in Chapter XIII, Sec. 30.)

17. Coercive Force. If we consider simply the "bulk" effects of magnetostriction upon the coercive force of a ferromagnet, we can see that the potential energy $H_c M_s$ of a volume element of the material subject to a field H_c (equal to the coercive force) would depend upon the strain energy density $\lambda \sigma_i$ in the element, where λ is the "isotropic" magnetostriction coefficient and σ_i is the "average" internal stress. It is shown in Becker and Döring [27] that $H_c M_s = C_1 \lambda \sigma_i$, so that the coercive force is given by

$$H_c = C_1 \lambda \sigma_i / M_s. \tag{17.1}$$

It can also be shown [28] that the initial susceptibility χ_0 of a ferromagnet depends upon the same quantities λ, σ_i, and M_s in the fashion

$$\chi_0 = C_2 M_s^2 / \lambda \sigma_i. \tag{17.2}$$

Elimination of $\lambda \sigma_i$ from Eq. (17.1) by the use of Eq. (17.2) gives the important relation

$$H_c = C M_s / \chi_0, \tag{17.3}$$

showing that the coercive force and the initial susceptibility are inversely proportional, at least in this first approximation.

Recent work in the theory of the coercive force has been directed toward deducing H_c from the detailed distribution and arrangements of dislocations in the ferromagnet. Vicena [29] and Kersten [30] have discussed this problem, and Vicena finds the relation

$$H_c = (2M_s I)^{-1} R \delta (\ln L/\delta)^{1/2} \varrho^{1/2}, \tag{17.4}$$

where M_s is the saturation magnetization, L is the average dimension of a ferromagnetic domain, R is the height of the potential barrier which must be overcome by the Bloch wall when passing over the dislocations, δ is the width of the Bloch wall, and ϱ is the density of dislocations. The calculation is based on the assumption that the Bloch wall does not change its shape or width when passing over the dislocations; the height of the potential barrier is calculated as the change in surface energy of the wall during passage over the dislocations.

Kersten [30] has given the relation

$$H_c = M_2^{-1}(KkT_C/a_0)^{1/2}n^{1/2}, \qquad (17.5)$$

where K is the coefficient of crystalline anisotropy, k is Boltzmann's constant, T_C is the Curie temperature, and a_0 is the lattice constant. This relationship is based on the assumption that the Bloch wall is pinned on the dislocation lines, and that it bends until its bent outside surfaces touch and unite. In a later paper [31] Kersten stated that his Eq. (17.5) above related primarily only to the annealed state. Dietrich and Kneller [32] had also given the formula

$$H_c = 1.5(\lambda/M_s)bG\varrho^{1/2}, \qquad (17.6)$$

in which λ is the coefficient of isotropic magnetostriction, b is the Burger's vector, and G is the modulus of elasticity. It can be shown that Eqs. (17.4) and (17.6) are essentially identical when expressed in terms of the same variables, except for a discrepancy represented by a numerical factor of the order of δ/L.

Measurements on the dependence of the coercive force upon plastic deformation by extension, which could be used for a comparison of Eqs. (17.4) and (17.5), were published independently by Dietrich and Kneller on single crystals of nickel [33], by Wotruba also on nickel [34], and by Malek on iron, nickel, and Fe–Ni alloys [35]. A comparison of the theoretical dependence in Eq. (17.4) above and that found experimentally was carried out for iron by Malek [35]. Good agreement was obtained both for the nature of the dependence of H_c on the several variables in Eq. (17.4), as well as in the absolute values of the coercive force H_c. On the other hand, Malek found that Kersten's relation, Eq. (17.5), leads to values which are two orders of magnitude greater than the measured values (see, however, Kersten [31]). Thus, it was found that Vicena's theory gave correct values for H_c in iron and nickel for dislocation densities in the range of 10^8–10^{11} dislocations per square centimeter.

In a later paper, Malek [36] altered Vicena's theory slightly to take into account the change in width of the Bloch wall with increasing magnetic anisotropy in the case of alloys of iron and nickel and was able to explain the behavior of H_c in these alloys. While the general agreement was quite satisfactory, several points should be noted. First, experimental verification of these theoretical formulations relating coercive force to dislocation density is, in fact, difficult to achieve. This difficulty arises because there is at present no absolute way to measure dislocation densities directly in bulk crystals. The comparison of Vicena's and Malek's theory with experiment rests upon the assumption that a given state of plastic deformation is uniquely associated with a certain density of dislocations. Estimates of dislocation densities in cold-worked metals have been based on measured values of the stored energy, and these estimates cannot be accurate without some knowledge of the dislocation model of the cold-worked metal. Second, Vicena's theory does not take into account interactions between dislocations, so that it cannot be expected to be correct for high densities of dislocations. Third, it is of interest that Eq. (17.3) above does not universally hold, as was found by Malek [36] in nickel–iron alloys; he observed that while H_c vs. ε (ϵ is the plastic strain) exhibits a maximum in some alloys, the curve of μ_i vs. ε for the same alloy does not necessarily exhibit a minimum. Thus, there appears to be some characteristic of these alloys which changes with deformation in such a way that it influences the coercive force H_c considerably more than it influences the initial permeability μ_0.

Seeger and Kronmüller [19a] consider the coercive force in nickel to depend mainly on the total number of dislocations and very much less on their arrangement. Their theory takes account of interactions between dislocations, and their results (data of Rieger, unpublished), shown in Fig. 1, indicate that below 700°K the coercive force H_c remains unaltered and that rearrangements only, not any annihilation of dislocations, takes place. Taking together the evidence contained in Fig. 1, Seeger and Kronmuller conclude that there are three separate temperature regions in which recovery takes place.

(1) below 700°K no annihilation of dislocations takes place. The long-range stress fields around dislocation groups are partially removed by dislocation rearrangements.

(2) In the temperature range between 700 and 900°K, a further reduction of the long-range stress fields goes hand in hand with the annihilation of dislocations.

(3) Above 900°K the distances between dislocations are larger than 300 Å. The long-range stress fields have largely disappeared. Further recovery in this temperature range is mainly due to the annihilation of dislocations.

Kronmüller et al. [20] have used measurements of coercive force H_c to follow recovery in nickel. They find that the sensitivity of H_c and that of high-field susceptibility χ_d to internal stresses can be used to study depleted zones in neutron-irradiated nickel. Their results indicate that at lower and intermediate temperatures these zones are accompanied by a measurable stress field. With the help of migrating vacancies, these zones are transformed into a configuration associated with much smaller stresses. Possible configurations are voids or stacking fault tetrahedra.

Of technological interest is the work of Stanley [37] on the effects of recovery in Fe + 3% Al alloys upon several magnetic properties, including coercive force, remanence, and "permeability." It should be mentioned that Stanley defines permeability simply as the ratio of the induction B at a specified field H to the field H. While this definition is of technological interest, it is not customarily used by physicists. Figure 4 shows schematic-

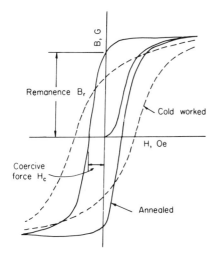

FIG. 4. Change in the hysteresis loop of cold-worked ferromagnetic metal after annealing.

ally the changes found in the B–H loop upon cold working and subsequent annealing. The coercive force and the remanence decrease upon annealing (recovery), and the permeability μ_{10} (B/H for $H = 10$ Oe) increases upon recovery. Stanley points out the technological implications and applications of the recovery process in the area of "soft" ferromagnetic materials.

(A more refined theory of the coercive force and related measurements on single crystals are given in Chapter XIII, Sec. 27.)

18. Hysteresis Loss. It has long been known that the magnetic properties of "soft" ferromagnetic materials may be significantly improved by annealing to relieve internal stresses. This is of special importance for materials used in transformers, for example, where it is desired to reduce hysteresis losses to a minimum. Besides eddy current losses, a ferromagnet exhibits magnetic losses ("hysteresis") which result from the irreversible changes in direction of magnetization. It is clear from the foregoing that the hysteresis losses should be affected by the density and arrangement of dislocations.

We shall discuss briefly the work of Dunn [38] on dislocation movements and their relation to magnetic recovery. Dunn attempted to correlate the restoration of hysteresis losses to the microstructure of flat single crystals of silicon–iron which had been deformed by bending. Tests were made on the crystals in the "as bent" condition and also after annealing for 1 hr at 650, 950, and 1100°C, respectively. The hysteresis losses and the value of H for $B = 15{,}000$ G were recorded and are shown in Fig. 5, plotted as a

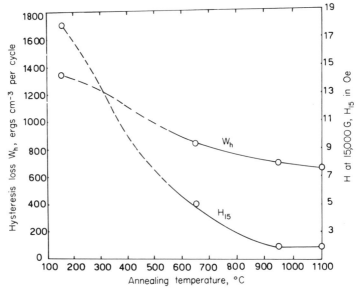

FIG. 5. Relation of hysteresis loss at 15,000 G and H at 15,000 G to annealing temperature in a plastically bent iron–silicon alloy.

function of the annealing temperature. Using etch pit techniques, Dunn associated the observed magnetic changes with the arrangement of dislocations. Below 650°C the dislocations were arranged along slip planes. The recovery of magnetic parameters was attributed to the relaxation of long-range residual stresses by the movement of dislocations on the glide planes. At temperatures above 650°C but not exceeding 950°C, the dislocations moved into walls perpendicular to the slip planes as polygonization occurred, and an additional change in the magnetic parameters was noted. In this temperature range some annihilation of dislocations was believed to have occurred but no recrystallization was noticed. Annealing above 950°C resulted in the growth of subgrains and additional small changes in the magnetic properties. Thus, the work and the suggestions of Dunn, in connection with hysteresis losses, is in essential agreement with that of later work of others associated with coercive force, differential susceptibility, initial permeability, and other magnetic parameters.

19. *Anisotropy Induced by Working of Metals: Torque Measurements.* The development of textures or preferred orientations in both cold-rolled and recrystallized metals is well known to metallurgists. As a result of these textures, ferromagnetic metals may exhibit a magnetic anisotropy [39]. The magnetic anisotropy of a rolled sheet may be measured by the torque

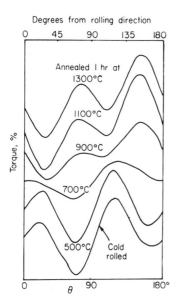

FIG. 6. Torque curve of silicon–iron reduced 40% by cold rolling and annealed for 1 hr at various temperatures.

exerted on a disk cut from the sheet when the disk is placed at various orientations relative to an external magnetic field. Typical torque curves for silicon–iron reduced 40% by cold rolling and then annealed for 1 hr at various temperatures are shown in Fig. 6 according to Tarasov [40]. These results show that the magnetic anisotropy of the cold-rolled material goes through a gradual change when it is annealed at increasingly higher temperatures. This suggests that the cold-rolled texture of silicon–iron changes upon recrystallization, and this conclusion has been confirmed by x-ray analysis. Since the texture of a cold-rolled metal is essentially unaltered by recovery, the use of magnetic anisotropy effects to distinguish between recovery and recrystallization phenomena is apparent. This technique is restricted to those ferromagnetic metals and alloys which exhibit a well-defined change in texture upon recrystallization. It has been applied to Ni and Ni–Fe alloys by Conradt et al. [41], whose results are indicated in Fig. 7. (An extensive treatment of textured materials is given in Chapter XV.)

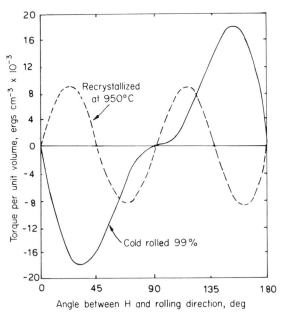

FIG. 7. Dependence of torque curves of a 93–7% nickel–iron alloy on heat treatment.

20. Magnetic Aftereffect. Many years ago, Ewing [5] observed that when a ferromagnetic material is placed in a magnetic field, a time delay is observed before the magnetization reaches the steady value associated with the applied field. The physical origin of the magnetic aftereffect may be looked at in the following way. Suppose that a ferromagnet is in a magnetic state such that changes in magnetization come from domain wall displacement, which means that the state is one below the "knee" of the virgin curve of M vs. H. Suppose now that the motion of the domain walls is such as to establish an energetically more favorable distribution of one or more kinds of lattice imperfections. This connection between domain walls and imperfections exists because of changes in the spatial distribution of stresses associated with these imperfections, the interactions being of a magnetoelastic nature. When the domain walls move in response to a change in the applied field, the distribution or "pattern" of the imperfections changes, or attempts to change, in such a way as to lower the overall energy of the ferromagnet. However, these "accommodating" changes in the pattern and arrangement of the lattice imperfections are not instantaneous, but rather are usually controlled by an activation energy so that there is a delay between the application of the applied field and the final equilibrium arrangement of the various imperfections.

The mechanism just described was originally proposed by Snoek [42] to explain magnetic aftereffects in iron containing carbon atoms in interstitial positions. In this case, the motion of domain walls upon sudden change of magnetization leads to a preferential directional distribution of the carbon atoms along the (100) axis parallel to the applied field, and the "accommodating" motions of the interstitial carbon atoms are determined by the thermally activated process of diffusion of these atoms. It should be mentioned that we are using the word "accommodating" here in a descriptive sense, and that there is another effect which was called "Disaccommodation" by Snoek [43] and which is related to but distinct from the usual magnetic aftereffect.

Néel [44] has extended and modified Snoek's concept by considering the effects of the carbon atom distribution upon the direct magnetic coupling between adjacent iron atoms, and has shown that these magnetic dipole effects may be considerably more sensitive to the presence of the carbon atoms than are the purely magnetoelastic effects treated by Snoek.

The importance of magnetic aftereffects in recovery processes is seen from the experimental work of Richter [45] and of Wittke [46], who found that the magnetic aftereffect in carbon–iron disappeared when the carbon was removed.

Biorci *et al.* [47] made a study of the magnetic aftereffect due to the motion of dislocations in iron. Here the effect is attributed to the relaxation of the magnetostrictive stress pattern of the dislocations as this pattern "follows" the changing positions of the domain walls. These authors use the concept of a "viscosity field" H_t, introduced by Néel, which represents the field which must be applied in order to move the domain walls when they have become fixed in certain positions by the diffusion of various crystalline defects. The general dependence of this viscosity field upon temperature was found to be in agreement with the dependence of internal friction upon temperature for the same specimens. An interpretation of the observations

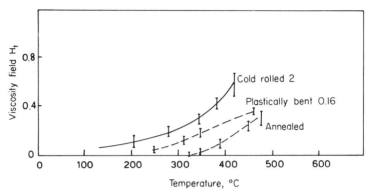

FIG. 8. Viscosity field versus temperature of an Armco iron specimen (large crystals); the frequency of the measuring field 30 Hz.

was given in terms of Vicena's theory of coercive force. The results of Biorci *et al.* on iron crystals are shown in Fig. 8. The values of the viscosity field are in the range 0–1 Oe, and the magnetic aftereffect was found to increase with increasing cold work and hence with increasing dislocation density. The application of aftereffect studies to diffusion problems is discussed in Chapter XVI.

Nonferromagnetic Metals and Alloys

21. The study of recovery and recrystallization of nonferromagnetic metals and alloys by means of magnetic techniques has been very limited. Because of the very great effects of minute traces of ferromagnetic inclusions, most, if not all, magnetic susceptibility studies are open to criticism concerning the interpretation of results. One cannot simply make "corrections" for the presence of ferromagnetic inclusions because the amount and nature of the latter are themselves dependent upon mechanical and thermal treatment of a specimen of metal even when the actual *total* amount of iron, for example, is fixed. During the heat treatment of a metal the iron may become internally oxidized, and the measured magnetization of the specimen may then depend upon the time and temperature of the heat treatment for reasons having little or nothing to do with recovery or recrystallization.

An indication of the difficulties involved may be seen from the work of Hutchinson and Reekie [48] and of McClelland [49] on pure copper. The results suggest that susceptibility measurements are not reliable in the study of recovery and recrystallization in nonferromagnetic materials. While it would seem that more research is called for in this area, it might be of more importance to study first the details of how magnetic inclusions are formed from iron or nickel impurities, as well as the effects of annealing atmosphere upon these inclusions.

One magnetic technique which is of importance in the study of recovery and recrystallization in nonferromagnetic metals and alloys is that of *nuclear magnetic resonance* (NMR). The application of NMR to such materials, as well as to cobalt, is discussed in Secs. 22-24.

22. Nuclear Magnetic Resonance (*NMR*). The theory and techniques of NMR have been described by many workers and the reader is referred to such standard works as Ingram [50] Andrew [51] and also Chapters III and XVII. In pure magnetic dipole resonance one of the fundamental observables, line position, is related to the local nuclear magnetic environment, and the other, linewidth, is related to relaxation mechanisms and inhomogeneities.

If the environment of a nucleus does not possess cubic symmetry, resulting in the presence of an electric field gradient, interaction between this and the nuclear quadrupole moment will produce a splitting of the nuclear spin levels even in the absence of a magnetic field. Transitions between these levels, induced by the rf *magnetic* field, give rise to pure quadrupole resonant absorption. In pure quadrupole resonance (NQR), a single crystal is, of course, not required. The occurrence of quadrupole interaction is particularly helpful in studying the distortion and annealing of crystals which ideally possess cubic symmetry. Quadrupole interaction occurs when $I > \frac{1}{2}$. For example, in a powder with $I = \frac{3}{2}$, first-order perturbation theory predicts three absorption lines—a central maximum and a pair of symmetrically located satellites. The separation between the satellites is [52]

$$\tfrac{1}{2}(e^2Q/h)\,\nabla V \qquad (22.1)$$

where Q is the nuclear quadrupole moment and ∇V is the electric field gradient.

In second approximation the central component ($m = \tfrac{1}{2}$ to $-\tfrac{1}{2}$) is influenced by quadrupolar interaction and is displaced towards lower frequencies. And again for $I = \tfrac{3}{2}$, the central component splits into two maxima the separation between which is

$$(25/64)(e^4Q^2/2h^2f_0)(\nabla V)^2 \qquad (22.2)$$

Nuclear magnetic resonance has been used to aid in the determination of the nature and density of metallic imperfections. It can be shown that quadrupole effects, which relate to the local electric field gradient, present when $I > \frac{1}{2}$, will predominate over magnetic dipole interactions. Important variables are the *position of the resonance line* and the *linewidth*. By observing the former, measurements on ^{57}Co have enabled the occurrence of various kinds of stacking faults to be studied. The intensity of the resonance in pure copper can be increased by annealing and decreased by cold work. Also in alloys like α-brass and Cu–Ag, the intensity is observed to decrease rapidly with increasing solute concentration. At 25% zinc, the copper resonance becomes so broad as to be unobservable. These effects are due to quadrupolar interaction and if suitably applied, can give information about the state or order, or disorder, of an alloy. Linewidth measurements in copper sheet have shown that an increase in width of up to 40% occurs with plastic deformation and there is evidence that the broadening is proportional to the stored energy. In aluminum after cold rolling, the linewidth increases by up to 120%, the broadening being proportional to the density of dislocations. It has been postulated that the density of dislocations as calculated from NMR is greater than that observed in aluminum by electron microscopy, and that the latter does not represent the true density.

23. Influence of Structural Defects on NMR. A detailed discussion of the response of NMR to crystalline defects has been given by Bloembergen [53]. The relevant effects will be briefly summarized.

Magnetic Dipole Effects. In the neighborhood of lattice dislocations the change of internuclear distance will usually be less than about 1%. It is therefore estimated that, in the absence of quadrupolar effects, dislocations in general will have negligible influence on the nuclear resonance. Vacancies and interstitials can have a pronounced temperature-dependent effect on the linewidth and nuclear relaxation times. The effect of diffusion motion has been studied by Torrey [54], Slichter and Norberg [55], and Gutowski [56].

Quadrupole Effects. The strain field around a dislocation line will produce a change in the gradient of the electric field at the nuclei near the dislocation. A general broadening of the components of the resonance will occur. In noncubic crystals, dislocations will broaden the satellites much more than the central component (only present if the spin is an odd multiple of $\frac{1}{2}$) to an extent which is inversely proportional to the magnetic field.

Consider now the effect of stress which will, for example, distort the cubic symmetry in an otherwise cubic crystal. Deformation of a single crystal or polycrystalline sample produces satellite lines which may become broadened to such an extent that they themselves become unobservable and only the central component remains. As pointed out by Faulkner [57], whose work will be described in the next section, for small strain it may be assumed that the electric field gradient ∇V is proportional to the value of the strain tensor at the site [58]; an NMR frequency shift Δf_Q also occurs, proportional to $\nabla_Z V_Z$, where Z is the direction of the applied magnetic field. With the stored energy, S, of a randomly strained isotropic material

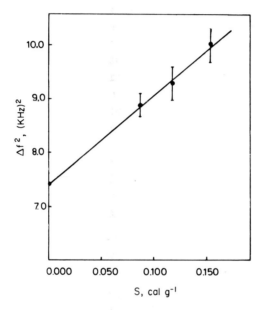

FIG. 9. Mean square linewidth in deformed and annealed copper sheet plotted against stored energy figures. The equation of the line is $\Delta f^2 = 17S + 7.4$. (From Faulkner [57].)

proportional to the mean square strain, $\langle (\nabla_Z V_Z)^2 \rangle_{av}$, and consequently $\langle (\Delta f_Q)^2 \rangle_{av}$ is proportional to the stored energy. The equation for mean square linewidth is therefore

$$\langle (\Delta f)^2 \rangle_{av} = \langle (\Delta f_Q)^2 \rangle_{av} + \langle (\Delta f_D)^2 \rangle_{av}$$
$$= kS + \langle (\Delta f_D)^2 \rangle_{av}, \qquad (23.1)$$

where D refers to the independent dipolar contribution present in the strain-free crystal. Figure 9 shows in fact that a plot of $\langle (\Delta f)^2 \rangle_{av}$ vs. S is linear, enabling S and $\langle (\Delta f_D)^2 \rangle_{av}$ to be determined.

24. *Application of NMR to the Study of Recovery and Related Phenomena.* Toth and Ravitz [59] have studied the ferromagnetic nuclear resonance in cobalt at room temperature. Resonances in cobalt occur at 221.0 MHz for the hexagonal close-packed phase and 213.1 MHz for the face-centered cubic phase. In analyzing the observed spectra in terms of the occurrence of various stacking fault patterns, it was assumed that the various hyperfine fields represented were simply related to differences in the c/a ratios. Since the interaction distances in the close packed planes are very nearly equal in both hcp and bcc phases, these ratios correspond to changes in the c axis. The resonance frequencies from nuclei in stacking faults were then calculated by comparing the average change in interplanar separation for a given faulted plane with the change in spacing between the fcc and hexagonal phases. Intensity comparisons indicated that intrinsic stacking faults are more abundant than the other types of faults or twins in accord with electron microscopy studies.

Work hardening in copper has been studied by Pavlovskaya and Stark [60] using powdered metal and by Faulkner [57, 61], using rolled foils. The Russian scientists used powdered samples of metal and measured NMR parameters after successive grindings in a mortar. No frequency shift was observed and only first-order quadrupolar interaction seemed to be present. The degree of lattice deformation ϵ was determined from the formula

$$\epsilon = (a^3/6\lambda)\, \nabla V \tag{24.1}$$

where a is the lattice constant, λ is an antishielding parameter (Bloembergen [53, p. 16]), and ∇V is obtainable from Eq. (22.1). They reported that the proportion of copper atoms in environments yielding the largest ϵ increased during the work hardening process, and at the highest state of work hardening achieved, the deformation was as high as 0.5%.

Faulkner [57] measured the mean square linewidth $\langle(\Delta f)^2\rangle_{av}$ and the "derivative maximum," G_{max}, during recrystallization by annealing of deformed samples. Specimens were prepared by rolling from an annealed copper strip. Since all samples needed to be thin, those of smallest deformation were prepared by a two-stage rolling process with an intermediate recrystallization. Deformation is designated by a percentage reduction in thickness; specimens whose deformations were 99, 75, and 25% were examined.

The deformed samples were heated at about $6°$ min^{-1} and G_{max} was obtained at suitable temperature intervals. It was observed that a pronounced increase in G_{max} occurred for each specimen in a specific temperature region. This temperature region was lower, the higher the degree of deformation (Fig. 10). This increase in G_{max} was attributed by Faulkner to recrystallization.

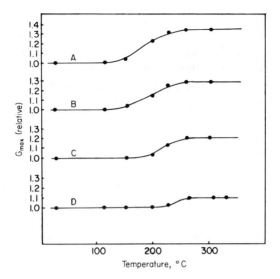

Fig. 10. Nuclear magnetic resonance in rolled sheet heated at 6°C min^{-1}. Deformation: (A) 99%; (B) 75%; (C) 50%; (D) 25%. (From Faulkner [57].)

The stored energy, derived from the results of work by Clarebrough *et al.* [62], was used to plot a graph of mean square linewidth versus stored energy, displaying the linear relationship of Fig. 9. Faulkner and Ham [63] have also studied the stored energy and density of dislocations in deformed aluminum. The linewidth was studied as a function of dislocation density, determined by electron transmission microscopy, and an approximately linear relationship was observed. It is interesting to note that these authors found increases of up to 120% in the mean square linewidth in heavily rolled aluminum whereas Rowland [64] found that filing had no effect on the linewidth presumably because of spontaneous recovery at room temperature. This is in contrast to the explanation proposed earlier by Bloembergen [53], who suggested that the dislocations in pure aluminum might be stacked in an ordered arrangement such that there is destructive interference of the stresses at large distances from the dislocation walls. Bloembergen also suggested that a self-annealing theory could be put to the test by conducting the whole experiment at liquid air temperature. It is, of course, likely that filing and rolling could produce different kinds of defects.

It is interesting to compare these experiments with some others of Rowland [64], who has also described the effect of annealing several aluminum alloys. One such sample was a powdered sample of Al–Mg (0.65 at. %)

alloy. It was prepared by filing and the absorption was observed. It was then heated for 87 hr at 250°C, cooled to room temperature, and again measured. Another specimen was measured after an anneal for 2 hr at 480°C. It was found that the NMR absorption steadily increased with heat treatment (Fig. 11), reaching a saturation value with the last specimen. It was concluded that the presence of the impurities inhibited the migration and recombination of dislocations so that the resonance signal increased as the dislocations were removed by annealing. The fact that the resonance in pure Al shows no effect of working by filing emphasizes the role that impurities play in trapping the dislocations.

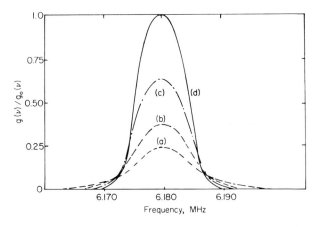

FIG. 11. Absorption curves obtained for 99.36–0.64% Al–Mg after the following treatments: (a) filed; (b) after a 250°C, 87-hr anneal; (c) after a 480°C, 2-hr anneal; (d) pure aluminum. (From Rowland [64].)

The difference between pure Al and copper at room temperature according to Seeger [65] can be ascribed to the fact that the activation energy needed to create a jog when two dislocation lines cross each other is 10 times lower in Al than in Cu.

Significance and Uses of Magnetic Analysis for Metallurgy

25. We have been concerned in this chapter with two aspects of the relations between magnetic properties and the metallurgical processes of recovery and recrystallization in metals and alloys. In the first place, the various processes of cold working alter many of the magnetic properties of metals and it is of fundamental as well as technological importance to know and understand how these properties "recover" during subsequent heat treatment. Second, it is of interest to know how observations on the magnetic properties may be used as a tool for the study of recovery and recrystallization in metals. In particular, metallurgists may well ask the following questions. What are the advantages of using magnetic methods for studying recovery and recrystallization? What are the limitations to these techniques? What additional information is obtained in this way that is not attainable with other methods?

In answer, one may begin with the categorical comment that the "standard" magnetic methods, apart from nuclear magnetic resonance, neutron diffraction, etc., are practically useless in the study of metallurgical processes in nonferromagnetic metals and alloys. Magnetic studies of nonferromagnetic metals are so thoroughly beset with difficulties arising mainly from the presence of minute amounts of ferromagnetic inclusions as to seem rather pointless. That of value here is a general investigation of the nature, formation, and distribution of such inclusions in nonferromagnetic metals. The statements below are therefore restricted to ferromagnetic materials, for which at least a good start has been made on the correlation between magnetic behavior and metallurgical state.

Pronounced changes in magnetic properties observed during the recovery and recrystallization processes in ferromagnetic metals and alloys make it possible to study the kinetics of these restoration processes nondestructively. Several magnetic parameters may be measured separately on the same specimen, and a distinction between recovery and recrystallization may be made by comparing the rates of restoration of these several parameters as the annealing progresses. In particular, Seeger and his co-workers have demonstrated how the measurement of magnetic susceptibility and of coercive force can yield information concerning the relaxation of long-range stresses and the annihilation of dislocations.

Similar work of this type, which utilizes a combination of more than one magnetic measurement, makes it possible to discern changes not easily detectable by other methods. The combined use of magnetic susceptibility, coercive force, and magnetic anisotropy allows one to separate recovery from recrystallization, provided that the texture changes upon recrystallization. The nondestructive nature of the magnetic method makes it possible to conduct numerous experiments upon one specimen as it is taken through various stages of annealing and/or cold working. Furthermore, use of magnetic techniques "at temperature" are straightforward, provided that one stays well below the Curie temperature of the material under investigation. A serious limitation of the ordinary magnetic methods is their insensitivity to the detection of the early states of recovery associated with the disappearance of vacancies; a very useful adjunct to magnetic methods in this case is the measurement of electrical resistivity, which is sensitive to the concentration of point defects. Finally, provided one can find reasonable correlations between magnetic properties and metallurgical state, the magnetic measurements have the virtue of being quite easy to perform under a wide variety of conditions of temperature, shape and size of specimen, application of stress, etc.

REFERENCES

1. S. Chikazumi, "Physics of Magnetism," Section 4.3, p. 72 ff. Wiley, New York, 1964.
2. W. F. Brown, Jr., *Phys. Rev.* **60**, 139 (1941).
3. L. Néel, *J. Radium* **9**, 184 (1948).
4. R. Becker and W. Döring, "Ferromagnetismus," English ed., Chapter III. Edwards Brothers, Ann Arbor, Michigan, 1943.
5. J. A. Ewing, *Proc. Roy. Soc. (London)* **46**, 269 (1889).
6. J. L. Snoek, *Physica* **5**, 663 (1938); see also Chikazumi [1] and Becker and Döring [4].
7. C. P. Slichter, "Principles of Magnetic Resonance." Harper, New York, 1963.
8. G. Masing and J. Raffelsieper, *Z. Metallk.* **41**, 65 (1950).
9. D. Kuhlmann, G. Masing, and J. Raffelsieper, *Z. Metallk.* **40**, 241 (1949).
10. T. Broom, *Advan. Phys.* **3**, 2 (1954).
11. N. F. Mott, *Phil. Mag.* **44**, 742 (1947).
12. D. Kuhlmann, *Z. Phys.* **124**, 468 (1947).
13. P. A. Beck, *Phil. Mag.* **3**, 256 (1954).
14. J. E. Bailey and P. B. Hirsch, *Phil. Mag.* **5**, 485 (1960).
15. J. C. Blade, J. W. H. Clare, and H. J. Lamb, *J. Inst. Metals* **88**, 365 (1959–60).
16. D. Reits, R. W. Starreveld, and H. J. De Wit, *Phys. Letters* **16**, 13 (1965).
17. R. Becker and W. Döring, "Ferromagnetismus," English ed., Chapter III, p. 171. Edward Brothers, Ann Arbor, Michigan, 1943.
18. E. Czerlinsky, *Ann. Physik* **13**, 80 (1932); see also Becker and Döring [4, pp. 172–173].
18a. H. Polley, *Ann. Physik* **36**, 625 (1939).
19. A. Seeger and H. Kronmüller, *J. Phys. Chem. Solids* **12**, 298 (1960); H. Kronmüller and A. Seeger, *ibid.* **18**, 93 (1961).
19a. A. Seeger and H. Kronmüller, *Phil. Mag.* **7**, 897 (1962).
20. H. Kronmüller, A. Seeger, H. Jäger, and H. Rieger, *Phys. Status Solidi* [5] **2**, K105 (1962).
21. H. Bumm and H. G. Müller, *Metallwirtschaft* **17**, 903 (1938).
22. M. Kersten, *Z. Angew. Phys.* **8**, 313 (1956).
23. M. Asanuma, *J. Phys. Soc. Japan* **15**, 1469 (1960).
24. D. Kuhlmann, *Z. Physik* **124**, 468 (1947).
25. W. Boas, Defects in crystalline solids (*Rept. Bristol Conf. Defects Crystalline Solids, July 1954*), p. 212, Phys. Soc. London (1955).
26. H. Träuble and A. Seeger, *Z. Angew. Phys.* **14**, 237 (1962).
27. R. Becker and W. Döring, "Ferromagnetismus," English ed., Chapter III, p. 214. Edward Brothers, Ann Arbor, Michigan, 1943.

28. R. Becker and W. Döring, "Ferromagnetismus," English ed., Chapter III, p. 155. Edward Brothers, Ann Arbor, Michigan, 1943.
29. F. Vicena, *Czech. J. Phys.* **5**, 480 (1955).
30. M. Kersten, *Z. Angew. Phys.* **8**, 496 (1956).
31. M. Kersten, *Z. Angew. Phys.* **9**, 280 (1957).
32. H. Dietrich and E. Kneller, *Z. Metallk.* **47**, 672 (1956).
33. H. Dietrich and E. Kneller, *Z. Metallk.* **47**, 716 (1956).
34. K. Wotruba, *Czech. J. Phys.* **6**, 498 (1956).
35. Z. Malek, *Czech. J. Phys.* **7**, 152 (1957).
36. Z. Malek, *Czech. J. Phys.* **9**, 613 (1959).
37. J. K. Stanley, *Trans. AIME Iron Steel Div.* **162**, 106 (1942).
38. C. G. Dunn, *Conf. Magnetism Magnetic Materials, 1st, 1955*, p. 126 (1955).
39. R. M. Bozorth, "Ferromagnetism," p. 586. Van Nostrand, Princeton, New Jersey, 1951.
40. L. P. Tarasov, *Trans. AIME* **135**, 353 (1959).
41. H. W. Conradt, O. Dahl, and K. J. Sixtus, *Z. Metallk.* **32**, 586 (1940).
42. J. L. Snoek, *Physica* **5**, 663 (1938).
43. S. Chikazumi, "Physics of Magnetism," p. 316. Wiley, New York, 1964.
44. L. Néel, *J. Radium* **13**, 249 (1952).
45. G. Richter, *Ann. Physik* **29**, 605 (1937).
46. H. Wittke, *Ann. Physik* **31**, 97 (1938).
47. G. Biorci, A. Ferro, and G. Montalenti, *Phys. Rev.* **119**, 653 (1960).
48. T. S. Hutchinson and J. Reekie, *Nature* **159**, 537 (1947).
49. J. D. McClelland, *Acta Met.* **2**, 406 (1954).
50. D. J. E. Ingram, "Spectroscopy at Radio and Microwave Frequencies." Butterworths, London, 1955.
51. E. R. Andrew, "Nuclear Magnetic Resonance," Cambridge Univ. Press, London and New York, 1956.
52. R. V. Pound, *Phys. Rev.* **79**, 685 (1950).
53. N. Bloembergen, *Rept. Bristol Conf. Defects Crystalline Solids, July 1954*, Phys. Soc. London (1955).
54. H. C. Torrey, *Phys. Rev.* **92**, 962 (1953).
55. C. P. Slichter and R. E. Norberg, *Phys. Rev.* **83**, 1074 (1951).
56. H. S. Gutowsky, *Phys. Rev.* **83**, 1073 (1951).
57. E. A. Faulkner, *Phil. Mag.* **5**, 843 (1960).
58. M. H. Cohen and F. Reif, *Solid State Phys.* **5**, 321 (1957).
59. L. E. Toth and S. F. Ravitz, *J. Phys. Chem. Solids* **24**, 1203 (1963).
60. V. S. Pavlovskaya and Yu. S. Stark, *Soviet Phys. Solid State (English Transl.)* **4**, 205 (1962).
61. E. A. Faulkner, *Nature* **183**, 1043 (1959).
62. L. M. Clarebrough, M. E. Hargreaves, and G. W. West, *Acta Met.* **5**, 738 (1959).
63. E. A. Faulkner and R. K. Ham, *Phil. Mag.* **7**, 279 (1962).
64. T. J. Rowland, *Acta Met.* **3**, 74 (1955).
65. A. Seeger, *Rept. Bristol Conf. Defects Crystalline Solids, July 1954*, Phys. Soc. London (1955).

CHAPTER XV

Textured Magnetic Materials

C. D. GRAHAM, Jr.
General Electric Research and Development Center
Schenectady, New York

INTRODUCTION	723
TEXTURES IN METALS	724
Specification of Textures	724
Determination of Textures	725
Sources of Texture	727
PERMANENT MAGNET MATERIALS	730
NICKEL–IRON ALLOYS	732
SILICON–IRON ALLOYS	734
The {110} ⟨100⟩ Texture	734
The {100} ⟨100⟩ Texture	739
MAGNETIC CONTROL OF TEXTURE	744
Field Applied during a Phase Change	744
Field Applied during Recrystallization	745
Field-Induced Recrystallization	745
References	745

Introduction

1. Single crystals of magnetic materials are almost always anisotropic; that is, they have different magnetic properties in different crystallographic directions. Therefore, a polycrystalline magnetic material whose grains are not oriented randomly will have anisotropic magnetic properties. This suggests the production of useful polycrystalline materials with superior magnetic properties in certain directions, by control of the crystallographic texture. The three principal applications of this idea are the directional-grain Alnico permanent magnets, the oriented 50–50% nickel–iron alloys, used in magnetic amplifiers, and the oriented silicon-iron alloys, almost universally used in large transformers. In these three materials the textures are obtained commercially by directional solidification, primary recrystallization, and by secondary recrystallization, respectively.

This section begins with a brief outline of the specification, determination, and origin of textures in metals, and continues with a discussion of the production and properties of textured magnetic materials.

Textures in Metals

Specification of Textures

2. The standard book in English on textures in metals is "Structure of Metals," 2nd ed., by C. S. Barrett [1]. A more extensive and recent book is "Texturen metallischer Werkstoffe," by G. Wassermann and J. Grewen [2]. Both of these give extensive references to the literature; Wassermann and Grewen have a section devoted specifically to magnetic materials.

The *texture* of a polycrystalline metal is the distribution of the orientations of the crystal axes of the individual grains with respect to some reference directions. A nonrandom texture is sometimes called a *preferred orientation*. A texture is specified with respect to the external dimensions of the material, that is, with respect to the axis of a wire, and the plane and edges of a sheet. The usual notation is to specify first the crystallographic plane which lies parallel to the plane of the sheet, and then crystallographic direction which lies parallel to the rolling direction of the sheet: $\{110\}\langle 100\rangle$. Frequently, the symbols meaning a particular plane and direction are used; in this case the direction specified must be one that lies in the plane specified: (110) [001]. If the texture is a strong one, and consists of a single component (that is, if each grain has approximately the same orientation as all the others), this method is adequate and satisfactory. If the texture is weak, or consists of a number of (possibly overlapping) components, some graphical representation of the data is required. The most common such representation is the *pole figure*, which is a two-dimensional plot of the orientations of the *poles*, or normals, of a family of crystallographic planes. It is usually drawn as a pole density plot, with contour lines of equal pole density drawn in (as in Fig. 5). Recently some alternative ways of representing this information have been proposed (for a summary, see Dunn and Walter [3]); however, these are not yet widely used.

Determination of Textures

3. Textures are usually determined by x-ray diffraction, and etch pit techniques have also been used. These methods are discussed in the books referred to above. Two additional methods are available in ferromagnetic materials, although both suffer from fairly severe limitations.

In the absence of an applied field, 180° domain walls lie parallel to an easy direction in each grain. If the crystallographic easy directions are known for the material in question (as they usually are), and if an easy direction lies approximately parallel to the surface observed, then the projection of the easy direction on the surface of a particular grain can usually be determined directly by looking at the domain structure. It is also possible to correlate the appearance of the domain pattern with the angle between the surface and an easy direction when the angle is small [4]. In a limited range of orientations, then, the observed domain pattern can give useful information about the crystal orientation of individual grains, and this method has been applied to help determine textures [5, 6].

A second method available in ferromagnetic materials is the measurement of torque, as discussed in the section on anisotropy measurements. The torque curve of a polycrystalline sample in sufficiently high fields will be the algebraic sum of the torque curves of the individual grains, and therefore

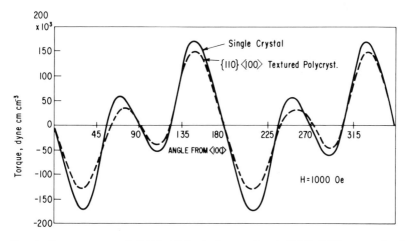

FIG. 1. Torque curves of {110} ⟨100⟩ single crystal and {110} ⟨100⟩ textured polycrystal.

the torque curve of a polycrystal with a strong single-component texture should approach the torque curve of a single crystal of the same orientation, as shown in Fig. 1. Since a torque curve is very much quicker and easier to determine than a complete x-ray diffraction pole figure, torque curves have been widely used in texture work on ferromagnetic materials. As long as the textures studied are simple and strong, the method is unobjectionable; in fact, it is superior to the usual x-ray methods in sensitivity and in the ability to average over a relatively large volume of coarse grained material. If the textures are weak, or contain many components, the torque method is of very limited value because under these conditions there is no unique relationship between the measured torque curve and the texture; many different textures can give similar or identical torque curves. A further difficulty is that the presence of a magnetic annealing anisotropy or a deformation anisotropy will give a contribution to the torque curve that cannot be easily separated from the contribution of the crystallographic texture. Several descriptions and discussions of the torque method of studying textures have appeared in the literature [7–12].

4. As the simplest example, consider a polycrystal with a $\{100\}\langle 100\rangle$ texture, and with deviations from a perfect texture occurring only as rotations about the normal to the surface, as in Fig. 2. This is a case of some

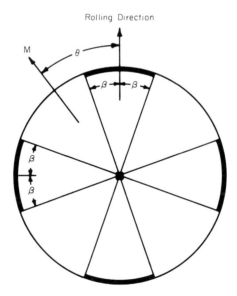

FIG. 2. $\{100\}$ pole figure of idealized $\{100\}\langle 100\rangle$ texture with deviations of $\pm\beta°$ rotation about the normal to the specimen surface.

importance, as discussed in Section 14. The magnetization in each grain rotates in a {100} plane as the torque curve is measured, and the total torque on an individual grain is given by

$$L = -(K_1/2)[\sin 4(\theta + \beta)]v$$

where K_1 is the first crystal anisotropy constant, θ is the angle from the magnetization to the rolling direction, β is the deviation of the $\langle 100 \rangle$ direction from the rolling direction, and v is the volume of the grain.

The torque curve of the polycrystalline sample is the sum of the torques on the individual grains:

$$L/V = \Sigma_i -(K_1/2)[\sin 4(\theta + \beta_i)]v_i/\Sigma_i v_i.$$

If we assume small grains of equal volume, uniformly distributed from $-\beta$ to $+\beta$ as in Fig. 2, the sums may be replaced by integrals over the angle β:

$$\begin{aligned}
L/V &= \int_{-\beta}^{+\beta} -(K_1/2) \sin 4(\theta + \beta) \, d\beta \Big/ \int_{-\beta}^{+\beta} d\beta \\
&= -(K_1/2) \cdot -\tfrac{1}{4}[\cos(4\theta + 4\beta)]_{-\beta}^{+\beta}/2\beta \\
&= -(K_1/2) \cdot (8\beta)^{-1}[\cos 4\theta \cos 4\beta - \sin 4\theta \sin 4\beta]_{-\beta}^{+\beta} \\
&= -(K_1/2) \sin 4\theta \, (\sin 4\beta)/4\beta.
\end{aligned}$$

The torque curve of the polycrystalline sample is therefore exactly the same shape as the torque curve of a single crystal {100}$\langle 100 \rangle$ disk, but reduced in magnitude by a factor $\sin 4\beta/4\beta < 1$. If the sample meets the assumed conditions, namely, that the only deviations from a perfect {100} $\langle 100 \rangle$ texture consist of rotations about the normal to the surface, then the ratio of the amplitude of the measured torque curve to that of a single crystal is numerically equal to $(\sin 4\beta)/4\beta$, and the value of β can be readily found.

In less simple cases, the results are much more complicated. Dunn and Walter [12] give general equations which can be applied to particular cases, and also emphasize the danger of trying to deduce information about texture from torque curves in cases where the torque is small compared to the single-crystal torque.

Sources of Texture

Textures in metals may arise from a number of sources; however, only three are important for this discussion. They are textures resulting from *solidification*, from *deformation*, and from *annealing after deformation*.

5. Solidification. If molten metal is poured into a mold with a cold bottom surface and heated or insulated side walls, solidification will begin at the cold bottom surface and will proceed upward through the metal. The resulting grain structure is usually *columnar*; that is, there are a number of parallel, pencillike grains with boundaries lying parallel to the direction of heat flow. In cubic metals, these columnar grains are found to have a $\langle 100 \rangle$ or cube edge direction parallel to the columnar axis [13]. The solidified ingot then has a $\langle 100 \rangle$ fiber texture, in which most or all of the grains have a $\langle 100 \rangle$ axis parallel to the direction of solidification.

6. Deformation. Metals deform in an anisotropic, crystallographic manner. As a result, if a random-textured polycrystalline material is plastically deformed, the resulting material will in general have some nonrandom texture. There is an extensive literature on this subject [1, 2], but thus far no useful theoretical predictions of deformation textures can be made, and only very limited empirical rules are apparent. For most magnetic applications, materials in the cold-worked state are not useful, and deformation textures are important only as a step toward an annealing texture.

7. Annealing. If a cold-worked metal is annealed at sufficiently high temperature, it will *recrystallize*; that is, the deformed grains will be replaced by a new set of approximately strain-free grains, with boundaries at new positions and with orientations usually differing from their predecessors. The mechanism of recrystallization is nucleation and growth: new grains nucleate or appear at various points in the material, and each of these nuclei grows until it is stopped by meeting other new grains. The driving force for this reaction is the reduction in strain energy stored in the material.

If the deformed material had a nonrandom texture, the recrystallized material also will have a texture. This may be the same texture as the deformation texture, but usually it is different. The recrystallization texture is sometimes stronger and sharper than its parent deformation texture. As in the case of deformation textures, there are no very useful theories or rules that can predict the texture that will be obtained in a particular material under particular deformation and annealing conditions. There is considerable evidence that very small changes in impurity concentrations can markedly affect recrystallization behavior [14].

If a fully recrystallized metal is annealed further, the usual result is a gradual increase in the average grain size. This occurs by the slow growth of larger grains at the expense of small ones, and often reaches a limit when the grain diameter is approximately equal to the sample thickness. The

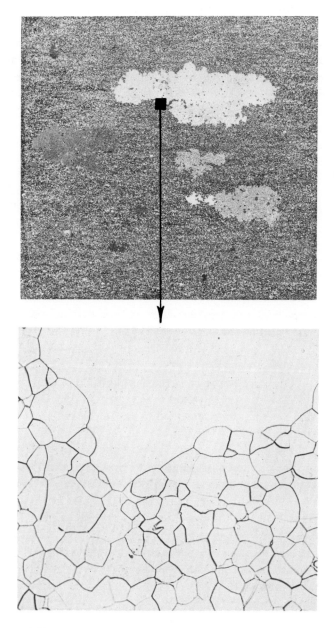

FIG. 3. Partially completed secondary recrystallization in commercial silicon–iron (top, 3×; bottom 250×).

driving force is the reduction of grain boundary area. Usually no significant changes in textures are associated with this normal grain growth.

If normal grain growth is prevented, prolonged annealing may result in the relatively sudden growth of a very few grains to occupy the entire volume of the sample. Although the driving force is again the reduction of grain boundary area, the results of this process are quite different from those of normal grain growth. The grain size increases very sharply, perhaps by several orders of magnitude, and the identification of the rapidly growing grain is unequivocal (see Fig. 3). The resulting texture is different from the recrystallization texture, and may be much sharper. This process is called by various names, including *secondary recrystallization*, *discontinuous grain growth*, and *exaggerated grain growth*; there are no agreed distinctions among the terms.

For secondary recrystallization to occur, normal grain growth must be prevented. Two mechanisms are known that can do this: the presence of second-phase inclusions at the grain boundaries (*inclusion* or *particle inhibition*), or the presence of a strong primary recrystallization texture (*texture inhibition*). (The boundaries between grains of similar orientation do not move easily.) The sample thickness effect mentioned above can also effectively limit normal grain growth.

Permanent Magnet Materials

8. The Alnico permanent magnet alloys are brittle, and are normally cast into their final shape. This precludes any texture development by working and recrystallization but still allows for orientation control by directional solidification. Such an approach seems to have been suggested independently in the United States and in England [15, 16]. Directional solidification favors the growth of columnar grains with $\langle 100 \rangle$ axes in Alnico, and this fortunately turns out to be the direction in which the best permanent magnet properties can be obtained.

Directional solidification is achieved by providing a mold whose bottom surface is a heavy, cold metal plate (which may be water cooled), and whose side walls are well insulated. Only straight magnets of substantially uniform cross section can be produced in this way, and a complete columnar structure is not obtained, as seen in the center photograph of Fig. 4. Only Alnico V is made in the columnar or directional grain form; the energy product is increased from about 5.5 million G Oe to about 7 million G Oe in commercial production [17]. Much higher energy products have been obtained

XV. TEXTURED MAGNETIC MATERIALS 731

FIG. 4. (Left) normal (random) Alnico V; (center) DG (directional grain) Alnico V; (right) UNI-80 supercolumnar Alnico V (Japanese manufacture) (approximately full size).

by variations in composition combined with careful directional solidification [18].

More recently, there has been interest in improved methods of casting oriented or single-crystal magnets of approximately the Alnico V and Alnico VIII compositions [19–21]; an example is shown in the right-hand photograph of Fig. 4. The use of a solid-state recrystallization method has also been suggested [22].

Texture in the workable permanent magnet alloys has been investigated only to a limited extent [23–25].

Nickel–Iron Alloys

9. The nickel–iron alloys near 75% nickel have very low values of magnetocrystalline anisotropy, so that there is normally no reason to develop texture in them. The alloys near 50% nickel, however, have a positive value of K_1, and a $\langle 100 \rangle$ easy direction. If the use of these higher iron content alloys is required because of their relatively high saturation magnetization, it may be worthwhile to produce a texture in them.

All of the face-centered cubic nickel–iron alloys develop a very sharp $\{100\} \langle 100 \rangle$ primary recrystallization texture when annealed after severe (95%) reduction by cold rolling, as shown in Fig. 5. This is a well-known but poorly understood metallurgical phenomenon, also found in copper, gold, aluminum, and certain other fcc alloys. The appearance of the cube texture in nickel–iron alloys was extensively studied in Germany and in The Netherlands in the 1930's, and it was recognized very early that the $\{100\} \langle 100 \rangle$ texture appears as a primary recrystallization texture after heavy cold rolling [1, 2]. Further annealing gives secondary recrystallization to a $\{120\} \langle 100 \rangle$ texture [26]. This is a standard case of secondary recrystallization by texture inhibition; however, the secondary recrystallization texture is usually less sharp than the primary texture [27].

Much of the early work on the 50–50% nickel–irons was done in connection with the development of Isoperm, a low-loss, constant-permeability material used for the cores of loading coils, which are high-frequency inductors used in telephone lines. Isoperm is quite different from the high-permeability nickel–iron alloys; it is made by cold rolling a cube-textured 50–50% nickel–iron. This leads to a strong anisotropy with its easy direction perpendicular to the rolling direction, and gives the desired properties [26, 28–31]. The anisotropy is a roll or deformation anisotropy, which is discussed in Chapter XII, Secs. 10–14.

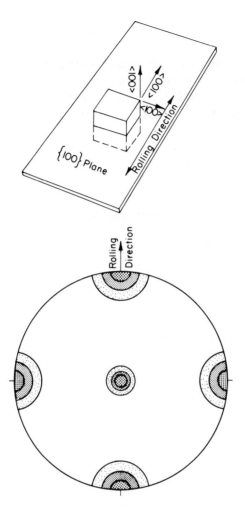

FIG. 5. The {100} ⟨100⟩ or cube texture: (top) schematic illustration; (bottom) {100} pole figure.

The *annealed* 50–50% nickel–iron alloy with {100} ⟨100⟩ texture was developed into an industrial material in Germany during World War II and this technology was adopted outside Germany after the war [32, 33]. Special precautions to avoid contamination during melting and annealing are required, and no mechanical strain can be tolerated after the final anneal. Good magnetic properties have been reported for material with the secondary recrystallization texture {120} ⟨100⟩ [34], but only the primary {100} ⟨100⟩ texture is available commercially. It is sold as Squaremu 49,

Orthonik, Deltamax, and Orthonol. Production methods are not generally revealed, but the essential steps are cold rolling 85–95%, followed by annealing at 900–1000°C in hydrogen. There is some evidence that too much cold rolling is detrimental [35].

Silicon–Iron Alloys

10. Unquestionably the most important textured magnetic materials in terms of volume and value of material used are the oriented silicon–iron alloys, sometimes called oriented silicon steels or electrical steels. The addition of silicon to iron is magnetically advantageous for a number of reasons, the most important of which are increased electrical resistivity (leading to lower eddy current losses) and the elimination of the high-temperature, face-centered cubic γ phase for silicon contents above about 2.2% silicon. The absence of γ-iron means that the alloys are body-centered cubic at all temperatures up to the melting point, and permits high-temperature heat treatments (to develop texture), which are impossible with pure iron. The iron–aluminum alloys are in many ways analogous to the iron–silicon alloys, and there have been attempts over the years to substitute aluminum for silicon in magnetic alloys. The substitution has not yet been commercially successful.

It was discovered about 1918 that the magnetic properties of single crystals of iron containing a few percent silicon are different in different crystallographic directions [36, 37]. These were, in fact, the first observations of magnetocrystalline anisotropy in metals. The easy direction in these alloys is the same as that in iron: $\langle 100 \rangle$. During the 1920's, it is probable that a number of industrial laboratories attacked the problem of finding a commercial method of developing a texture in silicon–iron alloys that would have $\langle 100 \rangle$ directions aligned parallel to the direction(s) in which the material would be magnetized in service [8, 38].

The $\{110\} \langle 100 \rangle$ Texture

11. *History.* The first successful process for making oriented silicon–iron sheet was that of Goss [39]. Briefly, his procedure consisted of starting with a hot-rolled sheet about 0.075 in. thick, reducing this to a final thickness of 0.012 or 0.014 in. in two stages of cold rolling, with an intermediate anneal at 800–900°C in a reducing atmosphere, and a final anneal at a temperature up to 1100°C in a nonoxidizing atmosphere. The resulting material had magnetic properties substantially superior to the hot-rolled sheets then standard in the electrical equipment business, if the properties were measured parallel to the rolling direction.

Goss used both torque and x-ray diffraction methods to study the effects of the various processing steps on his material, but he misinterpreted the x-ray results and believed that the final product had a random texture. However, this mistake was quickly discovered, and it was shown that the material had a {110} ⟨100⟩ texture [40], now sometimes known as the *cube-on-edge* or the Goss texture. This texture is illustrated in Fig. 6.

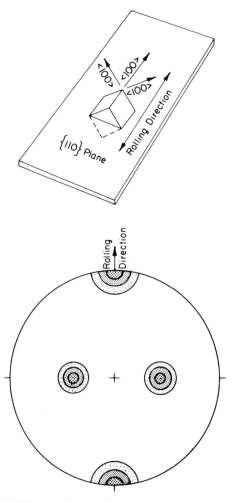

FIG. 6. The {110} ⟨100⟩ or cube-on-edge texture: (top) schematic illustration; (bottom) {100} pole figure.

A large number of modifications of the Goss process were patented: the cold rolling could be accomplished in three or more stages [41] or in one (instead of two) stages [42]; the final rolling was best done below room temperature [43], or else above room temperature [44]; the final anneal should be done in two stages [45], and then additionally there should be a very small cold-rolling reduction between the stages [46].

It seems that the $\{110\} \langle 100 \rangle$ texture arises under a fairly wide range of conditions. The "best" process for making oriented silicon–iron sheet depends on an assortment of related conditions, including the kind and amount of impurities in the starting material, and the nature of the rolling and annealing equipment available.

12. *Commercial Production.* Since no manufacturer publishes his production process, the following description of contemporary manufacturing methods is based largely on recent U.S. patents [47, 48]. The starting material is open-hearth steel containing approximately 3.2% Si, 0.06–0.10% Mn, 0.03% C, and 0.02% S as significant ingredients; other elements reported include about 0.007% P, 0.1% Cu, 0.01% Sn, and ˙0.01% Al. The cast ingots are hot rolled at temperatures near 1300°C to slabs 1–3 in. thick. The slabs may be reheated to 1300–1400°C before further rolling, or they may be rolled quickly before they have time to cool. Hot rolling ends with the material 0.060–0.100 in. thick; at this point it is called "hot-rolled band." The band may be given a relatively low-temperature anneal at 800–1000°C at this stage [49]. The oxide scale formed during hot rolling is removed with acid, and the strip is cold rolled to a final thickness of 0.010–0.014 in. in two approximately equal steps, with an intermediate anneal at 800–1000°C. The temperature during cold rolling may rise to 200°C or higher. A short decarburizing anneal in wet hydrogen at about 800°C follows the cold rolling. The final step is a high-temperature anneal in dry hydrogen at 1100–1200°C, which develops the $\{100\} \langle 100 \rangle$ texture. This final anneal is given to coils weighing 3 or 4 tons, and the total time in the furnace may be several days.

The decarburizing treatment reduces the carbon to about 0.003%, and probably also causes primary recrystallization. Higher carbon is bad because it precipitates slowly as iron carbide at the operating temperature of a transformer or generator and degrades the magnetic properties; this is known as "magnetic aging." The carbon could presumably be removed at any stage in the process, including the molten state; in practice, the removal is easiest and fastest when the sheet is thinnest, which is after the final rolling.

The sulfur level is reduced during the final high-temperature anneal, and may reach 0.002% if the final anneal is long and hot enough.

The final product is coarse-grained (1–5 mm grain size), and in the best material more than 90% of the grains will be in the {110} ⟨100⟩ orientation to within a few degrees [50]. The magnetic properties usually specified are the value of induction at an applied field of 10 Oe (or equivalently, the permeability at 10 Oe), and the rate of energy loss when the material is magnetized by an alternating field so that the induction varies sinusoidally with time and reaches a maximum value of 15,000 G. Lots of steel are tested at the mill, classified, and sold to a maximum-loss specification. The best commercial material currently has an induction of about 18,000 G at a field of 10 Oe, and total core losses of about 0.6 W/lb at 60 Hz when 0.012 in. thick. The losses in commercial material have decreased by nearly a factor of 2 in the past 25 years [51].

This material is sensitive to mechanical strain, and must be given a stress-relief anneal at 700–800°C after any cutting operation if the best magnetic properties are to be maintained.

This oriented magnetic sheet is the standard material for use in the cores of large and medium sized transformers, and of large motors and generators, operating at power frequencies. Since virtually all electrical power passes through several of these devices in its production, distribution, and use, small percentage power losses lead to very large cumulative losses, and small improvements in magnetic properties can mean large savings.

13. *Metallurgy*. The {110} ⟨100⟩ texture in silicon–iron alloys has been extensively studied for 30 years and much has been learned about the processes involved, although some important points remain obscure. The following discussion is fairly closely limited to the formation of commercially useful textures and does not deal with the general problem of texture formation.

Although it is not clear whether Goss's original process worked by primary or secondary recrystallization [52], the closely similar commercial process now used unquestionably makes use of secondary recrystallization [53].

Primary recrystallization in the commercial material occurs in a few minutes at temperatures around 800°C [53–55]; the primary recrystallization texture has been the subject of a number of investigations and a great deal of speculation. Usually a number of weak components are found, together with a strong background of randomly oriented material [53, 55–61]. On annealing at higher temperatures and for longer times, the characteristic secondary recrystallization process occurs, with a small number of grains growing to occupy the entire volume of the sample. These grains have the {110} ⟨100⟩ orientation [53, 54].

The importance of impurities in the development of the {110} ⟨100⟩ texture was suspected fairly early, and Mn and S were regarded as especially important [62, 63]. In 1956, it was reported that a purified silicon–iron sample, containing much less than the usual impurity levels, would not develop the {110} ⟨100⟩ texture unless 0.01–0.1% N was deliberately added [64]. The importance of Mn and S was shown in an important paper by May and Turnbull [65], and it is now believed that small particles of MnS act to pin the grain boundaries and prevent or greatly retard normal grain growth after primary recrystallization. The material therefore attains a temperature above 900°C with a relatively fine grain size, and a weak, multicomponent texture probably containing at least some grains with {110} ⟨100⟩ orientation. The details of what happens next are unclear, but the result is that grains with a {110} ⟨100⟩ orientation grow to the almost complete exclusion of other grains. It is usually assumed that the MnS goes into solution, thus releasing the {110} ⟨100⟩ grains [66], but measurements of the solubility product of MnS in Fe make this simple explanation seem unlikely [67]. Why only the {110} ⟨100⟩ grains grow is a major unsolved mystery, and even the role of the Mn and S is still in dispute.

Second-phase inclusions other than MnS can be used to inhibit normal grain growth in silicon–iron and produce the {110} ⟨100⟩ texture. Among the additions reported have been Ti + S, Cr + S [66], V + N [68], Ti + C [69], and V + C [70]. From the work of Fast [64], Si + N is presumably effective, and SiO_2 [71] is also a possibility. However, none of these has replaced the MnS, which occurs "naturally' in open-hearth steel.

The standard processing technique does not give good results for final thicknesses less than about 0.010 in. Thinner strips are needed for use in magnetic amplifiers, and to meet this need Littmann [72] devised a process in which conventionally produced strip of 0.010 in. thickness with the {110} ⟨100⟩ texture is further cold rolled to the desired thickness, and then annealed to obtain primary recrystallization. The recrystallization texture under these conditions is approximately {120} ⟨100⟩, which has an easy direction parallel to the rolling direction and gives reasonably good magnetic properties. More recently, it has been found possible to develop a strong {110} ⟨100⟩ texture in silicon–iron strip 0.002 in. thick by using vanadium nitride inclusions [73]. Furthermore, it has been shown that in silicon–iron containing MnS, the size of the particles can be controlled by the cooling rate from high temperatures prior to cold rolling, and that this MnS particle size is important in determining the completeness of {110}

⟨100⟩ texture formation [74]. By making the MnS particles small enough, the {110}⟨100⟩ texture can be developed by the conventional rolling and annealing methods in strip 0.004 in. thick [75]. In this thin material, it is necessary to guard against loss of S from the strip surface before the secondary recrystallization is complete.

The {100}⟨100⟩ Texture

14. The {110}⟨100⟩ or Goss texture silicon–iron sheet shows excellent magnetic properties in the rolling direction, but much inferior properties in other directions in the plane of the sheet. It is clear that for some applications, such as U-shaped transformer laminations, a sheet material with good magnetic properties both parallel and perpendicular to the rolling direction would be desirable. An attempt to make a material having two easy directions in the plane of the sheet was patented in 1936 by Bitter [76], who used a crossrolling technique that led to a primary recrystallization texture described as {100}⟨110⟩, with easy directions ±45° to the rolling direction. Only since 1957 has extensive work on the {100}⟨100⟩ or *cube texture*, with easy directions parallel and perpendicular to the rolling direction, appeared in the scientific and patent literature. It is now clear that most of the proposed methods depend on a new phenomenon: the control of texture by surface energy. This was apparently first suggested on theoretical grounds by Mullins [77] and was shown to apply to the case of silicon–iron by Detert [78] and by Walter and Dunn [79]. Walter and Dunn have carried out the most extensive investigation of the phenomenon [79–84]. As a result of this work, it is clearly established that if a sample of high-purity silicon–iron is heated near 1000°C in an atmosphere that prevents any surface contamination, grains with a {110} plane parallel to the sheet surface will grow at the expense of any other grains present. This growth is attributed to a difference in surface energy, of the order of 100 ergs cm^{-2} [79], between the {110} grains and grains of any other orientation. However, in the presence of small amounts of contaminants, either in the atmosphere or in the metal, the surface energy relationships can be changed so that grains with {100} parallel to the surface have the lowest energy. By control of the furnace atmosphere, {110} and {100} grains can be made to grow alternately in a sigle specimen. The contaminant that controls the process has been variously identified as oxygen [82] and as sulfur [85, 86]; presumably both could be effective.

Processes that make use of the surface energy phenomenon for producing a particular texture can usually be recognized even if the surface energy driving force is not known or slated explicitly. Such processes work best for very thin sheets, where the surface-to-volume ratio is high. The driving force is fairly small, so that grain boundary migration must not be impeded. This usually means high-purity starting materials and careful avoidance of contamination in melting and processing. Control of the atmosphere during the final anneal is critical; usually vacuum or hydrogen with a low dew point is specified. Various other procedures have been recommended: titanium in the furnace as a getter of oxygen; nickel, cobalt, or platinum to "promote the formation of atomic hydrogen from molecular hydrogen" [87]; and various etching treatments at intervals during the annealing treatment [88].

Material with the {100} ⟨100⟩ texture produced by the surface energy mechanism has a characteristic pole figure, in which the deviations of the {100} plane from the strip surface are very small (rarely exceeding 5°), but the deviations of the ⟨100⟩ directions from the rolling direction may be substantially larger (see Fig. 7). This suggests at once the major problem in

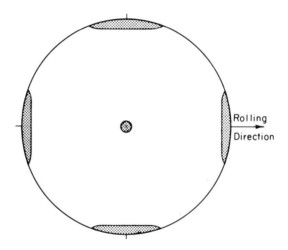

FIG. 7. {100} pole figure of cube texture sheet produced by surface energy driving force: the deviation of ⟨100⟩ directions away from the rolling direction is greater than deviations from sheet normal.

obtaining a good {100} ⟨100⟩ texture by the surface energy mechanism: the mechanism only acts to align a particular crystallographic plane parallel to the sheet surface and does not, by itself, establish any preferred direction in the sheet. The surface energy difference can only favor or retard the growth of existing grains, not create grains in orientations which did not previously exist. To obtain a $\{100\} \langle 100 \rangle$ texture, it is therefore necessary to insure that those grains in the primary (or secondary) texture that have a $\{100\}$ plane parallel to the sheet surface also have a $\langle 100 \rangle$ direction parallel to the rolling direction. One process for achieving this involves starting with a sheet material with the $\{110\} \langle 100 \rangle$ orientation; when rolled and (primary) recrystallized, this presumably gives some $\{100\} \langle 100 \rangle$ grains in addition to the $\{120\} \langle 100 \rangle$ grains as produced by the Littmann process. The $\{100\} \langle 100 \rangle$ primary grains can then be selected by surface energy mechanism and grow to occupy the entire sheet if the sample and atmosphere compositions are correctly chosen [89]. The effective means of obtaining the necessary $\{100\} \langle 100 \rangle$ nuclei in the other proposed processes are not clear [90–94].

Surface energy grain growth can occur after either primary or secondary recrystallization. When it occurs after secondary recrystallization, it can be called "tertiary recrystallization" [79]. Usually, however, the surface energy growth occurs after primary recrystallization and is then regarded as a kind of secondary recrystallization. It has been shown that for the most rapid growth of $\{100\} \langle 100 \rangle$ grains, the initial recrystallized grain size should be as small as possible, so that the reduction of grain boundary area can contribute to the driving force for boundary motion [83, 95]. The surface energy contribution is typically only about 10% of the total driving force for boundary motion, so that surface energy differences act primarily to *select* certain orientations for growth. Under these conditions, higher inclusion levels can be tolerated in the material than in the case where surface energy differences are the only driving force for boundary motion. Figure 8 shows the process of secondary grain growth by surface energy.

Other processes that do not or may not make use of the surface energy driving force have been reported for obtaining the $\{100\} \langle 100 \rangle$ texture in silicon–iron. In one case, an ingot is solidified to give a texture in the cast material, and subsequently rolled and recrystallized to the $\{100\} \langle 100 \rangle$ texture [96]. Other methods that probably do not depend on surface energy have been described [97–100]. The process of Möbius and Pawlek [101] does not obviously involve surface energy, but an examination of a sample of this material indicates a surface energy mechanism is operative [102].

The process described by Grenoble [103] calls for the initial hot rolling of an alloy containing sufficient Si + C to give some face-centered cubic γ-iron (austenite) at the rolling temperature. The hot-rolled material is decarburized, annealed, reduced to final thickness by warm rolling at about 200°C, and finally annealed at about 1200°C.

A number of authors have reported magnetic properties of $\{100\} \langle 100 \rangle$

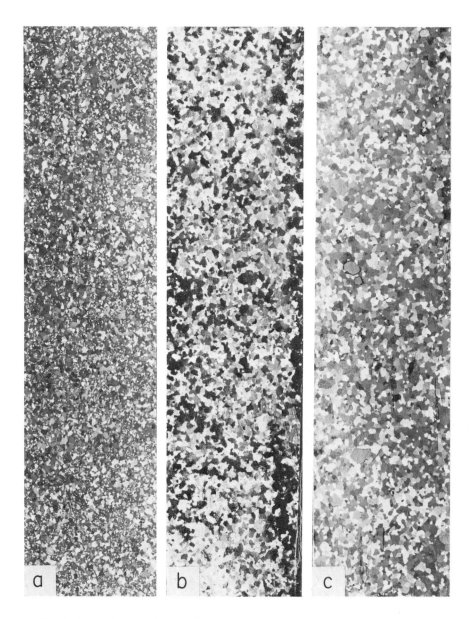

Fig. 8. Successive steps in secondary recrystallization by surface energy driving force in 5% silicon–iron (approximately 5×): (a) 10 sec; (b) 1 min; (c) 4 min; (d) 6 min; (e) 8 min; (f) 12 min.

XV. TEXTURED MAGNETIC MATERIALS

Fig. 8d–f.

textured silicon-iron sheets [91, 92, 104–107]. The properties of the best {100} ⟨100⟩ textured sheets measured in the rolling direction are substantially equivalent to the properties of good {110} ⟨100⟩ textured sheets measured in the rolling direction. The properties of the {100} ⟨100⟩ sheets measured perpendicular to the rolling direction are usually slightly inferior to the properties in the rolling direction but are very much better than the properties of the {110} ⟨100⟩ sheet in the perpendicular direction.

Since the surface energy mechanism is especially effective in thin sheets, it can be applied to the production of tapes for magnetic amplifier use. Both {110} ⟨100⟩ [108] and {100} ⟨100⟩* textures can be produced.

At this time (1964), {100} ⟨100⟩ textured silicon–iron is not a commercial material except for limited production of thin, tape-wound cores. The advantages the cube-textured material offers have not yet proved sufficient to offset the higher production cost as compared to the conventional {110} ⟨100⟩ texture in the thicknesses used at power frequencies.

Magnetic Control of Texture

In all the cases discussed above, the crystallographic texture is produced by energies or forces which are independent of the fact that the material is ferromagnetic. The magnetic anisotropy is only an incidental effect of the crystallographic texture, and is in no way responsible for the appearance of the texture. There are a few cases, however, in which the magnetic anisotropy can control or influence the crystallographic texture if a magnetic field is applied to the material at some step in the processing. It is important to distinguish these cases from the more common phenomenon of *magnetic annealing* (Chapter XI), in which a field applied during cooling introduces a magnetic anisotropy *without* a change in crystallographic texture.

Field Applied during a Phase Change

15. The only known example of crystallographic texture control by the application of a magnetic field during a phase change occurs in pure cobalt and in the nickel–cobalt alloys above 75% Co [109]. These materials transform from a high-temperature face-centered cubic structure to a low-temperature hexagonal structure at 400°C or below. Cooling through this transformation temperature in a strong magnetic field favors the appearance

---

* Westinghouse Electric Corporation product.

of hexagonal grains with their easy axes parallel to the applied field, and therefore tends to produce a crystallographic texture [110, 111]. The resulting magnetic anisotropy can be quite large ($\approx 10^6$ ergs cm^{-3}). For cobalt contents greater than about 95%, the sign of the magnetic anisotropy constant changes between the transformation temperature and room temperature. This corresponds to a 90° rotation of the eaxy axis, so that the anisotropy measured at room temperature has its easy axis *perpendicular* to the direction in which the field was applied during cooling.

Field Applied during Recrystallization

16. Control of texture by the application of a magnetic field during recrystallization has not been extensively studied. Measurable effects have been reported for 35–65% Co–Fe alloys [112, 113] but not for 50–50% Co–Fe [114]. Large but not entirely consistent results have been reported for pure iron [115]. The effects have been atributed to magnetostriction [112, 115] and to crystal anisotropy [113].

Field-Induced Recrystallization

17. Polycrystalline MnBi has been reported to recrystallize without prior deformation when heated between 270°C and the Curie point (350°C) in a magnetic field [116]. The resulting texture has the easy axis (*c* axis) parallel to the field applied during the anneal. In this case the magnetic anisotropy not only leads to the crystallographic texture, but also provides the driving force for recrystallization.

ACKNOWLEDGMENTS

J. L. Walter, H. C. Fiedler, H. E. Grenoble, J. J. Becker, and C. G. Dunn have all been generous with their time in reading the manuscript, correcting errors, suggesting improvements, and discussing in detail the subject and its presentation. J. L. Walter, H. C. Fiedler, and F. E. Luborsky kindly provided photographs and samples for the illustrations.

REFERENCES

1. C. S. Barrett, "Structure of Metals," 2nd ed. McGraw-Hill, New York, 1952.
2. G. Wassermann and J. Grewen, "Texturen metallischer Werkstoffe." Springer, Berlin, 1962.
3. C. G. Dunn and J. L. Walter, *J. Appl. Phys.* **31**, 827 (1960).
4. W. S. Paxton and T. G. Nilan, *J. Appl. Phys.* **26**, 994 (1955).

5. P. A. Albert, *Proc. AIEE Conf. Magnetism Magnetic Materials*, Boston, *1956*, p. 423.
6. G. W. Wiener, P. A. Albert, R. H. Trapp, and M. F. Littmann, *J. Appl. Phys.* **29**, 366 (1958).
7. N. Akulov and N. Bruchatov, *Ann. Physik* **15**, 741 (1932).
8. K. J. Sixtus, *Physics* **6**, 105 (1935).
9. L. P. Tarasov, *Trans. Am. Inst. Mining Met. Engrs.* **135**, 353 (1939).
10. L. R. Blake, *Brit. J. Appl. Phys.* **5**, 99 (1954).
11. G. I. Bunge, *Phys. Metals Metallog. (USSR) (English Transl.)* **13** [4], 30 (1962).
12. C. G. Dunn and J. L. Walter, *J. Appl. Phys.* **30**, 1067 (1959).
13. C. S. Barrett, "Structure of Metals," 2nd ed., p. 511. McGraw-Hill, New York, 1952.
14. W. G. Burgers, *in* "The Art and Science of Growing Single Crystals" (J. Gilman, ed.), p. 435 ff. Wiley, New York, 1963.
15. D. G. Ebeling, U.S. Pat. 2,295,082 (1951); D. G. Ebeling and A. A. Burr, *Trans. Am. Inst. Mining Met. Engrs.* **197**, 537 (1953).
16. K. Hoselitz and M. McCaig, *Nature* **164**, 581 (1949); M. McCaig, *Proc. Phys. Soc. London* **B62**, 652 (1949).
17. J. F. Hinsley, *in* "Permanent Magnets and Magnetism," (D. Hadfield, ed.), p. 155. Iliffe Books, London, 1962, and Wiley, New York, 1962.
18. A. I. Luteijn and K. J. de Vos, *Philips Res. Rept.* **11**, 489 (1956); A. J. J. Koch, M. G. van der Steeg, and K. J. de Vos, *Proc. AIEE Conf. Magnetism Magnetic Materials, Boston, 1956*, p. 173, AIEE, New York.
19. N. Makino, *Cobalt*, p. 3 (Dec. 1962); p. 26 (March 1963).
20. M. McCaig, *J. Appl. Phys.* **35**, 958 (1964).
21. J. E. Gould, *Cobalt*, p. 82 (June 1964).
22. E. Steinort, E. R. Cronk, S. J. Garvin, and H. Tiderman, *J. Appl. Phys.* **33**, 1310 (1962).
23. W. R. Hibbard, Jr., *Trans. Am. Inst. Mining Met. Engrs.* **206**, 962 (1956).
24. W. Baran, W. Brener, H. Fahlenbrach, and K. Janssen, *Tech. Mitt. Krupp* **18**, 81 (1960).
25. Y. Kimura, *Trans. Japan. Inst. Metals* **4**, 22 (1963).
26. F. Pawlek, *Z. Metallk.* **27**, 160 (1935).
27. G. W. Rathenau and J. F. H. Custers, *Philips Res. Rept.* **4**, 241 (1959).
28. W. Six, J. L. Snoeck, and W. G. Burgers, *Ingenieur (Utrecht)* **49**, 195 (1934).
29. J. L. Snoek, *Physica* **2**, 403 (1935).
30. O. Dahl and J. Pfaffenberger, *Metallwirtschaft* **13**, 527, 543, 559 (1934); **14**, 25 (1935).
31. H. W. Conradt, O. Dahl, and K. J. Sixtus, *Z. Metallk.* **32**, 231 (1940).
32. J. H. Crede and J. P. Martin, *J. Appl. Phys.* **20**, 966 (1949).
33. H. H. Scholefield, *J. Sci. Instr.* **26**, 207 (1949).
34. M. F. Littmann, E. S. Harris, and C. E. Ward, *J. Appl. Phys.* **33**, 1229 (1962).
35. S. Spachner and W. Rostoker, *Trans. Am. Inst. Mining Met. Engrs.* **203**, 921 (1955); M. F. Littmann, *ibid.* **206**, 593 (1956).
36. K. Beck, *Vierteljahrssch. Naturforsch. Ges. Zurich* **63**, 116 (1918); quoted by L. W. McKeehan, *Trans. Am. Inst. Mining Met. Engrs.* **111**, 11 (1934).
37. W. E. Ruder, *Trans. Am. Soc. Steel Treating* **8**, 23 (1925).
38. W. E. Ruder, *Trans. Am. Soc. Metals* **22**, 1120 (1934); see discussion.

39. N. P. Goss, U.S. Pat. 1,965,559 (1934); *Trans. Am. Soc. Metals* **23**, 511 (1935).
40. R. M. Bozorth, *Trans. Am. Soc. Metals* **23**, 1107 (1935).
41. A. A. Frey and F. Bitter, U.S. Pat. 2,112,084 (1938).
42. V. W. Carpenter, U.S. Pat. 2,287,466 (1942).
43. W. Morrill, U.S. Pat. 2,279,762 (1942).
44. G. H. Cole, R. L. Davidson, and V. W. Carpenter, U.S. Pat. 2,307,391 (1943).
45. W. Morrill, C. G. Dunn, and R. Ward, U.S. Pat. 2,534,141 (1950).
46. J. M. Jackson, U.S. Pat. 2,535,420 (1950).
47. M. F. Littmann and J. E. Heck, U.S. Pat. 2,599,340 (1952).
48. J. H. Crede, R. H. Henke, E. L. Pulaski, and C. H. Stroble, U.S. Pat. 2,867,557 (1959).
49. G. H. Cole and R. L. Davidson, U.S. Pat. 2,158,065 (1939).
50. R. H. Pry, *J. Appl. Phys.* **30**, 189S (1959).
51. W. E. Ruder, U.S. Pat. 2,264,859 (1941).
52. C. G. Dunn, *Trans. Inst. Mining Met. Engrs.* **224**, 612 (1962).
53. C. G. Dunn, *in* "Cold Working of Metals," p. 113. Am. Soc. Metals, Cleveland, Ohio, 1949.
54. P. K. Koh, *Trans. Am. Inst. Mining Met. Engrs.* **215**, 1043 (1959).
55. P. K. Koh and C. G. Dunn, *Trans. Am. Inst. Mining Met. Engrs.* **218**, 65 (1960).
56. C. G. Dunn and C. J. McHargue, *J. Appl. Phys.* **31**, 1767 (1960).
57. G. Wiener and R. Corcoran, *Trans. Am. Inst. Mining Met. Engrs.* **206**, 901 (1956).
58. J. R. Brown, *J. Appl. Phys.* **29**, 359 (1958).
59. H. C. Fiedler, *J. Appl. Phys.* **29**, 361 (1958).
60. H. Stablein and H. Moller, *Arch. Eisenhuttenw.* **29**, 433 (1958).
61. K. Detert, *Metall* **12**, 817 (1958).
62. G. H. Cole and R. L. Davidson, U.S. Pat. 2,158,065 (1939).
63. W. Morrill, *Metal Progr.* **54**, 675 (1948).
64. J. D. Fast, *Philips Res. Rept.* **11**, 490 (1956).
65. J. E. May and D. Turnbull, *Trans. Am. Inst. Mining Met. Engrs.* **212**, 769 (1958).
66. J. E. May and D. Turnbull, *J. Appl. Phys.* **30**, 210S (1959).
67. N. G. Ainslie and A. U. Seybolt, Jr., *J. Iron Steel Inst.* **194**, 341 (1960).
68. H. C. Fiedler, *Trans. Am. Inst. Mining Met. Engrs.* **221**, 1201 (1961).
69. H. C. Fiedler and R. H. Pry, U.S. Pat. 2,939,810 (1960).
70. H. C. Fiedler, U.S. Pat. 3,096,222 (1963).
71. E. V. Walker and J. Howard, *Powder Met.* **4**, 32 (1959).
72. M. F. Littmann, U.S. Pat. 2,473,156 (1949).
73. H. C. Fiedler, *Trans. Am. Inst. Mining Met. Engrs.* **227**, 777 (1963).
74. H. C. Fiedler, *Trans. Am. Inst. Mining Met. Engrs.* **230**, 95 (1964).
75. H. C. Fiedler, *Trans. Am. Inst. Mining Met. Engrs.* **230**, 603 (1964).
76. F. Bitter, U.S. Pat. 2,046,717 (1936).
77. W. W. Mullins, *Acta Met.* **6**, 414 (1958).
78. K. Detert, *Acta Met.* **7**, 589 (1959).
79. J. L. Walter and C. G. Dunn, *Trans. Am. Inst. Mining Met. Engrs.* **215**, 465 (1959).
80. J. L. Walter, *Acta Met.* **7**, 424 (1959).
81. C. G. Dunn and J. L. Walter, *Trans. Am. Inst. Mining Met. Engrs.* **218**, 448 (1960).
82. J. L. Walter and C. G. Dunn, *Acta Met.* **8**, 497 (1960).
83. J. L. Walter and C. G. Dunn, *Trans. Am. Inst. Mining Met. Engrs.* **218**, 914 (1960).
84. J. L. Walter and C. G. Dunn, *Trans. Am. Inst. Mining Met. Engrs.* **218**, 1033 (1960).

85. D. Kohler, *J. Appl. Phys.* **31**, 408S (1960).
86. G. W. Wiener, *J. Appl. Phys.* **35**, 856 (1954).
87. F. Assmus, R. Boll, K. Detert, D. Ganz, G. Ibe, and F. Pfeifer, U.S. Pat. 2,992,952 (1961).
88. G. W. Wiener, U.S. Pat. 3,078,198 (1963).
89. R. G. Aspden, U.S. Pat. 2,992,951 (1961).
90. F. Assmus, K. Detert, and G. Ibe, *Z. Metallk.* **48**, 344 (1957).
91. G. Wiener, P. A. Albert, R. H. Trapp, and M. F. Littmann, *J. Appl. Phys.* **29**, 366 (1958).
92. J. L. Walter, W. R. Hibbard, H. C. Fiedler, H. E. Grenoble, R. H. Pry, and P. G. Frischmann, *J. Appl. Phys.* **29**, 363 (1958).
93. G. Baer, D. Ganz, and H. Thomas, *J. Appl. Phys.* **31**, 235S (1960).
94. G. N. Kadykova and V. V. Sosnin, *Phys. Metals Metallog. USSR (English Transl.)* [3]**11**, 59 (1961).
95. K. Foster, J. J. Kramer, and G. W. Wiener, *Trans. Am. Inst. Mining Met. Engrs.* **227**, 183 (1963).
96. W. R. Hibbard and J. L. Walter, Belgian Pat. 560,938 (1957); H. J. Fisher and J. L. Walter, *Trans. Am. Inst. Mining Met. Engrs.* **224**, 1271 (1962).
97. V. P. Varlakov and V. M. Romashov, *Phys. Metals Metallog. (USSR) (English Transl.)* [5]**13**, 34 (1962).
98. S. Taguchi *et al.*, Japan. Pat. Specification 1459/63 (1963); translated by Rotha Fullford Leopold and Associates, Beaumaris, Victoria, Australia.
99. Hsun Hu, U.S. Pat. 3,089,795 (1963).
100. O. Madono, Company Rept. Riken Piston Ring Co., Tokyo, Japan (1963).
101. H. E. Möbius and F. Pawlek, *Arch. Eisenhuttenw.* **29**, 423 (1958).
102. H. C. Fiedler, private communication (1964).
103. H. E. Grenoble, U.S. Pat. 3,147,157 (1964).
104. D. Ganz, *Z. Angew. Phys.* **14**, 313 (1962).
105. K. Foster and J. J. Kramer, *J. Appl. Phys.* **31**, 233S (1960).
106. F. Assmus, R. Boll, D. Ganz, and F. Pfeifer, *Z. Metallk.* **48**, 341 (1957).
107. Ya. N. Kunakov and B. G. Livshits, *Phys. Metals Metallog. (USSR) (English Transl.)* [1]**15**, 52 (1963).
108. J. L. Walter, *J. Appl. Phys.* **33**, 1230 (1962).
109. M. Takahashi and T. Kōno, *J. Phys. Soc. Japan* **15**, 936 (1960).
110. C. D. Graham, Jr., *J. Phys. Soc. Japan* **16**, 1481 (1961); *Suppl. B-I* **17**, 321 (1962).
111. T. Sambongi and T. Mitui, *J. Phys. Soc. Japan* **16**, 1478 (1961); **18**, 1253 (1963).
112. R. Smoluchowski and R. W. Turner, *J. Appl. Phys.* **20**, 745 (1949); *Physica* **16**, 397 (1950).
113. B. Sawyer and R. Smoluchowski, *J. Appl. Phys.* **28**, 1069 (1957).
114. A. H. Geisler, J. P. Martin, E. Both, and J. H. Crede, *Trans. Am. Inst. Mining Met. Engrs.* **203**, 985 (1955).
115. V. S. Bhandary and B. D. Cullity, *Trans. Met. Soc. Am. Inst. Mining Met. Engrs.* **224**, 1194 (1962).
116. O. L. Boothby, D. H. Wenny, and E. E. Thomas, *J. Appl. Phys.* **29**, 353 (1958).

CHAPTER XVI

Diffusion

G. W. RATHENAU
Philips Research Laboratories
N. V. Philips Gloeilampenfabrieken
Eindhoven, the Netherlands
and
Natuurkundig Laboratorium der
Universiteit van Amsterdam
Amsterdam, the Netherlands

G. DE VRIES
Natuurkundig Laboratorium der
Universiteit van Amsterdam
Amsterdam, the Netherlands

SURVEY OF DIFFUSION PHENOMENA	749
MAGNETIC DETECTION OF DIFFUSION USING DIRECTIONAL ORDER	751
Interstitial Ordering	751
Induced Energy of Anisotropy	754
Aftereffect of the Magnetostriction	757
Stress-Induced Magnetic Anisotropy	758
Specified Domain Walls	759
Unspecified Arrays of Domain Walls	770
Examples of Diffusion Measurement	778
DIRECTIONAL ORDER IN OTHER SYSTEMS, ESPECIALLY SUBSTITUTIONAL ALLOYS	785
Theoretical Considerations	785
Experimental Approaches	795
Experimental Results	796
CONCLUDING REMARKS	810
References	810

Survey of Diffusion Phenomena

1. *Diffusion over Macroscopic Distances.* Little use has been made in the past of magnetic methods for measuring diffusion in bulk materials, as occurs, for example, while annealing a binary system such as a nickel sheet plated with chromium [1]. Measurement of the saturation magnetization at varying depth, however, may certainly often be a reliable tool to determine penetration and the diffusion constants.

2. Precipitation and the Kinetics of Transformation. For diffusion over smaller distances magnetic measurements have been extensively used to study precipitation and the kinetics of transformation (see for review, Wohlfarth [2]). In this kind of work it is advantageous for magnetic measurements to be made easily on materials at temperatures other than room temperature. Generally the kinetics cannot be determined unambiguously, because neither the distance between the precipitating particles nor their chemical composition is well known, while the short-circuiting of the diffusion in bulk by lattice imperfections such as dislocations may occur to an appreciable degree [3]. The cases most amenable to a quantitative treatment of the diffusion process are those where precipitation of a known ferromagnetic precipitate is completed, and only growth or change of shape of the precipitated particles occurs. That the precipitation of a ferromagnetic phase from a nonmagnetic matrix is indeed terminated can be inferred, for example, from the invariance with annealing of the saturation magnetization. This has been done in the case of cobalt precipitating from a Co–Cu alloy [4]. Also the growth of the Co particles and the change of their shape can be followed by magnetic measurements [5].

3. *The Development of Long-Range Order*. The development of long-range order can have a profound influence on the magnetic properties of metallic materials containing elements with partially filled $3d$ or $4f$ shells, and magnetic measurements have served in the past to study the ordering kinetics [6].

4. *Diffusion over Distances of the Order of the Lattice Constant*. Diffusion over distances of the order of the lattice constant has, during the past 30 years, been amply studied by magnetic methods in two or more component systems. The principle of these investigations, which use the phenomenon of directional ordering, is the following. Due to the presence of more than one component in the matrix, the atomic configuration in the lattice may exhibit locally uniaxial symmetry. The local magnetization also defines a direction.

Generally an interaction energy exists which depends on the angle between the configurational axis and the magnetization vector. Therefore, by changing the direction of the magnetization, for example, by application of an external magnetic field, it is possible to induce a redistribution of atoms. This in turn induces changes of the macroscopic magnetic properties, which can be investigated as a function of time.

5. *Self-Diffusion.* Dynamic measurements are useful in the study of self diffusion in metals. Both of relaxation times T_1 and T_2 of the nuclear magnetization depend on the mean time of stay of an atom at a particular atomic site. Nuclear magnetic resonance techniques can be used to determine the relevant activation energies of diffusion [7].

Magnetic Detection of Diffusion Using Directional Order

6. A remakable survey article, which stresses the physics of directional ordering and deals with elementary situations that can be treated theoretically, has been written by Slonczewski [8]. He also reviews the corresponding phenomena in ferrites. A short review is that of Néel [9]. A review of the whole field of magnetic annealing with an extensive bibliography up to 1958 has been given by Graham [10]. Other reviews have been written by Brissonneau [11] and by Schreiber [12], who introduces Néel's theory of interstial directional order. Reviews by Rathenau [13] and Bosman and de Vries [14] also deal with directional order in interstitial solutions.

Interstitial Ordering

7. *Mechanism and Interaction Energy.* This case, one of monatomic directional ordering according to Slonczewski's terminology, is the simplest one to visualize, and will be treated first in some detail before moving on to substitutional alloys and other cases where directional ordering is found. It has long been known [13] that on applying a magnetic field to iron specimens containing small amounts of interstitial impurities, the induction may lag due to "magnetic viscosity" while the initial permeability after demagnetization decreases with time. This phenomenon is called "disaccommodation." Following the work of Richter [15], Snoek proved that, in body-centered cubic iron interstitial atoms such as carbon and nitrogen can cause these effects, as well as the mechanical aftereffects of which internal friction is one manifestation [16].

Starting from ideas of Gorski [17], Snoek considered the noncubic surrounding of isolated interstitials in the iron lattice as basic for the observed phenomena. It may be assumed [13] that the stable interstitial sites are of the type $(00\frac{1}{2})$ as shown in Fig. 1. Therefore, each interstitial

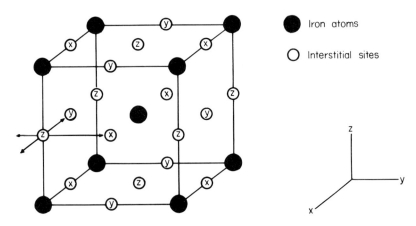

FIG. 1. Interstitial sites in the body centered structure. Possible jumps of an interstitial atom on a z site are indicated.

atom and its two nearest iron atom neighbors define an axis parallel to one of the cubic directions. The three types of site are named after this local axis: x, y, or z sites. An agent of directional character such as strain or magnetization may remove the equivalence of the three types of site. If initially the distribution of the interstitial atoms over the three types of site complies with thermal equilibrium and then suddenly the state of strain and magnetization is changed to another state, a new equilibrium distribution will be attained with a relaxation time τ. If only isolated interstitials of one species, e.g., carbon atoms, are present in the perfect lattice, τ is equal to $\frac{2}{3}\tau_m$. Here τ_m is the mean time an interstitial atom spends on a particular site. In order to derive this relation [18] between the mean time of stay and the macroscopically measurable relaxation time τ, it has been assumed that the new equilibrium distribution is obtained by jumps of the interstitial to the nearest available interstices, and that the mean time of stay is not much different for preferred and avoided sites. This is certainly true for the experiments performed. The relative difference between the times of stay is of the order of δ/kT, where δ is the energy difference between the sites. It will be shown later that the magnetization induces values δ of the order of 10^{-15} erg as does a stress of 1 kg mm^{-2}. The investigations have been carried out at temperatures exceeding 200°K, corresponding to 3×10^{-14} ergs for kT.

Snoek [16, 19], in order to explain mechanical aftereffects due to interstitials, considered the decrease of elastic energy when the interstitials move from sites which are compressed by the external stress to expanded sites. In order to deal quantitatively with the problem, he inserted the known cell deformation of tetragonal martensite into his formulas. His theory has been extended and refined by Polder [18]. Snoek assumed the magnetic aftereffects to be a consequence of mechanical aftereffects because of the lattice deformation brought about through magnetostriction.

Néel [20, 21], using Snoek's evidence, pointed out, however, that magnetostrictive coupling is too weak to explain the magnetic phenomena. He therefore introduced a phenomenological interaction between the occupation of a site and the magnetization vector. Néel's point of view was again verified experimentally by an investigation by Bosman et al. [22] of the iron–silicon system in which the magnetostriction constant λ_{100} for a concentration of 6 wt % passes a zero value while the aftereffects remain about constant. De Vries [23] incorporated both a magnetostrictive and a phenomenological "direct" interaction into a theory. As will be shown, experiments by de Vries et al. [24] allowed the complete separation of these different parts of the total interaction. The response of the interstitial solid solution to magnetic fields, which act through the magnetization vector and the stress, can be investigated theoretically by minimizing the free energy and with respect to those parameters free to adjust themselves. De Vries, extending the development given by Polder [18] for mechanical aftereffects, used

$$\begin{aligned} F = F_0 &+ \tfrac{1}{2}c_{11}(e_{xx}^2 + e_{yy}^2 + e_{zz}^2) + c_{12}(e_{xx}e_{yy} + e_{yy}e_{zz} + e_{zz}e_{xx}) \\ &+ \tfrac{1}{2}c_{44}(e_{xy}^2 + e_{yz}^2 + e_{zx}^2) + \epsilon(C_x e_{xx} + C_y e_{yy} + C_z e_{zz}) \\ &+ RT/2VC_0(C_x^2 + C_y^2 + C_z^2) + B_1[(\alpha_1^2 - \tfrac{1}{3})e_{xx} + (\alpha_2^2 - \tfrac{1}{3})e_{yy} \\ &+ (\alpha_3^2 - \tfrac{1}{3})e_{zz}] + B_2(\alpha_1\alpha_2 e_{xy} + \alpha_2\alpha_3 e_{yz} + \alpha_3\alpha_1 e_{zx}) \\ &+ D[(\alpha_1^2 - \tfrac{1}{3})C_x + (\alpha_2^2 - \tfrac{1}{3})C_y + (\alpha_3^2 - \tfrac{1}{3})C_z] \\ &+ K_1(\alpha_1^2\alpha_2^2 + \alpha_2^2\alpha_3^2 + \alpha_3^2\alpha_1^2) + \cdots . \end{aligned} \qquad (7.1)$$

Here the c_{ij} are the strain components, α_i are the direction cosines of the magnetization vector, C_i are the deviations of the concentrations of interstitials on sites i from C_0 (C_i and C_0 are supposed to be small), which is one-third of the total atomic concentration of interstitials in the iron lattice. The numerical values, as used by de Vries for temperatures around 250°K and partly arrived at by measurements of de Vries et al. and by Gersdorf [25], are reproduced here. All values are in joules per cubic meter, if not otherwise stated (1 J m^{-3} = 10 ergs cm^{-3}). The elastic constants are $c_{11} = 23.7 \times 10^{10}$, $c_{11} = 14.1 \times 10^{10}$, and $c_{44} = 11.6 \times 10^{10}$. The interaction constant between strain and population on one particular kind of site is $\epsilon = -9.9 \times 10^{10}$ for carbon and -9.2×10^{10} for nitrogen. For the entropy term, where T is the absolute temperature and V is the molar volume of iron (7.09×10^{-6} m^3), one inserts $R/V = 1.17 \times 10^6$ J (m³ °K)⁻¹. The magnetoelastic coupling constant representing magnetostriction are $B_1 = -3.3 \times 10^6$ and $B_2 = 7.3 \times 10^6$. The direct interactions between magnetization vector and interstitial population (dealt with in Secs. 5 and 6) are $D = 7.2 \times 10^6$ for carbon and 4.7×10^6 for nitrogen. The first magnetic crystal anisotropy constant is $K_1 = 42 \times 10^3$.

We observe that in the power development of the free energy the term in D_1 has cylindrical symmetry. The interstitial site actually has tetragonal symmetry. Only higher-order terms can show this lower symmetry.

In the following we will discuss the different consequences of diffusion which can be deduced from Eq. (7.1). First, we will deal with effects which are shown by the material in bulk. Subsequently, the rather spectacular behavior of domain walls will be considered.

Induced Energy of Anisotropy

8. In Eq. (7.1) the term with D and the two terms with ϵ and B_1 furnish interactions between the magnetization vector and the distributions of interstitials. These are the direct and the magnetostrictive interactions which have been discussed in the last section.

Now let us consider the following experiment [24] (Fig. 2). A flat disk

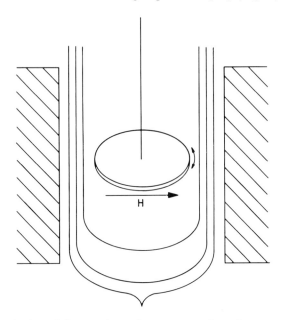

FIG. 2. Schematic view of the experimental arrangement for anisotropy measurements.

of an iron single crystal containing a known amount of a certain species of interstitials is brought between the poles of a magnet. At constant temperature it is subjected to a saturating field for a time long compared with the relaxation time (magnetic annealing). Then the interstitial distribution adapts itself to the direction of the magnetization vector, given by its direction cosines α_i. After the freezing-in operation an additional magnetic anisotropy energy is found that depends on the direction α_i during the magnetic anneal; this can be measured by an extra restoring torque. The extra anisotropy energy density connected with this torque can be deduced [22] for Eq. (7.1), to be

$$K_u = (VC_0/2RT)[D - \epsilon B_1/(c_{11} - c_{12})]^2 \{(\alpha_1^2 - \beta_1^2)^2 \\ + (\alpha_2^2 - \beta_2^2)^2 + (\alpha_3^2 - \beta_3^2)^2\}. \tag{8.1}$$

In this equation the β_i are the direction cosines of the magnetization vector during the measurement immediately after the magnetic anneal. Both types of interaction are found together.

$[D - \epsilon B_1/(c_{11} - c_{12})]^2$ appears in Eq. (8.1) as the square of the resultant interaction energy per interstitial atom times the number of iron atoms per unit volume. Neither the sign of the resultant interaction nor the magnitude of D can thus be determined from anisotropy measurements. The square in Eq. (8.1) can be understood as follows. The number of interstitials in excess on particular types of site is proportional to the ratio of the resultant interaction energy per interstitial atom and kT. Every interstitial atom in excess contributes the resultant interaction energy to the anisotropy energy.

The large natural magnetocrystalline anisotropy of iron makes it difficult to measure the additional anisotropy energy on single crystals with precision. In work by de Vries et al. [24] polycrystalline disks of iron containing C or N have been used which, due to weak texture, had two positions of easy magnetization in the plane of the disk normal to each other. These samples were investigated in the arrangement given schematically by Fig. 2. After subjecting a disk for a long time to a field of 1.7×10^6 A m^{-1} (21,000 Oe) in one of the stable positions, at time $t = 0$, the disk was suddenly turned through 90° around the axis. The period of small oscillations around the second stable position was measured as a function of time. This period decreases because in the course of time the interstitials redistribute themselves; they first contribute negatively and later, positively, to the restoring torque (Fig. 3). The stabilization by magnetic anneal and the anisotropy expressed in Eq. (8.1) will be approached exponentially with time [21] in the simple case which we have supposed to exist. The same will be true for the period of the oscillation if the restoring torque due to the magnetic anneal is small compared to the total torque. As seen in Fig. 3, the experiment directly gives the relaxation time τ, which is in quite good agreement with other measurements.

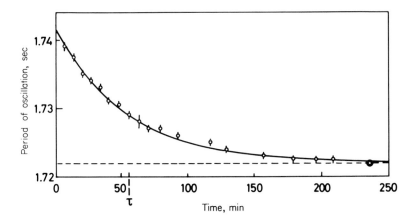

Fig. 3. Measurement of anisotropy energy due to magnetic annealing of interstitial carbon in iron in the arrangement of Fig. 2; carbon content of the sample, 0.015 wt %; temperature, 239°K.

The development of uniaxial anisotropy with time can also be followed by measurement of the area between the magnetization curves for the case in which an annealing magnetic field is applied parallel or normal to the measuring direction. Care should be taken that the measurement be rapid so as not to disturb the magnetic anneal, and that the temperature of anneal be well known.

In order to interpret the anisotropy measurements of polycrystalline material, one must take the average of Eq. (8.1), and keep in mind that the magnetostriction of a grain is partly suppressed by the surroundings. It can be calculated [24] that if the randomly oriented grains are small with respect to all dimensions of the sample, Eq. (8.1) can be replaced by

$$K_u' = (2VC_0/5RT)[D - \epsilon B_1'/(c_{11} - c_{12})]^2 \sin^2 \varphi \qquad (8.2)$$

with

$$B' = 0.77B_1 + 0.23B_2 \quad \text{and} \quad \epsilon B_1'/(c_{11} - c_{12}) = 0.75 \times 10^6 \ \text{J m}^{-3}.$$

Here φ is the angle between the direction of the magnetization vector during magnetic anneal and that during measurement. Torque measurements such as those represented in Fig. 3 can be summarized as follows:

For carbon at 239°K:

$$D - \epsilon B_1'/(c_{11} - c_{12}) = \pm 6.4 \times 10^6 \quad \text{J m}^{-3}$$
$$D = 7.15 \times 10^6 \quad \text{or} \quad -5.65 \times 10^6 \quad \text{J m}^{-3} \quad (8.3)$$

For nitrogen at $\pm 225°K$:

$$D - \epsilon B_1'/(c_{11} - c_{12}) = \pm 4.05 \times 10^6 \quad \text{J m}^{-3}$$
$$D = 4.8 \times 10^6 \quad \text{or} \quad -3.3 \times 10^6 \quad \text{J m}^{-3}. \quad (8.4)$$

Graham [10] measured the area between magnetization curves of polycrystalline material and estimated the effective anisotropy of iron-containing interstitials. The order of magnitude agrees with the values given above. The same is true for later work along similar lines [26] in which both the temperature of magnetic anneal and the interstitial concentration are not well known.

Aftereffect of the Magnetostriction

9. Let us now consider the following experiment [24]. An iron single crystal containing interstitials is held for a long time, compared to the relaxation time, in a strong magnetic field parallel to its [100] direction. Then the field is suddenly rotated 90° into a [010] direction of the crystal. One observes an immediate elongation due to magnetostriction and an aftereffect of the deformation. This aftereffect corresponds to the redistribution of the interstitials and would be positive, an elongation, if the interstitials preferred the sites on cube edges parallel to the magnetization vector.

Actually, as Fig. 4 shows, the observed aftereffect is negative. So, notwithstanding the magnetostrictive strain, which expands the lattice in the cube direction parallel to the magnetization vector, the interstitials avoid the corresponding sites. The aftereffect $\Delta\lambda$ of the magnetostriction λ_{100}, has been calculated to be [23]

$$\Delta\lambda/\lambda_{100} = (\epsilon V C_0/RTB_1)[\epsilon B_1/(c_{11} - c_{12}) - D]. \quad (9.1)$$

The experiment shows that for carbon and nitrogen in the temperature region investigated, the direct interaction D introduced in Eq. (7.1) is positive and larger than the magnetostrictive interaction. From experiments such as those reproduced in Fig. 4 the values of D can be calculated. These turn out to be in good agreement with the positive values of Eq. (8.4).

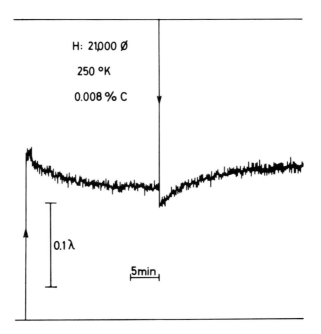

Fig. 4. A recording of the aftereffect of the magnetostriction induced by interstitial carbon.

At temperatures around 250°K, this procedure yields for carbon: $D = 7.3 \times 10^6$ J m^{-3} and for nitrogen $D = 4.6 \times 10^6$ J m^{-3}. The completion of the aftereffect again occurs exponentially. Thus, from the measurements such as those shown in Fig. 4, the relaxation time τ can be obtained, though not with very high precision since the effect is rather small. The physical origin of the interaction energy D is as yet speculative.

Stress-Induced Magnetic Anisotropy

10. The agent which introduces, in bulk material, magnetic anisotropy or an aftereffect of the elongation may be stress instead of a magnetic field which orients the magnetization. This is clear from Eq. (7.1).

In this case the aftereffect of the elongation is the well-known mechanical aftereffect measured on a magnetic material [27].

Stress-induced anisotropy has not been measured in interstitial solutions, but is well known for substitutional alloys (Sec. 28), and may serve to magnetically measure relaxation times for temperatures above the Curie temperature. The effects can be illustrated by considering interstitial solutions. In Eq. (7.1) external stresses are incorporated simply by adding

$-(P_{xx}e_{xx} + P_{yy}e_{yy} + P_{zz}e_{zz} + P_{yz}e_{yz} + P_{zx}e_{zx} + P_{xy}e_{xy})$. It is then easily found that with constant magnetization direction, or no magnetization at all, the changes in occupation due to the external forces are, for equilibrium, given by

$$C_i = [P_{ii} - \tfrac{1}{3}(P_{xx} + P_{yy} + P_{zz})][(\epsilon V C_0/RT) - (c_{11} - c_{12})^{-1}].$$

From this it follows that an extra anisotropy energy is introduced equal to

$$K_u = \epsilon' V' C_0/RT'(c'_{11} - c'_{12})^{-1})^{-1}[D - B_1/(c_{11} - c_{12})]$$
$$\times [P_{xx}(\tfrac{1}{3} - \alpha_1^2) + P_{yy}(\tfrac{1}{3} - \alpha_2^2) + P_{zz}(\tfrac{1}{3} - \alpha_3^2)]. \quad (10.1)$$

In this equation, as in some of the previous ones, $-\epsilon^2 V C_0/RT$ is neglected compared to $c_{11} - c_{12}$.

The primed constants refer to the conditions in which equilibrium is attained (while applying external pressure); the unprimed constants refer to the values of the parameters at the temperature where the magnetic anisotropy is measured.

T' is some high temperature at which the approach toward equilibrium is to be determined. After a definite stay at this temperature, the sample is quenched to T at which temperature the induced anisotropy can be determined conveniently.

Specified Domain Walls

11. It has been noted previously that the effects of interstitials on domain wall mobility can be particularly large. The magnetization vector **M** changes its direction within a wall. In sufficient time the equilibrium distribution of interstitials will adapt itself to the direction of **M** everywhere in the wall and in its surrounding. The local magnetization becomes stabilized (Sec. 9) and through this the position of the wall. The effect of this stabilization is particularly pronounced if the wall mobility is large in the absence of interstitials.

Néel [21] assumed only one kind of interaction which is constant everywhere in the material [D in Eq. (7.1)], and calculated the energy necessary to displace a stabilized wall over a distance u. The change of angle of the spins along the normal to the wall is assumed to obey the same laws that are valid in the absence of interstitials, a characteristic length being defined by $d = (A/K)^{1/2}$, where K is the crystalline anisotropy energy density and A is the exchange energy constant [28]. Passing along the normal through a wall of unit area, which has been displaced from the position in which the distribution of interstitials is relaxed, one adds everywhere the locally induced anisotropy energies as given by Eq. (8.1), with $\epsilon = 0$. The results of the calculation for a $\{100\}$ 90° wall and a $\{100\}$ 180° wall (which in iron is almost the same as two such 90° walls in series) is represented in Fig. 5. It is evident that after a large displacement of a 180° wall, the energy reaches a constant limiting value. This occurs because there are only two small regions of constant total volume where the interstitial distribution and the direction of the magnetization vector do not concur. These are the regions where the wall was initially at rest and that to which it has been

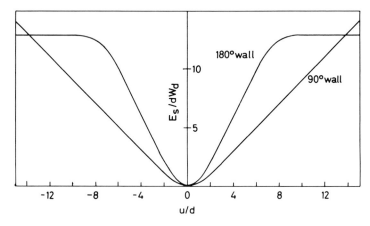

FIG. 5. Stabilization energy per unit area, E_s, as a function of the displacement u, for a $\{100\}$ 180° wall and a $\{100\}$ 90° wall $d = (A/K)^{1/2}$ and $W_d = (VC_0/RT)D^2$. (After Néel [21].)

displaced. For the volume in between, no stabilization energy is required since here the spins have only been rotated 180°, which makes no difference in the anisotropy energy density [Eq. (8.1)]. For the 90° wall, on the contrary, the volume where the interstitial distribution and the new direction of magnetization do not concur continues to increase with displacement.

The pressures P of a restoring character acting upon relaxed walls which are displaced over a distance u are found by differentiation of curves such as of Fig. 5, as is shown in Fig. 6. In the absence of other restoring forces, equilibrium is attained between the displacing magnetic field and P according to

$$\mathbf{H}(\mathbf{M}_1 - \mathbf{M}_2) + P(u) = 0. \tag{11.1}$$

The stabilization field of a wall is defined as that value of the external mag-

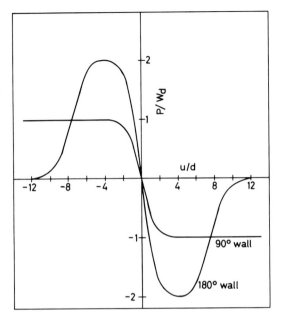

FIG. 6. Stabilizing pressure P acting upon a relaxed wall as a function of the displacement u, for a {100} 180° wall and a {100} 90° wall.

netic field, along one of the directions of magnetization separated by the wall, which balances P at a displacement u.

Equation (11.1) then becomes

$$H_{s'90°}(u) = P(u)/M_s \quad \text{and} \quad H_{s'180°}(u) = P(u)/2M_s. \tag{11.2}$$

For the two walls represented in Figs. 5 and 6 the maximum stabilization fields are equal.

Up to now the magnetostrictive interaction has been neglected compared to the direct interaction D. This may have to be taken into account in order to determine the absolute values of the curves representing the stabilization as a function of displacement. Moreover, the shape of the curves will also be altered if the relative importance of the two interactions changes along the normal to the wall. The dimensional changes within the wall parallel to its surface are determined by the strains in the large adjacent Weiss domains. Along the normal to a {100} 180° wall one finds almost everywhere the same strain, i.e., the magnetostrictive strain of the two adjacent domains. This demonstrates that it is indeed the interaction D alone, which in this case determines the restoring pressure P, as depicted in Fig. 6.

Within 180° walls that are not parallel to a {100} plane the strains generally differ from the strains in the adjacent domains. The restoring pressures for small displacement dP/du therefore do not depend only on D, as shown in Fig. 7 [29].

For the most common 90° wall of type (110) the restoring pressure at large displacements u/d is found to be [23]

$$P = (VC_0/RT)[D - \epsilon B_1/(c_{11} - c_{12})]^2,$$

which can be compared with the asymptotic value of P for a {100} 90° wall in Fig. 6. In the centre of the {110} 90° wall, however, the strains which

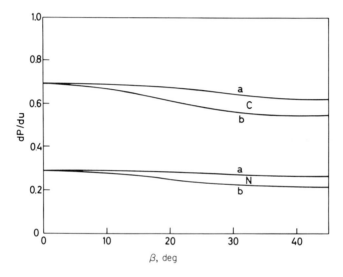

FIG. 7. Restoring pressure (dP/du) acting on 180° walls of which the normal direction is in a (100) plane making an angle β with a [010] direction: curve a, calculated with neglect of ϵ; curve b, calculated considering both D and ϵ.

counteract the direct interaction D cannot fully develop. Therefore, for small displacements u of such walls the restoring pressure rises to larger values than W. A maximum appears in the curve P vs. u, as for 180° walls. This maximum, shown in Fig. 8, is relatively large for N, because for this element D is smaller than for C [Eqs. (8.3) and (8.4)]. Equation (11.1) disregards the presence of other restoring pressures, $R(u)$, due to non magnetic inclusions, strain gradients, etc. acting in addition to the stabilization pressure. Including these and taking into account that the development of P takes time, one has

$$R(u) + (\mathbf{H}, \mathbf{M}_1 - \mathbf{M}_2) + P(u, t) = 0. \quad (11.3)$$

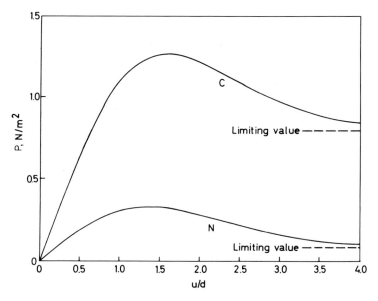

FIG. 8. Stabilizing pressure P acting upon a relaxed wall as function of the displacement u for a (110) 90° wall: curve C calculated for carbon with $\epsilon VC_0/RT = -0.0068$; curve N calculated for nitrogen with $\epsilon VC_0/RT = -0.0169$.

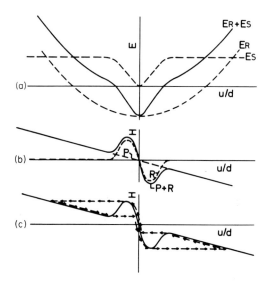

FIG. 9. A (100) 180° wall under the simultaneous influence of stabilization and other restoring pressures: (a) the energies; (b) the pressures; (c) the arrows indicate the displacement of the wall if to compensate $P + R$ an external alternating field is applied.

Figure 9 schematically represents a {100} 180° wall under the influence of P as well as a pressure R, where dR/du is a constant corresponding to the most simple potential trough. In Fig. 9a the energies are shown and in Fig. 9b are shown the pressures due to P and R. These can be balanced in a stable or unstable equilibrium by external magnetic fields H. Fig. 9c shows how the application of an alternating field H will make a relaxed wall go through a constricted hysteresis loop in which the inherent instability of the 180° wall is expressed. For a wall area of unit surface, namely, the magnetic moment obtained through the displacement u is $2M_s \cdot u$.

In an otherwise perfect lattice of pure iron containing one species of interstitial, the stabilization of a domain wall brought to, and held, at a well-defined new location, occurs exponentially with the same relaxation time τ introduced previously, which corresponds to an elementary interstitial jump

$$\frac{(P+R)_{t=\infty} - (P+R)_t}{(P+R)_{t=\infty} - (P+R)_{t=0}} = \frac{P_\infty - P_t}{P_\infty - P_0} = e^{-t/\tau}. \quad (11.4)$$

A very important case which we will consider in detail below (Sec. 13) is that of the stabilization of domain walls in the absence of dc fields after the specimen has been demagnetized. The phenomenon is called time decrease or time decay of initial permeability or disaccommodation. Superposed on this change a much slower one may be found.

It can be calculated that in the (110) 90° wall in which the spins turn from the x into the y direction, a density increase occurs, which depends on the position in the wall, through the magnetization direction [23]:

$$e_{xx} + e_{yy} + e_{zz} = (B_1 \alpha_3^2 - 2B_2 \alpha_1 \alpha_2)/(c_{11} + c_{12} + 2c_{44}). \quad (11.5)$$

This gives rise to a decrease of the concentration of interstitials. They move out of the wall into the adjoining domains with relaxation times of the order of $3 \times 10^5 \tau$. An additional stabilization pressure will develop by this migration, though not with an exponential time dependence [30]. The stabilization pressure for carbon with $VC_0/RT = -0.0068$ is shown in Fig. 10, which can be compared with Fig. 8, where the stabilization for the same value of the parameter due to carbon migration within the unit cell is given.

It is obvious that the stabilization by volume diffusion out of the wall drops to zero for displacements of the order of the wall thickness. Before discussing some observations on isolated or well-defined domain walls, we want to stress that the time decrease of initial permeability is a simple

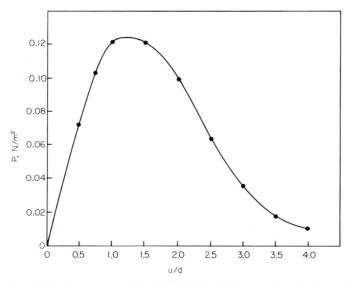

Fig. 10. Stabilization pressure of a {110} 90° wall due to the migration of carbon atoms out of the wall; $\epsilon V C_0 / RT = -0.0068$.

effect which measures the relaxation time τ of the underlying material diffusion Eq. (11.4) if one takes care that the measurement does not disturb the relaxation to be measured. This means, for example, that in a dc measurement the time during which the measuring field acts is short compared to all physical relaxation times, and that in an ac measurement this is true for half of the period.

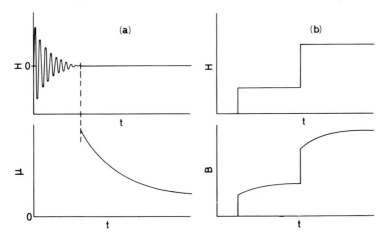

Fig. 11. (a) Time decrease of initial permeability after demagnetization, schematically. (b) Magnetic aftereffect or magnetic viscosity, schematically.

Magnetic viscosity or magnetic aftereffect, on the other hand, is a complicated effect even for a single domain wall. These effects are illustrated in Fig. 11. The connection between the two phenomena has been demonstrated by Snoek [31]. It can be summarized as follows. If a domain wall at $t = 0$ is brought to a fixed position and held there, the interstitials everywhere in the wall redistribute within a time given by τ, Eq. (11.4). If, however, walls are subjected to a constant field H, they are displaced over a distance u where the pressure exerted by the magnetic field is balanced by $P_m(u) + R(u)$. Immediately the distribution causing $P_m(u)$ starts to decay (while the distribution belonging to the new position starts to build up). The pressure opposing the displacement decreases and the wall moves on to a new position of equilibrium.

The effective relaxation time of magnetic viscosity will be larger than τ and increase with increasing ratio P_m/R. For $R = 0$ and $P_m(u) \neq 0$ at small fields, a wall will move with a constant velocity that is proportional to the exerted pressure, thus showing a viscous behavior that corresponds to an infinite relaxation time. Instead of developing the equations for one wall, e.g., a 90° wall, which is moved because a field H is applied at $t = 0$ parallel to the magnetization in one of the domains, we directly proceed to the magnetic viscosity in the presence of many arbitrary walls. We replace $P_m(u)/M_s$, by virtue of Eq. (11.2), by a maximum stabilization field H_{sm}, and this by $r_1 B$, introducing the reluctivity r_1. Analogously $R(u)/M_s$ is replaced by $r_0 B$, introducing the reluctivity r_0. The reluctivities are the inverses of permeabilities $r_i = \mu^{-1}$, a kind of magnetic resistivity. For a whole array of domain walls these resistivities are also assumed to be defined and additive, one due to the stabilization and one due to other restoring forces, respectively. For small fields in the region of initial permeability, r_0 and r_1 are found to be independent of the applied field. This is expected if the energy increase of the network of domain walls is proportional to the square of "its displacement." Corresponding to an exponential buildup of r, in the absence of a magnetic field, the equation for magnetic viscosity becomes

$$\Delta B/\Delta H = (r_0 + r_1)^{-1} + [r_0^{-1} - (r_0 + r_1)^{-1}]\{1 - \exp -[r_0/(r_0 + r_1)]^{t/\tau}\}. \tag{11.6}$$

We see that the effective relaxation time is

$$[(r_0 + r_1)/r_0]\tau > \tau. \tag{11.7}$$

The walls which are most mobile in the absence of stabilization ($r_1 \gg r_0$) give rise to the largest effective relaxation times of magnetic viscosity. Few experiments have been done on stabilization by interstitial atoms of well-defined wall structures. Bindels et al. [32] investigated single crystals of 3.25 wt % silicon–iron loaded with carbon. By making use of the arrangement and the crystals shown in Fig. 12 they evaluated the fields of (partial) stabilization Eq. (11.2) for 180° walls and 90° walls separately. Following Néel [20] and Brissonneau [33], who made similar measurements on

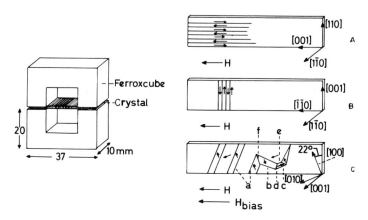

FIG. 12. Arrangement, dimensions, and domain structure of crystals used to measure stabilization of well-defined walls: (A) movement of 180° walls; (B) movement of 90° walls growing from closure domains near unmagnetic impurities; (C) with a biasing field of 4.7 Oe, the relative amount of wall length projected into the surface is: (a) 38%, (b) 51%, (c) 5%, (d) 1%, (e) 3%, (f) 2%, so 95% belonging to 90° walls.

polycrystalline material, they proceeded as follows. The magnetization curve is measured first for a material in which the interstitials are distributed as equally as possible over the three kinds of interstitial sites (unrelaxed curve). Then for the same material the magnetization curve is measured again after the interstitials have been given time to approach the equilibrium distribution corresponding to the magnetic structure (relaxed curve). The field strength at which a certain magnetization is obtained is larger for a relaxed curve than for the unrelaxed one. Let the difference be ΔH for a certain magnetization M, which corresponds to a certain wall displacement u if all walls in the relevant specimen behave alike. In that case, if the kind of walls and their number are known, the proportionality factor between u and B is known. From Eq. (11.3) it follows that the stabilization pressure is given by

$$\Delta \mathbf{H} \cdot (\mathbf{M}_1 - \mathbf{M}_2) = -P(u, t). \tag{11.8}$$

Thus, from a measurement of ΔH vs. M (or B), curves such as those

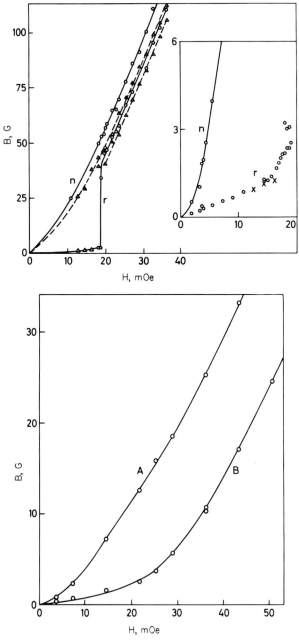

FIG. 13. (a) Magnetization curves (top left) and stabilization field ΔH (top right) for 180° walls of a {110} crystal (Fig. 12A). The points of the unrelaxed curve n were measured 10 sec after a demagnetization treatment. Every point of the relaxed curve r

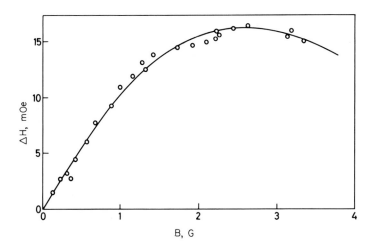

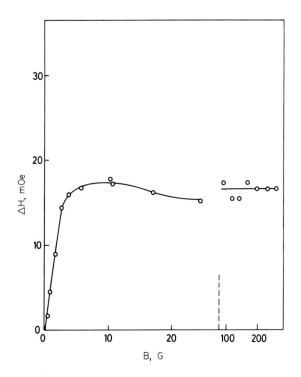

was measured by first demagnetizing (in a special way) and waiting for 10 min (250°K). (b) Magnetization curves (bottom left) and stabilization field ΔH (bottom right) for 90° walls of the crystal represented in Fig. 12B (250°K).

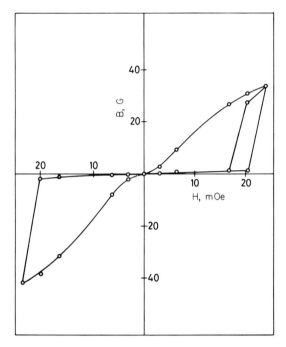

FIG. 14. (a) Constricted hysteresis loops due to 180° walls for the arrangement as in Fig. 12A; waiting time, 7.25 hr at 238°K.

given in Fig. 6 can be constructed. In the work referred to, the fundamental length d was determined in this way. The value of d was 165 Å ($+$10–20%) for 3.25 wt % SiFe. In Fig. 13a both a relaxed and an unrelaxed B–H curve are shown for 180° walls such as those represented in Fig. 12a, and also the $\varDelta H$ vs. B curve which was deduced from the values at low inductions. The intrinsic instability of the 180° walls is clearly seen and can be compared to that expected from Fig. 6. Likewise, Fig. 13b [32] shows the 90° wall behavior of a crystal represented in Fig. 12b. The $\varDelta H$ vs. B curve here is similar to that of the 90° wall curve in Fig. 6. Finally, constricted hysteresis loops for 180° and 90° walls are given in Fig. 14. In order to explain the small remanence of the 90° walls, one must consider the connections of the displaced walls with their original positions, e.g., by 90° wall spikes. The curve in Fig. 14a can be compared with the model of Fig. 9c.

Unspecified Arrays of Domain Walls

12. Stabilization. The development with time of stabilization fields and of constricted hysteresis loops connected with a special type of wall permits

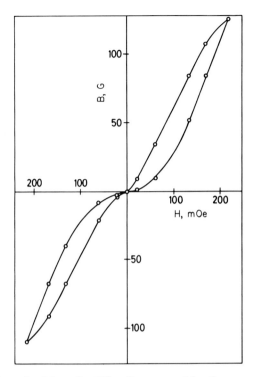

FIG. 14. (b) Constricted loop for 90° walls measured for the arrangement as in Fig. 12B; waiting time 6.25 hr at 238°K.

the measurement of relaxation times of diffusion. This section deals with arrays of unspecified domain walls. It will become clear that such arrays can serve the same purpose.

In Fig. 15 from the work of Brissonneau [33, 34] some magnetization curves of polycrystalline iron containing interstitial carbon are shown at different times after demagnetization. As shown in Fig. 13, in the course of stabilization larger fields become necessary to obtain the same values of the induction. The additional fields measured from the onset of stabilization at $t = 0$ are the fields of partial stabilization. Total stabilization is reached at $t = \infty$. In order to deal quantitatively with the stabilization process, the curves for $t = 0$ and $t = \infty$ should be reasonably well defined, which requires careful demagnetization.

Before discussing the time dependence of the field of partial stabilization, we represent a field of partial stabilization as a function of the magnetic induction in Fig. 16. As is seen, there is a maximum as a function of magnetization which is expected for isolated 180° walls (Figs. 13a and b). For higher field strengths the stabilization field drops to a rather constant value which is expected for 90° walls displaced large distances (Figs. 6 and 13b). On the basis of Néel's theory [21], Brissonneau [35] attributed the maximum of ΔH in experiments on polycrystalline material to the displacement of 180° walls out of their potential troughs.

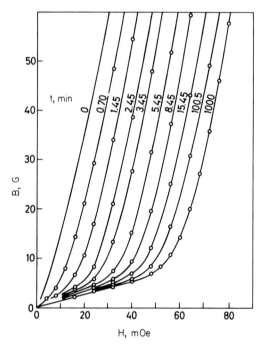

FIG. 15. Magnetization curves for iron containing interstitial carbon (4.6×10^{-3} wt % ±10%) at different stages of relaxation measured at 251.9°K. The time of relaxation, counted such that the onset of relaxation upon demagnetization in defined as $t = 0$, is indicated. The curve for $t = 0$ is obtained by extrapolation. (After Brissonneau [34].)

This view has been experimentally confirmed by him and by Bindels et al. [32], who compared measurements on well-defined walls with those on an arbitrary array of walls of different kinds. It should be kept in mind, however, that for pure iron with interstitials, isolated 90° walls of some kind can show a maximum stabilization field themselves.

It has been stressed by Brissonneau that it is dfficult to ascribe a clear physical meaning to the ΔH values measured on an unspecified array of walls. If the walls are not a coupled entity, this is certainly impossible for field strengths for which certain 180° walls have left the stabilization trough and are very mobile, while all of the 90° walls still experience relaxation.

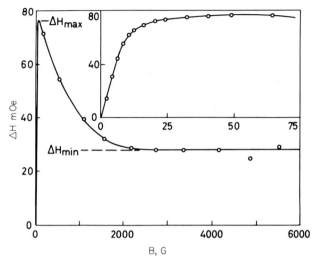

FIG. 16. Field of partial stabilization $\Delta H = H\,(M = 1000\text{ min}) - H\,(M = 0.7\text{ min})$ as determined from measurements such as those represented in Fig. 15: carbon content, 7.5×10^{-3} wt %; temperature, 239.7°K. (After Brissonneau [34].)

From Eq. (11.2) and Figs. 6 and 7 we may infer that the maximum stabilization field ΔH_{max} in polycrystalline material will be of the order of

$$\Delta H_{max} \sim W_0/M_s \cong VC_0D^2/M_sRT. \qquad (12.1)$$

It is unreliable to draw more quantitative conclusions from measurements of ΔH_{max} in polycrystalline materials, as has been done in the literature.

13. Establishment of the Stabilization Field: Time Decrease of the Initial Permeability. Even if the stabilization of every particular wall occurred with one well-defined relaxation time, this time would not generally be measured for the unspecified array of walls. This is partly due to the trivial fact that at fixed time t and external field H a spread exists of the ratio P/R in Eq. (11.3). This will be explained in the next section. In order to be able to generalize, we first return to an isolated wall. In Sec. 7 the stabilization of one wall with one relaxation time τ only was described by Eq. (11.4):

$$(P_\infty - P_t)/(P_\infty - P_0) = e^{-t/\tau} \quad [= 1 - g(t)]. \qquad (13.1)$$

Considering Eq. (11.2) this is equivalent to a stabilization field increasing from zero at $t = 0$ as

$$\Delta H(t)/\Delta H(\infty) = H_s(t)/H_s(\infty) = 1 - e^{-t/\tau} \quad [= g(t)]. \qquad (13.2)$$

If more than one relaxation time must be taken into account, $g(t)$ in Eqs. (12.1) and (13.1) has to be generalized by a more complicated function.

This function increases monotonically from 0 to 1 in the course of time. It represents all relaxation processes. An approximation which is rather crude, but is adequate for the description of many situations, is that the crystal can be subdivided into small parts [21], with one well-defined relaxation time being valid in each of them. The probability of finding this between τ and $\tau + d\tau$ is $p(\tau)$, and the normalization is such that

$$\int_0^\infty p(\tau)\, d\tau = 1. \qquad (13.3)$$

With this definition we arrive at the following representation of the relaxation function $g(t)$:

$$g(t) = \int_0^\infty p(t)(1 - e^{-t/\tau})\, d\tau. \qquad (13.4)$$

A further assumption regarding the distribution of time constants was made by Néel [21] and earlier by Richter [36]. It is treated in textbooks [37, 38] as the logarithmic law

$$p(\tau) = \tau^{-1}/(\log \tau_{max} - \log \tau_{min}), \quad \tau_{min} < \tau < \tau_{max}$$
$$p(\tau) = 0, \quad \tau_{max} < \tau \text{ and } \tau < \tau_{min}. \qquad (13.5)$$

If the spread in the relaxation times is due to a spread in activation energies for diffusion Q, where Q is defined by

$$\tau = \tau_0 e^{Q/RT}, \qquad (13.6)$$

where τ_0 is a constant, and if the probability of finding an activation energy were a finite constant in the interval $Q_{min} < Q < Q_{max}$ and zero outside this interval, one would expect Eq. (13.5) to hold strictly. The approximate validity of Eq. (13.4) does not indicate, of course, that a spread of Q is really at the basis of the spread of relaxation times.

Substitution of Eq. (13.5) into Eq. (13.4) leads to

$$g(t) = [\log(\tau_{max}/\tau_{min})]^{-1} \int_{\tau_{min}}^{\tau_{max}} (1 - e^{-t/\tau})\tau^{-1}\, d\tau, \qquad (13.7)$$

which reduces for $\tau_{max} = \tau_{min}$ to $1 - e^{-t/\tau}$.

In order to get an impression of the dispersion of time constant in a measurement, one plots $g(t)$ from Eq. (13.7), which is a well-known function [38], with $k^2 = \tau_{max}/\tau_{min}$ as a parameter. To introduce measuring points of the stabilization field according to Eq. (13.2), we observe that $g(t) = 0.575$ corresponds to $t = 0.85(\tau_{max}\tau_{min})^{1/2}$ almost independent of the value of k, as is seen in Fig. 17. Thus, measurements can easily be plotted as functions of the reduced time $t/(\tau_{max}\tau_{min})^{1/2}$.

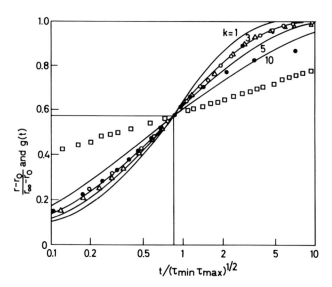

FIG. 17. $g(t)$ from Eq. (13.7), where $k^2 = \tau_{max}\tau_{min}$ interstitial (○) carbon in iron and (△) nitrogen in iron (measurements by Bosman [28]); (●) carbon in iron (measurements by Brissonneau [33, 39]); (□) carbon in 7 wt % SiFe, where the slope of this last curve corresponds to $k \simeq 450$ (measurements by Bosman [28]).

Equation (13.2) with the relaxation function $g(t)$ instead of the simple exponential can also be applied to an unspecified array of domain walls to evaluate the spread of the time constants and the value of the characteristic time constant $(\tau_{max}\tau_{min})^{1/2}$. This has been done for Brissonneau's measurements of Fig. 15 which are represented in Fig. 17. The course of the measured $g(t)$ curve is independent of the induction in the interval $7 < B < 56$ G. Also shown are Bosman's measurements on Fe; these lack the large spread at long relaxation times. These measurements have been deduced from the time decrease of initial permeability (Fig. 11a) of polycrystalline material, i.e., from measurements for very small values of B. We write for time t after demagnetization

$$\lim_{B \to 0} [\mu(t)]^{-1} = \lim_{B \to 0} [H(0) + \Delta H(t)]/B = r_0 + \Delta H(\infty)G(t)/B = r_0 + r_1 G(t). \tag{13.8}$$

In deriving Eq. (13.5) use has been made of Eq. (13.2) by generalizing it for the array of walls. The reluctivities r_0 and r_1 have been introduced above in Sec. 11. Summarizing Fig. 17, it can be stated that a large spread of relaxation times can be obtained by alloying.

In the experiments just discussed the development of the stabilization field with time has been treated. It is rather difficult to deal with the disappearance of the stabilized state in ac fields. Brissonneau made observations on the change with time of constricted hysteresis loop [33, 40] and attempted a quantitative interpretation for the case of the general array of walls [41].

14. *Time Dependence of Magnetic Viscosity: Loss Measurements.* In the stabilization experiments described in the last paragraph, care must be taken that during the accommodation process of interstitial atoms, wall motion in dc fields cannot disturb the diffusion, which is mainly of very short range, namely, within the unit cell. Likewise, the measurement must not influence the diffusion. If dc fields are used, they must act in a time short compared with all relaxation times. Under these conditions such stabilization experiments are relatively easy to interpret. As has been discussed in Sec. 11, magnetic viscosity is a more complicated process. The effective time constants are increased, as shown by Eq. (11.7) because the wall moves during the measurement.

Snoek [31] showed that in a dc measurement of permeability, there is an interrelation of time decrease of permeability and magnetic viscosity. This is qualitatively illustrated in Fig. 18, which shows a measurement of the permeability at different temperatures.

The specimen is demagnetized at $t = 0$ and kept field free for 3 min. After this a dc field of $H = 0.01$ Oe is applied and $\mu = B/H$ is measured with a ballistic galvanometer having an oscillation time of 27 sec. Let us assume that one well-defined relaxation time exists for the interstitial jumps. At very low temperatures this relaxation time is so long that no time decrease at all occurs in 3 min. At rising temperatures relaxation does occur within 3 min. At the temperature indicated by T_w the relaxation time τ coincides approximately with the waiting time of 3 min. At still higher temperatures, on the rising branch, $\tau \ll 3$ min so that stabilization is complete at the beginning of the measuring operation. Now, however, the oscillation time of the galvanometer is no longer small compared with the relaxation time.

When the field is switched on, the walls begin to move and continue to do so through magnetic viscosity. The displacement during about one-fourth of the oscillation time of the galvanometer is recorded. At high enough

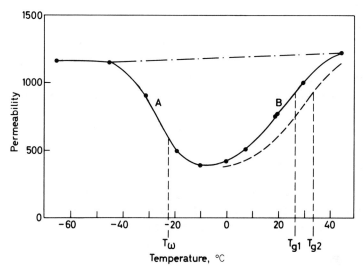

FIG. 18. Permeability, at $H = 0.01$ Oe, of iron with interstitials, measurements being taken each time 3 min after demagnetization. Measurements taken with a ballistic galvanometer with a free period of 27 sec are given by x, with a ballistic galvanometer of free period 10 sec by the dashed line. The point dashed line schematically gives the permeability in the absence of interstitials. (After Snoek [16].)

temperatures magnetic viscosity will allow the walls to move just as far as the restoring forces other than those from stabilization allow them to go. At T_g the relaxation time τ, or according to Eq. (11.6) $\tau(r_0 + r_1)/r_0$, will approximately coincide with one-fourth of the oscillation time of the galvanometer.

If the material is subjected to small ac magnetic fields, there will be a phase shift $\tan \delta$ between B and H, which in accordance with Eq. (11.6) has a maximum for that angular frequency ω of the field, for which

$$\omega\tau(1 + r_1/r_2)^{1/2} = 1. \tag{14.1}$$

If $r_1/r_0 \ll 1$, the maximum of $\tan \delta$ occurs at that temperature for which $\omega = \tau^{-1}$. In that case

$$\tan \delta = 2\omega\tau/[(\omega\tau)^2 + 1](\tan \delta)_{\max}. \tag{14.2}$$

Equation (14.2) assumes that there is only one relaxation time and moreover that the diffusion lag is the predominant reason for the phase shift. If we look at Fig. 19, which has been taken from Bosman [29], we see that these assumptions are reasonably true for the Fe–C alloy but not for the Fe–Si–C alloys. For the 3.25 wt % and the 4.5 wt % Si alloys the original Fe–C peak is still visible but not for the 7 wt % Si alloy. In the neighborhood of 145°C there are other peaks for the Si alloys. These are not very well defined because of large hysteresis losses which occur concurrently with large permeabilities at elevated temperatures.

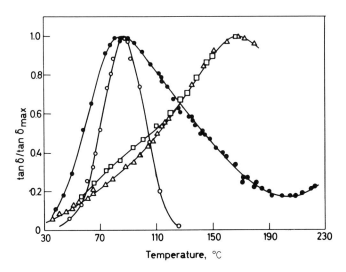

FIG. 19. Relaxation losses for iron and iron–silicon alloys, containing carbon, as a function of temperature (30 Hz): (○) pure iron; (●) 3.25 wt % Si; (□) 4.5 wt % Si; (△) 7.0 wt % Si.

The relaxation phenomena in alternating magnetic fields, especially at larger amplitudes, are very complicated and difficult to interpret, and we will not venture to go into details. Relevant calculations have been made by Néel [21], and experiments have been reported by Brissonneau [33, 40] and Schreiber [42].

We will also refrain from discussing the attempts to separate ac losses of different origin in magnetic systems; these have been developed with great ingenuity. In particular, Feldtkeller [43] and his school [38] must be mentioned in this connection.

Examples of Diffusion Measurement

15. Activation Energy without Measurement of Relaxation Times. Figure 20 shows the time decrease of permeability as measured by Bosman *et al.* [44] for iron to which nitrogen had been deliberately added. Curves such as this can be measured at different temperatures and analyzed into exponentials with different time constants, as discussed in Sec. 16.

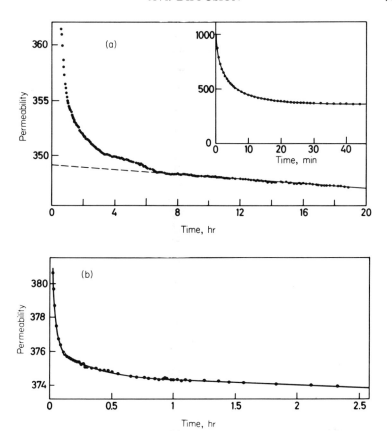

FIG. 20. Time decrease of permeability for iron to which nitrogen had been added (0.006 wt %): (a) $T = -39.11°C$; (b) $T = -9.46°C$.

In order to see whether there is one definite activation energy and to derive its value, one can proceed simply as follows. We assume that the time scale of the phenomenon responsible for the observed decrease of the permeability varies with temperature as $e^{Q/RT}$. For Eq. (13.8) we may then write

$$r = r_0 + r_1 g(t e^{-Q/RT}) \quad (15.1)$$

if r_0 and r_1 do not depend on temperature. In this case the relation

$$\mu = \mu_0 - \mu_1 g^*(t e^{-Q/RT}) \quad (15.2)$$

holds just as well.

For fixed values of g or of g^* corresponding to a constant value of $t e^{-Q/RT}$, $\ln t$ is plotted as a function of T^{-1} (Fig. 21). The slope determines Q/R.

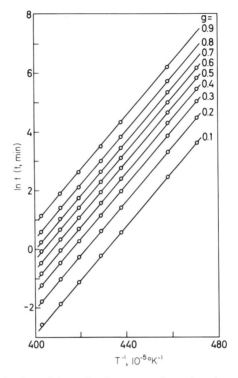

FIG. 21. Determination of the activation energy from time decrease of permeability, analyzed according to Eq. (15.1).

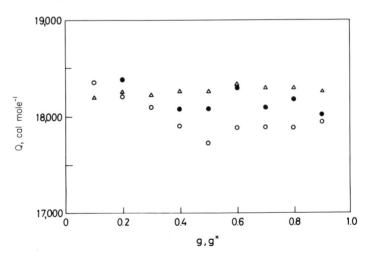

FIG. 22. Activation energies determined for different values of g or g^*: ($\triangle$) g (Fig. 21), 0.006 wt % N in Fe; ($\bigcirc$) g^*, using the same measured data; ($\bullet$) g^*, 0.002 wt % N in Fe.

Actually r_0 and r_1 depend slightly on temperature (respectively 0.1 and 0.3% per °C). This has as a consequence that equal values of g or of g^* do not necessarily correspond to the same relative progress of the atomic process. The uncertainty however is not large, as shown by Fig. 22, in which the Q values are plotted as a function of g (Fig. 21) and g^*. The difference is small, and the value 18,000 cal mole^{-1} (± 200) thus determined is in agreement with volume diffusion and internal friction measurements [45, 46].

16. Relaxation Times. It has been shown in Bosman et al. [44] that the time decrease can be decomposed into contributions with different relaxation times τ_k by writing

$$r = r_0 + \sum_{k=1}^{n} r_{1k}(1 - e^{-t/\tau_k}). \tag{16.1}$$

This decomposition is appropriate only if the τ_k are sufficiently different and do not fill a continuous spectrum. This was approximately the case for rather pure iron deliberately loaded with nitrogen. The procedure of decomposition is as follows. If the τ_k are sufficiently different, as assumed, the behavior of r at large time t is determined by the largest τ_k, say τ_1. A plot of $\log\{(r_\infty - r_0)/(r_\infty - r)\}$ vs. time will therefore show for large t a straight line relationship that gives r_{11} and τ. For large times, namely,

$$\log\{(r_\infty - r_0)/(r_\infty - r)\} \sim \log \sum_{k=1}^{n} (r_{1k}/r_{11}) + (t/\tau_1). \tag{16.2}$$

The slope of the straight line relationship of Eq. (16.2) with time gives τ_1, while r_{11} is given by the ordinate at $t = 0$ since $r_\infty - r_0 = \sum_{k=1}^{n} r_{1k}$ is known by straightforward measurement. If $r_{11}(1 - e^{-t/\tau_1})$ is then subtracted from r, one can proceed in a similar manner to find τ_2, r_{12}, etc.

In the actual case of pure iron loaded with nitrogen, a very weak relaxation τ_0, r_{10} corresponding to very long times is at first subtracted from the other relaxation phenomena. This is done by using for the sum of those phenomena a value of r_∞ obtained by extrapolating the straight line in Fig. 20 to $t = 0$. One then finds the $\log\{(r_\infty - r_0)/(r_\infty - r)\}$ values represented in Fig. 23 as a function of time. The slope of the asymptotic part yields $\tau_1 = 102$ min and extrapolation to $t = 0$ gives $r_{11}/\sum_{k=1}^{n} r_{1k} = 0.040$. Correcting the curve in Fig. 23 for τ_1, r_{11} yields the curve of Fig. 24, the straight part of which gives $\tau_2 = 11.0$ min and $r_{12}/\sum_{1}^{n} r_{1k} = 0.704$. Continuing in the same manner one gets the values given in Table I.

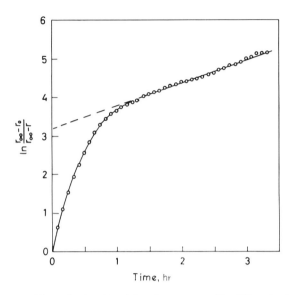

FIG. 23. Decomposition of relaxation into phenomena with different relaxation times.

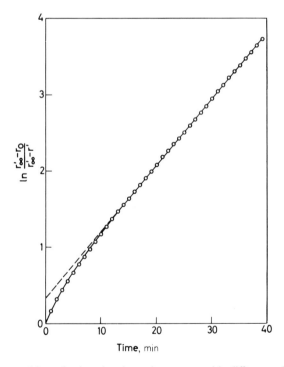

FIG. 24. Decomposition of relaxation into phenomena with different relaxation times.

TABLE I
DECOMPOSITION OF THE TIME DECREASE OF PERMEABILITY INTO DIFFERENT RELAXATION PROCESSES

τ_k (min)	Relaxation strength	Responsible center
$\tau_0 = ?$	$r_{10}(1 - \exp[-t/\tau_0]) / \Sigma_{k=1}^{4} r_{1k} \simeq 0.01$	?
$\tau_1 = 102$	$r_{11} / \Sigma_{k=1}^{4} r_{1k} = 0.040$	C
$\tau_2 = 11$	$r_{12} / \Sigma_{k=1}^{4} r_{1k} = 0.704$	N
$\tau_3 = 3.5$	$r_{13} / \Sigma_{k=1}^{4} r_{1k} = 0.218$	N
$\tau_4 = 0.5$–1.0	$r_{14} / \Sigma_{k=1}^{4} r_{1k} = 0.038$	N

Note that there is a carbon relaxation where the time τ_1 corresponds to that known from internal friction; the impurity probably was introduced during heat treatment. The quantity of carbon can be estimated from the relaxation to be less than 10^{-4} wt %. The relaxation time τ_2 apparently corresponds to the elementary jump of nitrogen as an interstitial. This time is in the range of those measured from the same alloys in internal friction. τ_3 and τ_4 are considered to be sattelites of τ_2, having little significance of their own. If there is a spread in the mobilities of different walls in the absence of stabilization, stabilization will quickly reduce the relative contribution of the most mobile ones [44]. This results in short apparent relaxation times which are represented in τ_3 and τ_4. From the analysis in Table I and calculations it appears [29] that the spread is of the order of 100 in the wall mobility of the nonrelaxed material used. It is not quite clear which physical interpretation should be attached to the weak relaxation τ_0 Later measurements, which were made for carbon in the same iron sample [47], showed that in this case the relaxation at long times is about proportional to the square of the carbon concentration. The explanation proposed is that the corresponding relaxation is due to reorienting pairs of carbon atoms. Pairs of interstitials at much larger concentration (O in Ta) are well known from the work of Powers and Doyle [48], while the pair interaction of C in α-iron was theoretically treated by Fisher [49]. The measured effect [47] approximately agrees with this calculation. Perhaps the long time effect of interstitial solutions of N in Fe must also be regarded as being due to nitrogen pairs. This question deserves further attention.

The dependence on temperature of τ_2 gives the activation energy of 18,200 cal mole^{-1} $\pm$ 500, in good agreement with literature and the values reported in Sec. 15.

17. Measurement of a Small Relative Change of the Relaxation Time.
We will presently treat some experiments in which, through a change in the standard conditions, a small change in relaxation time results. For instance, instead of atmospheric pressure, a high hydrostatic pressure is applied. The change in relaxation time τ_2 must be determined since this is the basic relaxation time. The analysis of the preceding paragraph permitted only an accuracy in τ_2 of about 4%. It will be shown, however, that relative changes in τ_2 can be determined with higher accuracy if the circumstances are approximately those indicated in Table I. We therefore restrict our attention to the interval of 0.8–$2.2\tau_2$. The contribution to relaxation process 4 is completely negligible in this time interval. At the beginning of this interval, relaxation process 3 still contributes 1.75% of the total relaxation, and at the end practically nothing. Relaxation process 1 decreases in almost linear fashion from 3.7 to 3.15%. The relaxation of interest decreases from 31.6 to 7.8%. As the other relaxations are only a small fraction of the total effect in the interval considered, their change in value due to the hydrostatic pressure can be neglected in a first approximation. The procedure now is straightforward. At a reasonably long time, e.g., $7\tau_2$, relaxation process 2 may be considered to be complete. The measured relaxation at this time is taken as the end value. Then, for both curves, with and without pressure, the effect of the long relaxation is first subtracted if this effect is important. Except for a small correction due to τ_3, the resulting curves can be represented by an exponential: $c \exp - t/\tau_2$ and $c^* \exp - t/\tau_2^*$, where c, τ_2 and c^*, τ_2^* refer to the situations with and without external pressure. Now we plot the ratio $(c/c^*) \exp - t[\tau_2^{-1} - (\tau_2^*)^{-1}] \simeq (c/c^*)[1 - (\Delta\tau/\tau_2^2)t]$ as a function of time. It should be noted that in this way the correction due to τ_3 drops out in first approximation; $\Delta\tau = \tau_2^* - \tau_2$ is immediately determined from the slope. A similar formalism has been applied to the change with hydrostatic pressure of the mean time of stay of nitrogen [50] and carbon [51] as interstitials in iron (see also Bosman [29]).

Figure 25 shows the ratio $(c/c^*)(1 - \Delta\tau \cdot t/\tau_2)$ as a function of time when applying a pressure of 3000 kg cm^{-2} at constant temperature to a solution of carbon in iron. A similar procedure in which the difference in experimental circumstances was only a small difference in temperature led to the correct change in relaxation time. The change of the mean time of stay with pressure is exceedingly small in these cases. This makes the so-called activation volume which has been defined by Nachtrieb et al. [52] as

$$\Delta V = (\delta \Delta G/\partial p)_T, \qquad (17.1)$$

where ΔG, the activation Gibbs free energy of the diffusion process, is

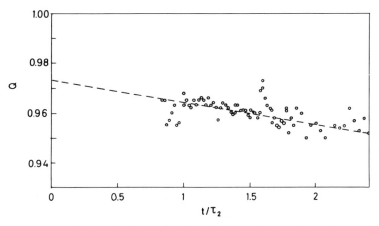

FIG. 25. Determination of a change in relaxation time of carbon due to the application of high pressure, where $T = -34.34°C$, and $\Delta p = 3000$ atm. The straight line corresponds to $\Delta \tau/\tau = 0.009$.

very small. If a is the lattice constant, approximately

$$\Delta V = RT(\tau^{-1}\,\partial\tau/\partial p - a^{-1}\,\partial a/\partial p)/[(1 - RT)/2\,\Delta G]. \qquad (17.2)$$

For nitrogen in iron $RT\tau^{-1}\,\partial\tau/\partial p = (3.3 \pm 3.3)\,10^{-2}$ cm^3 mole^{-1} at 234°K and $\Delta V = (3.7 \pm 3.3) \times 10^{-2}$ cm^3 mole^{-1}. For carbon in iron $RT\tau^{-1}\,\partial\tau/\partial p \simeq -5 \times 10^{-2}$ cm^3 mole^{-1} at 250°K and $\Delta V = (-4.1 \pm 3.5) \times 10^{-2}$ cm^3 mole^{-1}. A very small activation volume for the diffusion of interstitial carbon in iron is confirmed by internal friction measurements [53].

The same procedure has also been applied to study the dependence of the mean time of stay of interstitials on mass. Within the limit of error the simple $m^{1/2}$ relation has been confirmed by replacing 99% ^{12}C as interstitial in iron by 95% ^{13}C [22, 29]. One has $\tau(^{13}C)/\tau(^{12}C) = 1.044 \pm 0.005$ vs. 1.041 for the $m^{1/2}$ relationship.

Directional Order in Other Systems, Especially Substitutional Alloys

Theoretical Considerations

18. *Generalization with Respect to the Symmetry of the Ordering Configuration.* In the foregoing, the effects of interstitial diffusion in iron on the magnetic and elastic properties were discussed most extensively as in this case the mechanism is well understood. As stated before, the same phenomena can in principle be caused by any configuration of defects, or dissolved atoms, that locally introduces a lower symmetry than that of the lattice as a whole. We may think in this connection of divacancies, interstitialcies, pairs of dissolved atoms, combination of a point defect with a dissolved atom, etc.

In discussing these phenomena we will restrict our attention to cubic lattices. The effect of asymmetric centres obviously is most marked in structures of high symmetry. The experimental data relevant to this subject deal almost exclusively with cubic metals and alloys.

The considerations of the first part must be generalized with respect to the possible symmetries of the anisotropy centres: In the case of interstially dissolved atoms in iron, we had centres of tetragonal symmetry which could be labeled by indicating a cube edge direction. We then distinguish (100), (010), and (001) sites. Pairs of nearest neighbors in a bcc lattice show trigonal symmetry and can be labeled by the corresponding body diagonal. This yields (111), ($1\bar{1}\bar{1}$), ($\bar{1}1\bar{1}$), and ($\bar{1}\bar{1}1$) type sites. Pairs of nearest neighbors in fcc lattices show orthorhombic symmetry. We get six types, indicated by (011), (01$\bar{1}$), ($\bar{1}$01), (101), (110), and (1$\bar{1}$0). It is not difficult to imagine centers of still lower symmetry. The types mentioned, however, are most commonly considered for the analysis of experimental results.

Again as in the preceding part, a magnetic anisotropy energy caused by the anisotropy center is introduced, considering only the lowest-order terms, quadratic in the direction cosines. These are

For (100): $w_{100}(\alpha_1^2 - \tfrac{1}{3})$

For (111): $w_{111}^{2/3}(\alpha_2\alpha_3 + \alpha_3\alpha_1 + \alpha_1\alpha_2)$ (18.1)

For (110): $w_{110}^{(1)}(\alpha_1\alpha_2 - \tfrac{1}{2}\alpha_3^2 + \tfrac{1}{6}) + w_{110}^{(2)}(\alpha_1\alpha_2 + \tfrac{3}{2}\alpha_3^2 - \tfrac{1}{2})$.

In all these cases we have a term of the form $w(\cos^2\phi - \tfrac{1}{3})$, where ϕ is the angle between the indicated direction and the magnetization vector. For the (110) case, a term in $w_{110}^{(2)}$ is added to describe the deviation from cylindrical symmetry around the $\langle 110 \rangle$ axis [8].

The lattice deformation caused by the anisotropy centers can be described with expressions of the same form as those used for the magnetic anisotropy energy.

An expression for the free energy, similar to Eq. (7.1), introduced for interstitial atoms or interstices in iron, would suffice for the prediction of all measurable quantities (supposing questions concerning domain configuration, etc. to have been settled). Generally, however, these expressions are much more complicated than Eq. (7.1). A general discussion of the possible symmetries of defects in different lattices has been given by Nowick [54]. Most experiments deal with concentrated alloys.

Therefore, simplifications which are valid for low concentrations are not allowed; ordinary order–disorder theory must be incorporated. The temperature depence of the coefficients, neglected until now, may play an important role in these experiments. This effect is partially due to a decrease in magnetization and may be derived with the use of statistical arguments, but an intrinsic temperature dependence cannot be excluded *a priori*.

The large number of unknown parameters makes it much more difficult to obtain quantitative information on diffusion by magnetic measurements of the type that proved so useful in the case of interstitial atoms in iron. In the following we will try to evaluate the experimental and theoretical situation with this question in mind.

19. *Pair Ordering.* Of all anisotropy centers that can be imagined, apart from the interstitial atoms, the pair of neighboring like atoms in an alloy is most extensively discussed in the literature. Atomic rearrangement is at the base of the magnetic annealing effects found in binary alloys that are heat treated at about 400°C. The mechanism of pair ordering in its present form was introduced into the theory by Néel [55, 56] and by Taniguchi and Yamamoto [57, 58]. Originally, however, it was proposed by Chikazumi [59], who sought to explain the magnetic annealing effects that he observed in NiFe alloys [60]. In his picture the number of like atom pairs in the direction of magnetization would differ from the number in the perpendicular direction. Due to a difference in bond length, which can be related to the change in volume upon ordering, this anisotropic distribution causes a strain in the preferred direction. Due to the magnetostriction this strain is equivalent to an induced magnetic anisotropy. He found that in order to explain the observed magnitude of the induced anisotropies, the extra strain should be about 40 times the normal magnetostriction. In later theories the idea of pair ordering is retained, but magnetostriction plays only a secondary role. The main effect is thought to be a direct interaction with the magnetization direction through an energy of the form of Eq. (18.1), as later also was proposed by Chikazumi and Oomura [61]. The energy of a pair of nearest neighbors is in these theories written as

$$K = w(\cos^2 \phi - \tfrac{1}{3}) \qquad (19.1)$$

in which ϕ is the angle between the direction of magnetization and the line joining the two atoms considered. In the theory of Néel the term considered here is only one of a set of terms in which the interaction between neighboring pairs of atoms can be developed as a function of the distance between them and the direction ϕ. He showed how the different terms were related to the values for crystalline anisotropy, the magnetostriction, ordering parameters, etc. The term, mentioned explicitly, turns out to be responsible for the magnetic annealing effects.

This theory yields very good results qualitatively and leads to the right order of magnitude for the energies involved. However, it is not sufficient for a satisfactory quantitative description. In an effort to give a complete account of the behavior of alloys, based on interaction between near atoms, Ferguson [62] set up a formalism for a binary alloy that should have included some 90 adjustable constants, extending the development only to second-nearest neighbors. This number of constants seems much too large to give a useful reference frame for the analysis of the experimental data, even if the physical basis were more certain.

In the work of Taniguchi and Yamamoto only the term in $(\cos^2 \phi - \frac{1}{3})$ was introduced. Its magnitude was estimated using the theoretical arguments that Van Vleck used in his treatment of the ordinary ferromagnetic anisotropy energy [63].

As already mentioned, the effects of ordinary order must be included from the start. This was recognized by Néel. Iwata [64–66] later gave a somewhat more refined theory using the quasi-chemical method for the ordering problems.

For the magnetic interaction he also took an expression of the form of Eq. (19.1), taking into account nearest neighbor interaction only. As will be seen below, the assumption of only nearest neighbor interactions cannot lead to quantitative agreement with experiment, at least not in the bcc structures. Still the simple theories seem to yield good physical insight. The more involved theories have not led to a better understanding of the observed phenomena. We will not try to give a detailed account of the theories, but will restrict our attention to some general features common to all theories and, as far as possible, indicate the specific results obtained and the inherent limitations of the approaches used.

20. *The Anisotropy of the Magnetic Annealing Effect.* The type of anisotropy center determines the form of the magnetic anisotropy energy that can be introduced by magnetic annealing. In the case of interstitial atoms in iron this form was given by Eq. (8.1) as

$$K_u = \text{const}[(\alpha_1^2 - \beta_1^2)^2 + (\alpha_2^2 - \beta_2^2)^2 + (\alpha_3^2 - \beta_3^2)^2]$$

or, if we consider only the part depending on both α and β

$$K_u = \text{const}(\alpha_1^2\beta_1^2 + \alpha_2^2\beta_2^2 + \alpha_3^2\beta_3^2). \tag{20.1}$$

It is immediately seem from Eq. (18.1) that for any type (100) center the anisotropy must have the form

$$K_u = C_1\alpha_1^2 + C_2\alpha_2^2 + C_3\alpha_3^2, \tag{20.2}$$

where C_1, C_2, and C_3 are the numbers of the (100), (010), and (001) centers multiplied by w_{100}. If the concentration is sufficiently low, the numbers are determined by the Boltzmann factors during the treatment with field direction β:

$$C_i = \tfrac{1}{3}nw_{100}\{1 - (w_{100}/kT)(\beta_i^2 - \tfrac{1}{3})\} \tag{20.3}$$

in which n is the total number of centers. Therefore

$$K_u^{100} = -\tfrac{1}{3}(nw_{100}^2/3kT)(\alpha_1^2\beta_1^2 + \alpha_2^2\beta_2^2 + \alpha_3^2\beta_3^2) \tag{20.4}$$

In a completely similar manner it follows that for (111) centers in general

$$K_u^{111} = C_1\alpha_2\alpha_3 + C_2\alpha_3\alpha_1 + C_3\alpha_1\alpha_2 \tag{20.5}$$

and, when a Boltzmann distribution is established for field direction β

$$K_u^{111} = -(4nw_{111}^2/9kT)(\alpha_1\alpha_2\beta_1\beta_2 + \alpha_2\alpha_3\beta_2\beta_3 + \alpha_3\alpha_1\beta_3\beta_1). \tag{20.6}$$

By straightforward symmetry considerations [8] it can be shown that an expansion of the induced anisotropy in ascending powers of α and β has, for cubic crystals, as lowest-order terms, expressions of the form of Eqs. (20.4) and (20.6):

$$K_u = k_1(\alpha_1^2\beta_1^2 + \alpha_2^2\beta_2^2 + \alpha_3^2\beta_3^2) + k_2(\alpha_1\alpha_2\beta_1\beta_2 + \alpha_2\alpha_3\beta_2\beta_3 + \alpha_3\alpha_1\beta_3\beta_1)\cdots. \tag{20.7}$$

If a body-centered cubic crystal pairs of neighboring atoms are thought to be the important symmetry centers, k_1 should be zero, as follows from a comparison of Eqs. (20.6) and (20.7). Only the interaction beyond nearest neighbors could lead to nonvanishing k_1.

In face-centered cubic crystals matters are somewhat more complicated. In the usual pair theories only the term in $w_{110}^{(1)}$ is retained. In that case the same reasoning that led to Eq. (20.4) and Eq. (20.6) leads to

$$K_u^{110} = -(nw^2/kT)[(1/12)(\alpha_1^2\beta_1^2 + \alpha_2^2\beta_2^2 + \alpha_3^2\beta_3^2) + \tfrac{1}{3}(\alpha_1\alpha_2\beta_1\beta_2 + \alpha_2\alpha_3\beta_2\beta_3 + \alpha_3\alpha_1\beta_3\beta_1)]. \tag{20.8}$$

Therefore, for this case $k_2/k_1 = 4$. For concentrated alloys this ratio is still 4 as long as, in the absence of directional order the numbers of pairs in the different directions are given by the *a priori* probabilities [65, 66].

In the presence of an energy term depending on order this ratio may be altered. For instance, if like pairs are energetically favored, sets of three mutually neighboring like atoms will occur more often than in a random distribution. Such a set of atoms however shows trigonal symmetry [67]. From Eq. (18.1) it follows that for a triplet in a plane normal to a (111) direction the magnetic anisotropy energy would be $-w_{110}^{(1)}(\alpha_1\alpha_2 + \alpha_2\alpha_3 + \alpha_3\alpha_3)$, still neglecting $w_{110}^{(2)}$. These sets will give a contribution to the induced anisotropy energy, which adds only to the second term of Eq. (20.7). In that case we would expect $k_2/k_1 < 4$. If the ordering energy has the opposite sign, $k_2/k_1 < 4$. For fcc lattices Iwata has computed the k_2/k_1 values as a function of composition, temperature, and the ordering energy [66] in the absence of long-range order, and for the composition A_3B, also taking long-range order into account (see Fig. 26 [65]). Next nearest neigh-

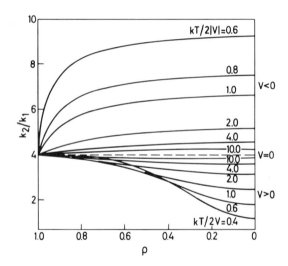

FIG. 26. Ratio of k_2/k_1 [Eq. (20.7)] as a function of alloy composition and ordering energy V, in fcc cubic alloys, as computed by Iwata [66]; $\varrho = 0$ corresponds to equal amounts of the two constituents.

bor interaction in these alloys would lead to an increase in k_1 only, or to a decrease in the k_2/k_1 ratio. As already stated, however, on general grounds both $w_{110}^{(1)}$ and $w_{110}^{(2)}$ may differ from zero in which case Iwata's predictions need not hold even for nearest neighbor interaction only. The expression similar to Eqs. (20.4) and (20.6) in that case is

$$K_u = -(n/kT)(1/12)(w_{110}^{(1)} - 3w_{110}^{(2)})^2(\alpha_1^2\beta_1^2 + \alpha_2^2\beta_2^2 + \alpha_3^2\beta_3^2)$$
$$- \tfrac{1}{3}(w_{110}^{(1)} + w_{110}^{(2)})^2(\alpha_1\alpha_2\beta_1\beta_2 + \alpha_2\alpha_3\beta_2\beta_3 + \alpha_3\alpha_1\beta_3\beta_1). \qquad (20.9)$$

From this expression any positive value for k_2/k_1 may be obtained.

21. Temperature Dependence of the Ordering Energy. The temperatures at which the magnetic annealing experiments can be performed are usually above about 400°C, a temperature region where the saturation becomes notably temperature dependent. Therefore, the constant w giving the magnitude of the magnetic anisotropy energy caused by an anisotropy center may also be expected to be temperature dependent. The experiments in which the induced magnetic anisotropy is investigated usually are conducted in such a way that the external field is kept in the direction specified by the direction cosines β_i for a sufficiently long time, while the specimen is at the temperature T'; the actual measurement of the induced anisotropy is carried out at room temperature T. In that case the equations of the preceding section should be modified in such a way that everywhere w^2/T is replaced by $w(T)w(T')/T'$.

Quenching from different temperatures T' determines $w(T')$ [naturally assuming that the conditions for the validity of Eqs. (20.4), (20.6), and (20.9) are satisfied]. Similar to Zener's [68] theory for the crystalline anisotropy, a theoretical temperature dependence may be derived in the following way (Ferguson [62]): a picture of the macroscopically observed magnetization is obtained by assuming that locally the magnitude of the magnetization is independent of temperature; the direction, however, fluctuates in time, with an amplitude that increases with increasing temperature. If the angle by which the local magnetization deviates from the macroscopically observed magnetization $M(T)$ is indicated by δ, the magnitude of M is given by $M(T) = M(0)\langle\cos\delta\rangle_{av}$. If now it is supposed that relations of the type of Eq. (19.1) hold on a microscopic scale, the expression $(\cos^2\phi - \tfrac{1}{3})$ must be averaged over all δ. It turns out that the only effect is a multiplication of the constant w_0 that obtains for perfect alignment of the local magnetization vectors by the average of $\tfrac{3}{2}(\cos^2\phi - \tfrac{1}{3})$. From this expression it is seen that for small deviations δ the constant w decreases proportional to the third power of the relative magnetization. A simple way to show this is as follows. Suppose that in a small fraction of the volume the magnetization is uniformly rotated by a small angle from the direction of the macroscopic magnetization. Due to this rotation the contribution to the magnetization of this region is decreased by a relative amount $(1 - \cos \delta)$. The effective anisotropy of some center in this region in decreased by a relative amount $1 - \frac{3}{2}(\cos^2 \delta - \frac{1}{3})$. For small δ these expressions can be approximated by $\frac{1}{2}\delta^2$ and by $\frac{3}{2}\delta^2$ respectively, or $\Delta M/M = \frac{1}{3}\Delta W/W_0$.

At high temperatures, where the magnetization drops considerably below the value at zero temperature, δ no longer is small and the distribution function for δ must be introduced. With the random walk function that Zener [68] originally introduced to explain the temperature dependence of the crystalline anisotropy, the M^3 dependence is found for all temperatures. A Boltzmann distribution found for independent spins in the molecular field) at high temperatures would lead to a M^2 dependence. Also Taniguchi and Yamamoto [57, 58] proposed a M^2 dependence for pair interaction. In their model the interaction energy arises from a pseudodipolar interaction between neighboring spins, different for different types of neighbors. Taniguchi averages this energy for independent deviations of the neighboring spins. This certainly is not applicable at low temperatures but at high temperatures it leads to the same dependence on M as a Boltzmann distribution with perfect correlation between the spins on the atom pair.

No clear-cut statement can be made about the temperature dependence of the anisotropy energy on these statistical grounds as there is no theory available that predicts the distribution of δ's. At low temperatures, where the theory is on firmer ground, the predictions become somewhat doubtful if we consider that in many cases the results of analogous theories on the temperature dependence of crystalline anisotropy and magnetostriction proved to be wrong. Instead of a 10th power in $M/M(0)$ for the constant K_1 in cubic crystals, in nickel something like a 50th and in iron a 5th power is found,* while in iron and iron alloys the magnetostriction constants also do not fit in with these views [70].

22. The Interference between Directional Order and Ordering.

It has been mentioned previously that ordinary order can have a large influence on the anisotropy of directional order. It also affects the magnitude of these effects. To make this clear it is necessary to discuss pair ordering in somewhat greater detail. As stated before, the energy of a pair of atoms is taken to depend on the magnetization direction as $l_{AA}(\cos^2 \phi - \frac{1}{3})$. Here the constant l_{AA} is introduced to indicate that the pair considered consists of two similar atoms of type A. In the alloy AB we must consider similar expressions in which l_{AA} is replaced by l_{BB} or l_{AB}. The resulting anisotropy is found by

---

* Values are tabulated in the "American Institute of Physics Handbook" and Bozorth, "Ferromagnetism" [69].

summing over all atom pairs. In cubic crystals the result is zero unless the numbers of pairs in the different crystallographic directions are different.

As in the ordinary order theories, if in some specified direction the number of AA bonds is increased by one, when interchanging two atoms, the number of BB bonds also increases by one, and obviously the number AB bonds decreases by two. Thus, the contribution of these pairs to the anisotropy energy increases by $(l_{AA} + l_{BB} - 2l_{AB})(\cos^2 \phi_k - \tfrac{1}{3}) \equiv w(\cos^2 \phi_k - \tfrac{1}{3})$ for every AA pair added to the number of AA pairs in the direction specified by k. It is therefore only necessary to count the pairs of one kind and to attribute to this kind the energy $w(\cos^2 \phi_k - \tfrac{1}{3})$, the form introduced before. For example, consider a body-centered cubic alloy 50–50% A–B. In the disordered state one-quarter of all bonds is of the type AA and relatively large differences between the numbers in different directions may be found in thermal equilibrium, when a magnetic field is present. With order, however, the number of AA pairs may go to zero and therefore also the magnitude of the magnetic ordering effects. Iwata [64, 65] has discussed these effects in detail for complete thermal equilibrium. In actual experiments short-range order is established much faster than long-range order. A metastable situation is then reached. For some of these cases Iwata's calculations also apply.

In special cases long-range order may considerably enhance the effects of magnetic annealing. If ordering occurs in a fcc alloy on alternate $\langle 100 \rangle$ planes, the resulting structure is anisotropic. In the presence of a magnetic field, domains in which the preferred direction of the magnetization is parallel to the magnetic field will grow at the expense of less favorably oriented domains. Obviously this may be a very slow process, as only diffusion at the domain boundaries is important if the specimen is completely ordered. Field cooling from above the ordering temperature could lead to large effects in short times.

23. Time Dependence. We saw that with interstitial atoms in iron the field-induced anisotropy was established with an almost exponential time dependence. Because the atomic concentration is only a few times 10^{-4} in that case, usually no other interstitials are found in the neighborhood of a given interstitial atom. All are similarly surrounded by other atoms and the probability for jumping is the same. This results in an exponential approach to equilibrium. In the experiments on pair ordering the concentrations are much higher. Therefore, the individual pairs may differ considerably in neighboring configuration, and a spread in relaxation times should be expected. In the limit of low concentration the spread should be small again and the relaxation time would correspond to the average time during which a pair exists in a given direction. This may be the time during which two atoms form a pair before the combination is broken up. However, if there is a simple way of rearranging the pair without breaking it up, the time found will depend also on the probability for this process. The time constants found need not be related in a simple way to the time constant deduced from diffusion measurements on a macroscopic scale.

Similar to the case of the diffusion of interstitials in iron, the time dependence usually changes with temperature as given by Eq. (13.6): $\tau = \tau_0 e^{Q/RT}$. The activation energy, Q, is the sum of the formation energy of a vacancy and the energy needed for exchange of one of the atoms of the pair with the vacancy. The constant τ_0 usually is of the order of 10^{-14} sec but is sometimes considerably shorter, apparently due to the factor $e^{-\Delta S/k}$ which is part of the theoretical expression for τ_0 [71], where ΔS is the "activation entropy."

24. *Relation between Magnetic and Elastic Aftereffects.* In the case of interstitial atoms in iron we saw that the distribution over the different sites depends both on the direction of magnetization and on the strain of the lattice. The same relations can in principle be expected for any type anisotropy center. In fact the idea of pair ordering was originally proposed by Zener [72] to explain the elastic aftereffects in alloys like CuZn.

Whether or not the elastic effects are important for the magnetic properties depends on the value of the strain upon ordering and on the magnitude of the magnetostriction. This may differ considerably among the different alloy systems. Even for different concentrations within the same system, relations may change due to the concentration dependence of the magnetostriction constants. A complication may arise because in practice different types of neighbors are often active. The effects can only be separated by measurements on single crystals.

25. *Other Causes of Aftereffects Due to Diffusion apart from Interstitial Foreign Atoms and Pair Ordering.* In most studies of the magnetic annealing effects directional ordering is invoked to explain the observed phenomena. A number of other causes can be imagined and these have also been considered in the literature.

An effect expected to play a role, especially in pure metals, is creep under the action of magnetostrictive strain [73, 74]. In a disaccommodation experiment a decrease of initial permeability can result from this effect. Also a preferred direction of magnetization in this way can be introduced into polycrystalline material if the magnetostriction is anisotropic.

XVI. DIFFUSION

In ordering systems a magnetic aftereffect may be caused by interphase boundaries that have preferred orientation with respect to the magnetization direction [74a].

An effect that can show up only in measurements in the mobility of Bloch walls is the diffusion into or out of these walls of vacancies or of one of the constituents of the alloy [23, 30]. Thus, long-range diffusion can affect the initial permeability. Annealing in a magnetic field possibly could have a small effect if diffusion into or out of differently oriented grains is considered.

Interstitials [75, 76] and double vacancies [77] in deformed or irradiated nickel have been proposed as a cause of aftereffects found at (very) low temperatures [78]. Complexes of vacancies with dissolved atoms also could play a role. This effect would depend strongly on the nature of the alloy system. The temperature at which the last-mentioned mechanisms operate are very much lower than in the case of pair ordering.

Experimental Approaches

The most direct way to study the effects of magnetic annealing is the measurement of the magnetic anisotropy induced in single crystals by an anneal at sufficiently high temperature in the presence of a saturating magnetic field. Some examples of the experimental setup can be found in the literature [61, 62, 79]. Ideally this should be combined with measurement of the aftereffect of the magnetostriction [23] and a determination of the purely mechanical aftereffects. These data then should be obtained as a function of composition, annealing temperature and duration, and the measuring temperature. If the aim of the investigation is to attain a better insight into the diffusion process, these results should be compared with the results of more conventional diffusion experiments.

26. The induced magnetic anisotropy can be determined most directly by a torque measurement; however, measurement of the stabilization field in suitable single crystals can be used to obtain the same information [80]. Measurement of the aftereffect of the magnetostriction can be replaced by the measurement of strain-induced magnetic anisotropy [81, 81a]. On single crystals no experiments of this sort have been performed.

Often the only interest is in time constants. In that case measurements on polycrystalline material may serve just as well. Again in this case measurement of the induced anisotropy is preferred as no unknown factors depending upon domain configuration are involved. The anisotropy can be shown to be given by

$$K_u = (\tfrac{2}{5}k_1 + \tfrac{3}{10}k_2)(\cos^2 \phi - \tfrac{1}{3}) \tag{26.1}$$

if Eq. (20.7) held for single crystals and strain effects could be neglected. Here ϕ is the angle between the magnetization and the direction of the magnetic field in which the sample is annealed. Often in the literature less direct experiments are described, a quantitative evaluation of which is somewhat less certain, for example, a change in the magnetization curve caused by annealing in or without a magnetic field, the occurrence of constricted hysteresis loops when the specimen is heat treated in zero field, and a change in magnetic properties if the sample is annealed in a magnetic field or quenched from high temperatures. Related to these effects is the difference in magnetic work done in magnetizing the sample when the relative orientation of annealing and magnetizing field is altered. This method may be expected to yield the right order of magnitude of the induced anisotropy energy but in the usual experimental conditions the exact value depends upon the unknown initial domain configuration. Measurement of the decrease of initial permeability is a very convenient method for determining the time constants of the diffusion mechanism if, as is the case for interstitial atoms in iron, the induced anisotropy is small compared to the crystalline anisotropy. In that case the diffusion process has only a negligible influence upon the width of the Bloch wall. The presence of the interstitial atoms gives rise to a contribution to K_1 if measured under conditions in which the distribution is always in thermal equilibrium [22]. The magnitude of this term is easily shown to be $nw^2/3kT$, equal to the coefficient in Eq. (20.4). In the more general case of Eq. (20.7) the correction is found to be $(-k_1 + \tfrac{1}{2}k_2)$. If this is not negligible, the wall thickness changes during the diffusion process and the stiffness, or reluctivity, reaches its final value with a time constant that is longer than the time constant for the underlying process. Related to disaccommodation measurement is the measurement of the stabilization field at higher induction [31–33, 80].

Experimental Results

In this section we will discuss data on some alloy systems, in order to illustrate the preceding sections, and we will outline the experimental situation.

27. The Iron–Silicon System. This system seems to be the first in which magnetic annealing effects have been observed [82]. High-temperature annealing in a (alternating) magnetic field resulted in an increased maximum permeability and remanence.

The effect is easily understood on the basis of the theory given. Of the three originally equivalent directions of easy magnetization the one nearest to the magnetizing field is preferred. Many of the later papers deal in more or less systematic fashion with room temperature magnetic properties. The first direct measurement of induced anisotropy is due to Brissonneau and Moser [80] who in single-crystal "picture frames" measured the stabilization field as a function of the magnetization. The terms in k_1 and k_2 supposedly due to second- and first-neighbor ordering, respectively, lead to different restoring forces on the wall when it is displaced from the equilibrium position.

We will not give a full discussion of this experiment here. It was found that for 3.25 wt % Si the constant k_2 is about 100 ergs cm^{-3} at 400°C and k_1 is smaller by a factor of 8.7. A very careful, direct measurement of these anisotropies was made by Brommer and 't Hooft [81b] using a torque method. The anisotropy introduced by an anneal at 400°C in the [100] direction showed k_1 to be about 140 ergs cm^{-3}. Annealing in the [110] direction led to a slightly lower anisotropy, with k_2 about 250 ergs cm^{-3}. These values are rather small and no magnetostrictive effects were taken into consideration in the analysis.

Any discrepancy between the two sets of values may possibly be due to this. In direct measurements the induced anisotropy seemed to scale approximately with the crystalline anisotropy constant K_1, in agreement with

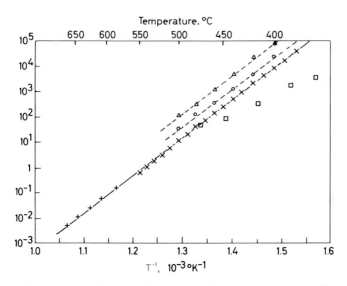

FIG. 27. Time constants for the disaccommodation of SiFe polycrystalline samples (the values of τ are about a factor of 5 larger than would be found when comparing with a pair ordering theory): (△) 0.95 wt %; (○) 2.2 wt %; (×) 3 wt % Si in Fe (after Balthesen [85]); (□) 3.5 wt % Si (from Biorci et al. [83]).

results reported by Ferro *et al*. [81c]. Disaccommodation, the time decrease of initial permeability, has been observed by different authors [80, 84, 85]. Figure 27 shows Balthesen's results. They were obtained with due regard for the importance of the experimental parameters like the magnitude and the frequency of the measuring field. In this figure $\tau(s)$ is the time constant describing the disaccommodation curve.

In discussing the annealing effects in silicon–iron most authors assume the mechanism of pair ordering to be at the base of the observed effects. Dietze [30], however, holds that the diffusion of vacancies into or out of the wall is the cause of disaccommodation. In our opinion this assumption is false.

The agreement Dietze claims between calculated and observed time de-

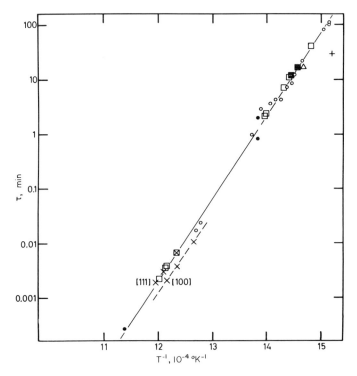

FIG. 28. Relaxation time measured in iron–silicon as a function of temperature: (□, ○) time dependence of induced anisotropy measured by a torque method, (Brommer *et al*. [81b]); (○) a 4% Si polycrystalline sample; (□) a 3.9% Si–Fe single crystal [100] (001); (+) torque measurement (taken from a published curve by Lutes *et al*. [85a]); (●) disaccommodations (taken from published curves of Balthesen [85]); (△) curve with a correction for the temperature dependence of the effect (from Biorci [83]); (■) stabilization field measurements (from Brissonneau *et al*. [81]); (×) internal friction (Zener damping) (Boesono *et al*. [85b]).

pendence can also be found with a reasonable spread in relaxation times. The results presented in Fig. 28 show that the relaxation times found for disaccommodation and for the development of induced anisotropy in single-domain monocrystalline samples agree very well. There clearly is no need for different mechanisms. From the combined measurements the value $\tau_0 \approx 10^{-19}$ sec and an activation energy of 69,000 cal mole^{-1} can be deduced.

Classical diffusion measurements in Fe with 2.3–3.7 wt % Si yielded the value 48,000 cal mole^{-1}, with $D_0 = 0.44$ cm^2 sec^{-1} [86] corresponding to a τ_0 for the breakup of a pair of 4×10^{-16} sec, if the binding energy can be neglected.

Obviously the agreement between the classically obtained numbers and the values deduced from the magnetic measurements is not very good. Careful work on this problem is needed, both from a theoretical and experimental standpoint.

28. The Iron–Aluminum System. As in iron–silicon, the room temperature magnetic properties of iron–aluminum can be influenced by heat treatment in a magnetic field. These effects were investigated by Sugihara [87]. For the induced anisotropies much higher values have been reported than those for iron–silicon alloys. The results of some experiments on single crystals are represented in Table II. It is seen that k_1 and k_2 are about equal.

TABLE II

INDUCED ANISOTROPY IN FeAl SINGLE CRYSTALS

Al (at. %):	17.5	20	23	25	29.5	18.7	22	25.2	25[a]
k_1:	5000	3000	0	0	0	8000	2200	0	2900
k_2:	5000	4000	0	0	0	11,000	2900	0	9800
Ref.:	88	88	88	88	88	90	90	90	89

[a] Chikazumi [88] quoting the results of Suzuki [89] gives a value of 22.5 at. % for this alloy.

Therefore, the anisotropy induced by annealing with the magnetic field in the [100] direction is twice as large as with the field in the [111] direction. Here bcc lattice nearest neighbor interaction is apparently less important than the other interactions. Near the composition Fe$_3$Al long-range ordering occurs, and directional ordering is suppressed.

This effect was found by Birkenbeil and Cahn [81] in a measurement of mechanically induced magnetic anisotropy in a fairly narrow region of concentration (Fig. 29). Note in this figure the large value of the induced anisotropy, as well as the very strong influence of strain on directional ordering. The stress of 14 kg mm^{-2} put on the sample is much larger than the magnetostrictive stresses expected, but at lower concentrations the strain-induced anisotropies also are much larger than the field-induced anisotropies. The interference between the two mechanisms for directional ordering can be expected to have a large influence on the observed effects of magnetic annealing as the magnetostriction constant λ_{100} and the "Zener" relaxation strength are large in this alloy system. This is shown in Fig. 30 where several results on internal friction are plotted [91]. Also the same concentration dependence is found here as that for the magnetic effects, demonstrating the common cause.

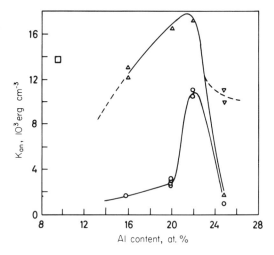

FIG. 29. Induced anisotropy in FeAl alloys: (△) slow cooled under stress; (▽) quenched under stress; (○) slow cooled in field; (□) biaxial anisotropy, stress induced. (After Birkenbeil and Cahn [81].)

It would be very interesting to systematically investigate the relation between magnetization-induced and stress-induced anisotropy and their connection through magnetostriction. The relevant experiments should be performed on single crystals, making use of materials of very high purity. The occurrence of field-induced anisotropy of a higher-order than ($\cos^2 \phi - \frac{1}{3}$), found by many investigators [68, 74], may be caused by long-range order [92].

Measurement of the time constant by magnetic methods is still very incomplete. Biorci et al. who measured disaccommodation remark that

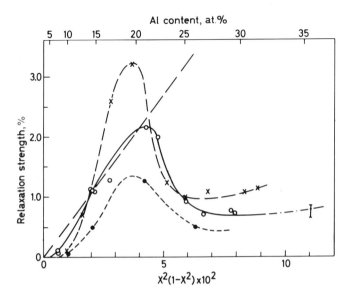

FIG. 30. Internal friction measured in FeAl alloys. Different curves represent data obtained by different authors. The straight line represents a proportionality to $x^2(1 - x^2)$ (when x is the atomic fraction aluminum) expected for a random alloy. (After Fishbach [91].)

the activation energy of the diffusion process controlling the effect is about the same as that found for 3% silicon–iron. Note, however, the influence of vacancies measured by Steinert [92a]. The time constants for the related effect of Zener relaxation has been investigated by several investigators. Fishbach [91] reports for an alloy with 14 at. % Al, $\tau_0 = 2 \times 10^{-17}$ sec and $Q = 58.9$ kcal mole^{-1}; for the 17% alloy these values are 8×10^{-17} sec, and 56 kcal mole^{-1}, respectively. He reported similar results for the higher concentration, where ordering sets in, but these were considered less certain because of interference with the ordering effects. Also it is difficult in these alloys to reconcile the findings of the different investigators, especially with respect to the magnitude of the effects. The difference between the experiments on single crystals and on polycrystalline material possibly may be connected with magnetostrictive effects.

The small size of k_2 is probably connected with the low probability of finding two aluminum atoms at neighboring positions. This agrees with the strong tendency to form an ordered phase of the type FeAl.

A rapid decrease of the effects with decreasing concentration agrees with the calculations of Iwata. At low concentrations it is easy for the aluminum atoms to avoid each other but at higher concentrations they are forced together and an increase in these effects is much more rapid than to the square of the product of the concentrations. This reasoning would apply uot only to nearest but also to next nearest neighbors. Experimentally this is supported by the observations of Vlasova and Veronova [93]. Using x-rays, they found that in a 15 at. % AlFe alloy the probability of Al atoms being nearest or next nearest neighbors was less than the random probability.

The hypothesis of pair interaction of the simple kind we considered need not necessarily lead to quantitatively correct results, as can be inferred from the neutron diffraction data on the ordered phases of Fe_3Al [94] and Fe_3Si [95]; these showed that the magnetic moments of the iron atoms on the different positions in this lattice differ from each other. Therefore, it is very likely that interaction with a neighboring atom would also depend on the other atoms surrounding this pair. In order to obtain information that can be compared with diffusion data, it would be most useful to make the measurements at as small a concentration as possible. Even then the results are not directly comparable. The time effects considered here are always connected with the formation of pairs. The activation energy and entropy can be expected to be quite different from those determined for classical diffusion that may proceed, at any rate in dilute alloys, mainly by jumps of single atoms.

29. The Iron–Cobalt System. Disaccommodation measurements in this system have been carried out by Ferro *et al.* [96] for alloys with less than 20 at. % Co in Fe. An activation energy of the order of 50,000–60,000 cal mole^{-1} was reported, with a τ_0 of approximately 10^{-16} sec. Ferguson [62] made a more thorough study of the induced anisotropy for alloys with 40 and 50% Co in Fe. In these alloys there is a very strong tendency toward long-range order. His object was to investigate the interference between the two types of order. The time constants related to the establishment of directional order proved to be very much spread out. It was impossible to reach true thermodynamic equilibrium within a reasonably short time (hundreds of hours). Ordinary long-range order, on the other hand, seems to be established in much shorter time periods. Possibly the effects of magnetic annealing are not caused only by simple pair ordering, as it is hard to see how one can account for the very long time constants in this model. It would not seem unreasonable to assume that rearrangements of much larger groups of atoms play an important role in this alloys system. Again, for the study of diffusion phenomena the study of more dilute alloys would be most useful.

30. *The Nickel–Iron System.* In this system the magnetic crystalline anisotropy becomes very small, and the effects of directional order on the room temperature properties is therefore very noticeable. Much of the older literature is concerned with these effects. We mention here only the more recent experiments that can give some information on the kinetics and the relevant interaction constants of the pair ordering mechanism.

Data have been reported by Chikazumi [67], Aoyagi [97], Yamamoto *et al.* [98], Takahashi [99], and Takahashi *et al.* [100, 101] on single crystals. The values of the induced anisotropies are on the order of 600 ergs cm^{-3} in an alloy with 17% Fe in Ni and about 3000 ergs cm^{-3} in an alloy of 46% Fe in Ni. As in the previously discussed alloys, the ratio k_2/k_1 differs considerably from the value predicted by the simple theory. In Fig. 31 the results of Takahashi *et al.* are reproduced for NiFe without long-range order.

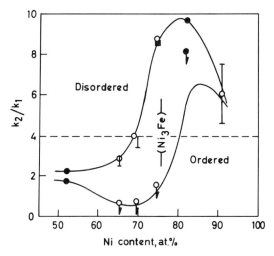

FIG. 31. Ratio of k_2/k_1 measured in FeNi single-crystal samples. (After Takahashi *et al.* [100].)

Comparison with the theoretical curves of Iwata (Fig. 26) shows that the experimental values of k_2/k_1 are larger by at least a factor of 2 than can be accounted for by the theory for ordering alloys. From Eq. (20.9), however, it is found where $w^{(2)}_{110}/w^{(1)}_{110} = 0.079$ or 0.744, this factor can be explained. The first value in particular seems very reasonable. Iwata's calculations are not modified except by this factor when the constant $w^{(2)}_{100}$ is introduced. The best, but far from perfect, fit would be obtained for $kT/2V = 0.4$. This value again is reasonable in view of Iida's [102] value for the transformation energy in FeNi$_3$ of 888 cal mole^{-1}.

Measurements of the induced anisotropy were performed by Ferguson [62] with polycrystalline material on fcc alloys of several concentrations. He also determined the induced anisotropy as a function of the annealing temperature. Within measuring accuracy a linear temperature dependence is found near the Curie temperature. He determined the time constants for the approach to equilibrium in two alloys. Without any claim to great accurancy he quotes values for the activation energy Q/R of 30,800°K and 35,200°K for the 70% Ni and 60% Ni alloys, respectively. The preexponential factor has the very small value of 5×10^{-18} sec.

In one case he made a careful determination of the time dependence and compared this with a theoretical curve deduced with the same logarithmical spread in relaxation times employed in Eqs. (13.5) and (13.7). The results are reproduced in Fig. 32. In these experiments, where a bulk effect is meas-

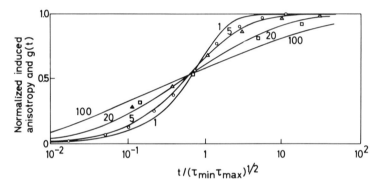

FIG. 32. Development of induced anisotropy with time compared with $g(t)$ curves [Eq. (13.7)]. The parameter k is indicated, where $k^2 = \tau_{max}/\tau_{min}$. (After Ferguson [62].)

ured, a definite spread in relaxation times is seen to exist. Disaccommodation measurements have been reported by Fahlenbrach [103] and by Gerstner and Kneller [104]. The latter authors made a very careful study of the effects of different heat treatments on the kinetics of the directional ordering phenomena. One of their figures is reproduced in Fig. 33. For alloys with 81.5 and 60 wt % Ni in Fe, respectively, activation energies of 54 ± 2 kcal mole^{-1} and 63 ± 2 kcal mole^{-1} were found, and values for τ_0 of 3×10^{-14} sec and 0.6×10^{-15} sec. The spread in relaxation times found for the 60% alloy involved a factor of 20 between the highest and lowest value, which is not far from the value of about 12 found by Ferguson.

An interesting observation was made on the influence of quenching on the kinetics of the ordering process. One of the experiments is shown in Fig. 34. Curve 3 shows the low initial susceptibility that results when the

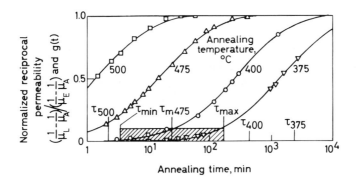

FIG. 33. Disaccommodation in 19.5–81.5% Fe–Ni, compared with $g(t)$ curves for $\tau_{max}/\tau_{min} = 50$. (After Gerstner and Kneller [104].)

sample (60% Ni) is slowly cooled without external field. By this treatment the local direction of magnetization is stabilized everywhere and the sample behaves like uniaxial material with randomly oriented grains. Curve 1 is a case where the stabilization is suppressed by an air quench to room temperature. In the case of curve 2 this is achieved by alternately applying,

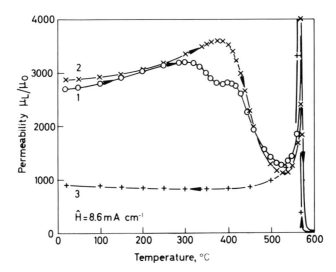

FIG. 34. Permeability of a 40–60% Fe–Ni alloys measured after different heat treatments [104]. Curve 1, measured with increasing temperature (400°C hr^{-1}) after air quench from 650°C (20°C sec^{-1}). Curve 2, measured as curve 1, but after slow cooling in a saturating magnetic field applied in the measuring direction, and in a direction perpendicular to it, changing direction every 40 sec. Curve 3, measured while cooling from 650°C (400°C hr^{-1}).

during the slow cooling, a field parallel and perpendicular to the measuring direction. Both, therefore, exhibit a large value of the initial susceptibility at room temperature.

Succeeding slow heating reveals the difference between the two treatments. The air quenched sample starts to anneal some 100°C below the temperature where the slowly cooled sample drops in permeability. Quenched-in vacancies probably are responsible for this difference in behavior. Gerstner and Kneller made this plausible by estimating the effects using known activation energies for vacancy generation and mobility in nickel. A very small vacancy concentration of only 10^{-7} could explain the described effect.

They also investigated systematically the effects that are found on a very large time scale. This is shown for the composition 75% Ni in Fig. 35. The effects of decrease of initial susceptibility that develops when continu-

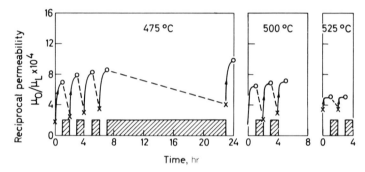

FIG. 35. Repeated measurements of the disaccommodation in 25–75% Fe–Ni. During the times indicated by the shaded rectangles a saturating field alternating in direction (Fig. 34, curve 2) was applied [104].

ously alternating transverse and longitudinal fields are applied are found below the critical temperature for long-range ordering, but not above. Gerstner and Kneller correlated these observation (also found in 60% Ni alloy) with observations of the increase in magnetic saturation and change in resistivity. Although not completely consistent, the results make the connection with the establishment of long-range order very probable.

In this system small amounts of impurities may play a large role. Heidenreich and Nesbitt [105–108] did magnetic annealing experiments on a number of Fe–Ni–Co alloys, and Fe–Ni–Mo (2%) and Ni–Fe alloys. With electronmicroscopic and magnetic means it was found that very small amounts of oxygen (0.007%) resulted in an inhomogeneous structure, probably ⟨111⟩ stacking faults more or less regularly spaced. The distribution over the different directions could be influenced by a magnetic anneal, thus giving rise to magnetic anisotropy. They even found that deoxidizing the sample resulted in no magnetic annealing effect by their standard field treatment: heating to 750°C and cooling in the field at 4° min^{-1}. Similar low-oxygen samples also showed no stacking fault inhomogeneities. Unpublished experiments of Graham failed to duplicate these findings. The reason for the discrepancies is not clear and would merit more study.

31. The Cobalt–Nickel System. In this system the absolute value of the ordinary ordering energy is low as no precipitation or long-range order has been observed with certainty. It therefore should be an ideal case for the study of magnetic annealing effects. Yamamoto et al. [98] measured the induced anisotropy in polycrystalline material with 10, 20, 30, 40, 50, and 60% Co in Ni; this was also done on two single-crystal specimens with 12 and 20% Co in Ni, respectively. All of these alloys are face-centered cubic. Dependence on annealing temperature was checked in the 30% alloy; it agreed well with the expected proportionality with $M(T)^2$. The dependence on measuring temperature was checked on the 12% single-crystal specimen; again a M^2 dependence was found, in both cases assuming M to be given by a Brillouin curve for $S = \frac{1}{2}$. The concentration dependence deviated considerably from the expected behavior for an ideal solution. Comparison with different theories would lead to a very strong tendency to precipitation. At present it is not at all clear why there is such a large difference between

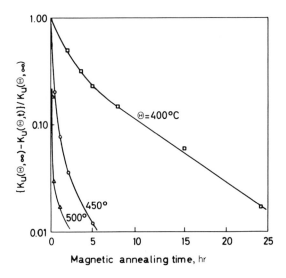

FIG. 36. The development of the induced magnetic anisotropy with time in a 30.84 wt % Co–Ni alloys for different annealing temperatures. (After Yamamoto et al. [79].)

theory and experiment. The assumption of independent pair interactions could be too crude in this system; the tendency to form like pairs could be more pronounced for second-nearest neighbors than for nearest neighbors, which should lead to precipitation of a $CoNi_3$ phase at much lower temperatures (an indication for this last possibility is found in the low value of k_2/k_1); indetermined impurities (O) may also play a role here. Iwata [66] when comparing his theory with the experiments considers yet another possibility. He states that the temperature dependence found for the magnetic ordering energy is proportional to M^2 only in the 12% Co in Ni alloy. In the 20 and 33% alloys large deviations are said to exist from this law. As the experimental results used to check the concentration dependence had to be corrected for the temperature dependence, the resulting uncertainty may give sufficient room for agreement with theory. In any case a thorough study of this system seems to be indicated. The dependence upon time for the approach the equilibrium is shown for different annealing temperatures in Fig. 36. Before being held at the indicated temperatures the specimen was annealed for 30 min at 750°C. The magnetic field was applied 6 min after the temperature Θ was reached. That it is dangerous to give values for the activation energies involved is illustrated in Fig. 37, where at one annealing temperature the approach to

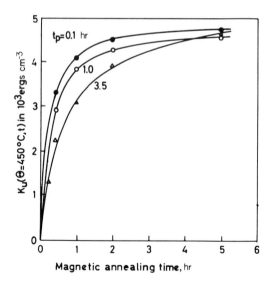

Fig. 37. The development of the induced magnetic anisotropy with time in a 30.84 wt % Co–Ni alloy at 450°C after keeping the sample at 450°C during the indicated time without an applied magnetic field [79].

equilibrium is shown for three different instances before application of the magnetic field. Possibly quenched-in vacancies are again responsible for the observed effect.

32. Other Systems and Pair Ordering in Nonequilibrium Conditions. In some other alloy systems measurements on magnetic annealing effects have been performed [96, 108a]. In all ferromagnetic alloys magnetic annealing effects can be expected if the Curie point is sufficiently high; while the sample is in the ferromagnetic state, atomic diffusion should still be possible. Even if this is not the case, magnetic annealing effects can be brought about in many cases when the diffusion is artificially enhanced by quenching in vacancies or by irradiation with fast particles or gamma rays. The defects produced in this way cause atomic displacements when they wander through the crystal lattice but may just as well exhibit their own anisotropic structure and in this way give rise to magnetic aftereffects and related phenomena, as will be discussed in the next section. In Sec. 30 an example of the effect of quenched-in vacancies was discussed for the Ni–Fe system. Disaccommodation was measured in silicon–iron by Dietze and Balthesen [85, 109] while the sample was exposed to a flux of fast neutrons. For technical reasons, only the decrease of the permeability immediately after demagnetization was determined. Above 450°C the temperature dependence was the same as that for the nonirradiated case. After a sufficiently long time in a neutron flux of about 10^{12} neutrons cm^{-2} sec^{-1}, a temperature-independent effect was found between 400 and 250°C. In this region the decrease in mobility of the defects apparently is compensated for by a corresponding increase in density. Unexplained however is the insensitivity of this curve to the actual neutron flux at the sample, which was reported to be changed by a factor of at least 10 without noticeable influence upon the measured curve. In this type of experiment changes in domain configuration and mobility of Bloch walls due to clustered defects that may be caused by the irradiation can be expected to modify the experimental results. Measurements of bulk effects, as the induced uniaxial anisotropy are preferred where feasable. Paulevé et al. [110, 111, 112] experimented on 50–50% Ni–Fe. A very large anisotropy resulted under the influence of both fast electron irradiation and neutron irradiation. This was attributed to the formation of a tetragonal ordered phase in which {100} planes are occupied alternately by Ni and Fe. When a magnetic field is present, regions where the preferred direction is aligned with the field grow fastest and a very large anisotropy results.

Néel et al. also studied the growth of tetragonal domains of FeNi in a single crystal of the alloy under isotropic conditions [113].

33. Magnetic Aftereffects Due to Other Causes. An interstitialcy in nickel can be thought to consist of two atoms around one atomic position, with the line joining these atoms along a cube direction. Interstitials of this type can cause magnetic aftereffects just as interstitial carbon atoms cause these effects in iron. Seeger *et al.* [75] proposed this mechanism for the disaccommodation measured in cold-worked and neutron-irradiated nickel. The anisotropy centers were observed in both magnetic measurements and internal friction experiments [76, 77]. The activation energy for reorientation of the defect was found to be 0.81 eV. Studies of the annealing behavior of this effect led to an activation energy of 1.02 eV for the disappearance of these centers.

Peretto *et al.* [78] irradiated Ni at 27°K with fast neutrons. In this case large magnetic effects are also found, even at these low temperatures. The annealing behavior observed with continuous warming up of the sample is very complex, but could partially be correlated with the annealing of the electrical resistance after irradiation. As yet there is no complete understanding of the observed phenomena, but without doubt the combination of magnetic and other measurements possible in ferromagnetic materials is a powerful tool for the clarification of the mechanisms of radiation damage. In principle, mechanical measurements can give the same type of information, but often matters are complicated by motion of dislocations; these are not expected to play such a large role in the magnetic measurements.

Concluding Remarks

34. In this chapter we have tried to indicate the potential usefulness of magnetic methods in diffusion problems. For interstitial atoms in iron the contribution from this side indeed has been important. In the binary alloys the method has not been pushed to its potential limits by any means. This is mainly due to the complexities of the problems in this case. In dilute alloys where conceptual difficulties are much fewer, the measurements become far from simple. In any case the magnetic methods constitute only one of several possible techniques that should be used in combination for the study of diffusion in alloy systems. This chapter was submitted to the editors in October, 1964.

REFERENCES

1. P. Chevenard and X. Waché, *Rev. Met.* **41**, 353, 389 (1944); W. Köster and J. Raffelseiper, *Z. Metallk.* **42**, 387 (1951).
2. E. P. Wohlfarth, Hard magnetic materials, *Phil. Mag. Suppl.* **8**, 87 (1959).

3. D. Turnbull, *Rept. Conf. Defects Crystalline Solids*, p. 203. The Physical Society, London (1955).
4. J. J. Becker, *Trans. AIME* **209**, 59 (1957); **212**, 138 (1958); J. D. Livingston and J. J. Becker, *ibid.* **212**, 316 (1958).
5. L. Weil and L. Gruner, *Compt. Rend.* **243**, 1629 (1956); J. D. Livingston, *Trans. AIME* **215**, 566 (1959); J. D. Livingston and C. P. Bean, *J. Appl. Phys.* **30**, 318S (1959); C. W. Berghout, *J. Phys. Chem. Solids* **24**, 507 (1963); R. Hahn and E. Kneller, *Z. Metallk.* **49**, 426, 480 (1958); A. E. Berkowitz and P. J. Flanders, *J. Appl. Phys.* **30**, 111S (1959).
6. S. Kaya and A. Kussman, *Z. Physik* **72**, 293 (1931); N. J. Goldman and R. Smoluchowski, *Phys. Rev.* **75**, 140 (1949); **76**, 471 (1949); R. M. Bozorth, *Rev. Mod. Phys.* **25**, 42 (1953).
7. C. P. Slichter, *Rept. Conf. Defects Crystalline Solids*, p. 52. The Physical Society, London (1955); D. F. Holcomb and R. E. Norberg, *Phys. Rev.* **93**, 919 (1954); **98**, 1074 (1955); H. C. Torrey, *ibid.* **92**, 962 (1953).
8. J. C. Slonczewski, Magnetic Annealing, *in* "Magnetism" (G. T. Rado and H. Suhl, eds.), Vol. I. Academic Press, New York, 1964.
9. L. Néel, *J. Appl. Phys.* **30**, 3S (1959).
10. C. D. Graham, Jr., *in* "Magnetic Properties of Metals and Alloys" (R. Bozorth, ed.). Am. Soc. Metals, Cleveland, 1959.
11. P. Brissonneau, *J. Phys. Radium* **19**, 490 (1958).
12. F. Schreiber, *Z. Angew. Phys.* **9**, 203 (1957).
13. G. W. Rathenau, *in* "Magnetic Properties of Metals and Alloys" (R. Bozorth, ed.). Am. Soc. Metals, Cleveland, 1959; *J. Appl. Phys.* **29**, 239 (1958); *Metaux (Corrosion Ind.)* **37**, 158 (1961).
14. A. J. Bosman and G. de Vries, *Ned. Tijdschr. Natuurk.* (*in Dutch*) **27**, 65 (1961).
15. G. Richter, *Ann. Physik* **26**, 605 (1937); *Ann. Physik* **32**, 683 (1938).
16. J. L. Snoek, *Physica* **6**, 161, 591 (1939); **8**, 711 (1941); "New Developments in Ferromagnetic Materials." Elsevier, New York and Amsterdam, 1947.
17. W. S. Gorski, *Phys. Z. Soviet Union* **8**, 457 (1935).
18. D. Polder, *Philips Res. Rept.* **1**, 5 (1945).
19. J. L. Snoek and F. K. du Pré, *Philips Tech. Rev.* (*in Dutch*) **8**, 57 (1946); J. L. Snoek, *Physica* **9**, 862 (1942); *Ned. Tijdschr. Natuurk.* (*in Dutch*) **8**, 177 (1941).
20. L. Néel, *J. Phys. Radium* **12**, 339 (1951).
21. L. Néel, *J. Phys. Radium* **13**, 249 (1952).
22. A. J. Bosman, P. E. Brommer, and G. W. Rathenau, *J. Phys. Radium* **20**, 241 (1959).
23. G. de Vries, *Physica* **25**, 1211 (1959).
24. G. de Vries, D. W. van Geest, R. Gersdorf, and G. W. Rathenau, *Physica* **25**, 1131 (1959).
25. R. Gersdorf, Magnetostriction of iron and iron alloys, thesis, Univ. of Amsterdam, 1961; R. Gersdorf, J. H. M. Stoelinga, and G. W. Rathenau, *Proc. Intern. Conf. Magnetism Crystallography, Kyoto, 1961*; *J. Phys. Soc. Japan Suppl. B-I*, **17**, 342 (1962).
26. G. Biorci, A. Ferro, and G. Montalenti, *Nuovo Cimento* **18**, 229 (1960).
27. C. Zener, "Elasticity and anelasticity of metals." The Univ. of Chicago Press, Chicago, 1948.
28. C. Kittel and J. K. Galt, *Solid State Phys.* **3**, 437 (1956).

29. A. J. Bosman, Investigation on magnetic aftereffects due to interstitials, thesis, Univ. of Amsterdam, 1960.
30. H. D. Dietze, *Tech. Mitt. Krupp* **17**, 67 (1959); *Phys. Status Solidi* **3**, 2309 (1963).
31. J. L. Snoek, *Physica* **5**, 663 (1938).
32. J. Bindels, J. Bijvoet, and G. W. Rathenau, *Physica* **26**, 163 (1960).
33. P. Brissonneau, *J. Phys. Chem. Solids* **7**, 22 (1958).
34. P. Brissonneau, *Compt. Rend.* **244**, 1341 (1957).
35. P. Brissonneau, *Compt. Rend.* **239**, 346 (1954).
36. G. Richter, *Ann. Physik* **29**, 605 (1937).
37. R. Becker and W. Döring, "Ferromagnetismus." Springer, Berlin, 1939.
38. E. Kneller, "Ferromagnetismus." Springer, Berlin, 1962.
39. P. Brissonneau, *Compt. Rend.* **244**, 1174 (1957).
40. P. Brissonneau, *Compt. Rend.* **244**, 868 (1957).
41. P. Brissonneau, *J. Appl. Phys.* **29**, 249 (1958).
42. F. Schreiber, *Z. Angew. Phys.* **8**, 539 (1956).
43. R. Feldtkerller, "Theorie der Spulen und Übertrager." Hirzel, Stuttgart, 1958 (see bibliography).
44. A. J. Bosman, P. E. Brommer, A. J. van Daal, and G. W. Rathenau, *Physica* **23**, 989 (1957).
45. J. D. Fast and H. B. Verrijp, *J. Iron Steel Inst.* **176**, 24 (1954).
46. G. A. Wert, *J. Appl. Phys.* **21**, 1196 (1950).
47. A. J. Bosman, P. E. Brommer, L. C. H. Eykelenboom, C. J. Schinkel, and G. W. Rathenau, *Acta Met.* **8**, 728 (1960).
48. R. W. Powers, M. V. Doyle, *Trans. AIME* **215**, 655 (1959).
49. J. C. Fisher, *Acta Met.* **6**, 13 (1958).
50. A. J. Bosman, P. E. Brommer, and G. W. Rathenau, *Physica* **23**, 1001 (1957).
51. A. J. Bosman, P. E. Brommer, L. C. H. Eykelenboom, C. J. Schinkel, and G. W. Rathenau, *Physica* **26**, 533 (1960).
52. N. H. Nachtrieb, J. A. Weil, E. Catalano, and W. A. Lawson, *J. Chem. Phys.* **20**, 1189 (1952).
53. J. Bass and D. Lazarus, *Phys. Chem. Solids* **23**, 1820 (1962).
54. A. S. Nowick and W. R. Heller, *Advan. Phys.* **14**, 101 (1965); A. S. Nowick, *ibid.* **16**, 1 (1967).
55. L. Néel, *Compt. Rend.* **237**, 1613 (1953).
56. L. Néel, *J. Phys. Radium* **15**, 225 (1954).
57. S. Taniguchi and M. Yamamoto, *Sci. Rept. Res. Inst. Tohoku Univ. Ser. A* **6**, 330 (1954).
58. S. Taniguchi, *Sci. Rept. Res. Inst. Tohoku Univ. Ser. A* **7**, 269 (1955).
59. S. Chikazumi, *J. Phys. Soc. Japan* **5**, 333 (1950).
60. S. Chikazumi, *J. Phys. Soc. Japan* **5**, 327 (1950).
61. S. Chikazumi and T. Oomura, *J. Phys. Soc. Japan* **10**, 842 (1955).
62. E. T. Ferguson, thesis, Univ. of Grenoble, 1962.
63. J. H. Van Vleck, *Phys. Rev.* **52**, 1178 (1937).
64. T. Iwata, *Sci. Rept. Res. Inst. Tohoku Univ. Ser. A* **10**, 34 (1958).
65. T. Iwata, *Sci. Rept. Res. Inst. Tohoku Univ. Ser. A* **13**, 337 (1961).
66. T. Iwata, *Sci. Rept. Res. Inst. Tohoku Univ. Ser. A* **13**, 356 (1961).
67. S. Chikazumi, *J. Phys. Soc. Japan* **11**, 551 (1956).
68. C. Zener, *Phys. Rev.* **96**, 1335 (1954).

69. "American Institute of Physics Handbook." McGraw-Hill, New York, 1963; R. M. Bozorth, "Ferromagnetism," Van Nostrand, Princeton, New Jersey, 1961.
70. R. Gersdorf, J. H. M. Stoelinga, and G. W. Rathenau, *J. Phys. Soc. Japan Suppl.* **17**, B-I, 343 (1962).
71. C. Zener and C. Wert, *Phys. Rev.* **76**, 1169 (1949).
72. C. Zener, *Phys. Rev.* **71**, 34 (1947).
73. R. M. Bozorth, *Phys. Rev.* **46**, 232 (1934).
74. G. Biorci, G. Montalenti, and A. Ferro, *Proc. Conf. Solid State Phys.* (*IUPAP*), *Brussels, 1958*, p. 235.
74a. H. Wagner and H. Gengnagel, *Phys. Status Solidi* **9**, 45 (1965); H. Dressel and H. Gengnagel, *Z. Angew. Phys.* **18**, 479 (1965).
75. A. Seeger, P. Schiller, and H. Kronmüller, *Phil. Mag.* **5**, 853 (1960).
76. M. V. Klein and H. Kronmüller, *J. Appl. Phys.* **33**, 2191 (1962).
77. F. J. Wagner, P. Schiller, and A. Seeger, *Phys. Status Solidi* **6**, K39 (1964).
78. P. Peretto, P. Moser, and D. Dautreppe, *Compt. Rend.* **258**, 499 (1964).
79. M. Yamamoto, S. Taniguchi, and K. Aoyagi, *Sci. Rept. Res. Inst. Tohoku Univ. Ser. A* **13**, 117 (1961).
80. P. Brissonneau and P. Moser, *J. Phys. Soc. Japan Suppl. B-I* **17**, 333 (1962).
81. H. J. Birkenbeil and R. W. Cahn, *J. Appl. Phys.* **32**, 3625 (1961).
81a. R. Vergne, thesis, Univ. of Grenoble, 1962.
81b. P. E. Brommer and H. A. 't Hooft, *Phys. Letters* **26A**, 52 (1967).
81c. A. Ferro, P. Mazzetti, and G. Montalenti, *Nuovo Cimento* **23**, 280 (1967).
82. H. Pender and R. L. Jones, *Phys. Rev.* **1**, 259 (1913).
83. G. Biorci, A. Ferro, and G. Montalenti, Tech. Rept. 3[a] Inst. Elettrotecnico Naz. "Galileo Ferraris" Centro Studi Elettrofisica C.N.R. (Nov. 1959); *J. Appl. Phys.* **31**, 2121 (1960); **32**, 630 (1961).
84. H. Fahlenbach and G. Sommerkorn, *Tech. Mitt. Krupp* **15**, 161 (1957).
85. E. Balthesen, *Phys. Status Solidi* **3**, 2321 (1963).
85a. O. S. Lutes and R. P. Ulmer, *J. Appl. Phys.* **38**, 1009 (1967).
85b. Boesono, G. J. Ernst, M. C. Lemmens, M. J. van Langen, and G. de Vries, *Phys. Status Solidi* **19**, 107 (1967).
86. W. Batz, H. W. Mead, and T. E. Birchenall, *J. Metals* **4**, 1070 (1952).
87. M. Sugihara, *J. Phys. Soc. Japan* **15**, 1456 (1960).
88. S. Chikazumi and T. Wakiyama, *J. Phys. Soc. Japan Suppl. B-I* **17**, 325 (1962).
89. K. Suzuki, *J. Phys. Soc. Japan* **13**, 756 (1958).
90. H. Gengnagel and H. Wagner, *Z. Angew. Phys.* **13**, 174 (1961).
91. D. B. Fishbach, *Acta Met.* **10**, 319 (1962).
92. S. Chikazumi, private communication, 1964.
92a. J. Steinert, *Phys. Status Solidi* **9**, K109 (1965).
93. Ye. V. Vlasova and V. I. Neronova, *Fiz. Metal. i Metalloved.* **15**, 254 (1963); *Phys. Met. Metallog.* (*USSR*) (*English Transl.*) **15**, 79 (1963).
94. R. Nathans, M. T. Pigott, and C. G. Shull, *J. Phys. Chem. Solids* **6**, 38 (1958).
95. P. Lecocq and A. Michet, *Compt. Rend.* **258**, 1817 (1964).
96. A. Ferro, P. Mazzetti, and G. Montalenti, *Nuovo Cimento Ser. X*, **23**, 280 (1962).
97. K. Aoyagi, *Sci. Rept. Res. Inst. Tohoku Univ. Ser. A* **13**, 137 (1960).
98. M. Yamamoto, S. Taniguchi, and K. Aoyagi, *J. Phys. Soc. Japan Suppl. B-I* **17**, 328 (1962).
99. M. Takahashi, *J. Phys. Soc. Japan* **18**, 734 (1963).

100. M. Takahashi, T. Sasakawa, and H. Fujimori, *J. Appl. Phys.* **35**, 869 (1964).
101. M. Takahashi, T. Sasakawa, and H. Fujimori, *J. Phys. Soc. Japan* **17**, 585 (1962).
102. S. Iida, *J. Phys. Soc. Japan* **10**, 9 (1955).
103. H. Fahlenbrach, *Tech. Mitt. Krupp* **19**, 219 (1961).
104. D. Gerstner and E. Kneller, *Z. Metallk.* **52**, 426 (1961); *J. Appl. Phys.* **32**, 364S (1961).
105. R. D. Heidenreich and E. A. Nesbitt, *Phys. Rev.* **105**, 1678 (1957).
105. R. D. Heidenreich, E. A. Nesbitt, and R. D. Burbank, *J. Appl. Phys.* **30**, 995 (1959).
107. E. A. Nesbitt and R. D. Heidenreich, *J. Appl. Phys.* **30**, 1000 (1959).
108. E. A. Nesbitt, R. D. Heidenreich, and A. J. Williams, *J. Appl. Phys.* **31**, 228S (1960).
108a. H. B. Aaron, M. Wuttig, and H. K. Birnbaum, *Acta Met.* **16**, 269 (1968).
109. H. D. Dietze and E. Balthesen, *Nukleonik* **3**, 93 (1961).
110. J. Paulevé, D. Dautreppe, J. Laugier, and L. Néel, *Compt. Rend.* **254**, 965 (1962).
111. W. Chambron, D. Dautreppe, L. Néel, and J. Paulevé, *Compt. Rend.* **255**, 132 (1962).
112. W. Chambron, D. Dautreppe, and J. Paulevé, *Compt. Rend.* **255**, 2037 (1962).
113. L. Néel, J. Paulevé, R. Pauthenet, J. Langier, and D. Dautreppe, *J. Appl. Phvs.* **35**, 873 (1964).

CHAPTER XVII

Interpretation of Magnetic Resonance Measurements in Metals

D. S. RODBELL

General Electric Research and Development Center
Schenectady, New York

FERROMAGNETIC RESONANCE	816
ANTIFERROMAGNETIC RESONANCE	831
NUCLEAR RESONANCE	831
References	837

1. Introduction. This chapter is concerned with the ways in which magnetic resonance may be used as a tool to study some properties of metals. The treatment given here is therefore restricted to metallic, magnetic systems. The properties discussed are, in some cases, of primarily magnetic significance while in other cases these properties have direct implications for the nonmagnetic properties of metals.

The features of a magnetic resonance and the experimental techniques in general use have been discussed by P. E. Seiden in Chapter III, Volume 1. Magnetic resonance may be conveniently divided into two subgroups; one is concerned with electronic magnetic moments and the other with nuclear magnetic moments, both being connected to spin angular momentum. Within the area of electronic spin resonance are two subareas: one is of disordered (or isolated) spin systems and the other is of ordered systems (e.g., ferromagnetic, ferrimagnetic, antiferromagnetic). While the former is of extreme importance in understanding microscopic properties, it is considered beyond the aim of this chapter to discuss it.

The frequency of resonance for either electronic or nuclear spins is sufficiently high that the penetration depth into metallic samples is a nontrivial problem to consider and has some important consequences both in experimental technique and in the analysis of observations made on metallic systems.

Ferromagnetic Resonance

2. The Condition for Resonance. A lossless system, in stable equilibrium, will oscillate when disturbed from that equilibrium. The frequency of free oscillation depends upon the restoring forces and inertial properties of the system. If a disturbing force of constant amplitude but variable frequency is applied to such an oscillatory system, the response (amplitude of excursion versus frequency) will exhibit a characteristic "resonant" peak when the applied frequency corresponds to the frequency of free oscillation. The presence of damping (energy loss due to absorption and dissipation) modifies the response by decreasing the amplitude of the peak and increasing the relative amplitude in the wings of the response curve. Damping will also effect the resonant frequency to a small degree. The "width" of the resonance line is often used to describe its sharpness, and for a single oscillator it is related to the losses that prevail.

The net magnetization per unit volume, **M**, in a ferromagnetic material is an oscillatory system. In recognizing the magnetization to be fundamentally connected with angular momentum, we may classically describe its motion as we would any gyroscope. This description assumes that the magnitude of **M** is a constant. Although this assumption is adequate for an initial description, it overlooks certain important physical situations referred to below. The equation of motion in the absence of damping is simply a statement that the time rate of change of angular momentum is equal to the torque:

$$d\mathbf{J}/dt = \mathbf{M} \times \mathbf{H}_{\text{eff}}, \tag{2.1}$$

where **J** is the angular momentum per unit volume and is related to **M** through the ratio, $\gamma = M/J$.* We rewrite Eq. (2.1) as

$$d\mathbf{M}/dt = \gamma \mathbf{M} \times \mathbf{H}_{\text{eff}}. \tag{2.2}$$

The solution of Eq. (2.2) for the natural frequency of **M** about $\mathbf{H}_{\text{eff}}$ estab-

---

* The quantity γ is sometimes called the magnetomechanical ratio.

lishes the "resonance condition,"

$$\omega_{\text{res}} = \gamma H_{\text{eff}}. \tag{2.3}$$

Specifically, a small oscillatory disturbing magnetic field transverse to the equilibrium orientation of **M** will result in a large precession of the magnetization about its equilibrium direction when the disturbing angular frequency satisfies Eq. (2.3). Landau and Lifshitz [1] considered this phenomenon theoretically in 1935. Several authors [2–6] have demonstrated that Eq. (2.3) may be satisfactorily derived from a quantum mechanical treatment. That this is reasonable is seen from the point of view of the correspondence principle, since in ferromagnetic solids, quantum numbers on the order of 10^{22} are involved. The original papers referenced contain the detailed assumptions of the quantum mechanical treatment; a particularly important one is that only a very small orbital contribution is mixed in with the dominant spin angular momentum.

3. The Significance of γ. The ratio γ is an important physical parameter and may be extracted from experimental observations of the resonance. To see its significance, we must note first that the net magnetization of a ferromagnetic sample is the sum of the contributions made from each atom in the sample. The moment per atom, m_a, may be written for a solid in terms of the spin-only angular momentum, S_a, as

$$m_a = g(e\hbar/2mc)S_a = \gamma\hbar S_a, \tag{3.1}$$

where quantity g is the spectroscopic splitting factor, and the other symbols have their standard meaning. For an electron with "spin-only" g has the value 2.002; deviations from this value are used to describe the orbital contribution to the angular momentum. It is worth remarking that in some of the ferrimagnetic insulators all magnetic atoms are not equivalent, and the apparent "g" value deduced from experiment must be interpreted with care, as described originally by Wangsness [7]. The usual ferromagnetic resonance experiment excites what is called the uniform precession in which the net magnetization retains its magnitude, M, a constant—each m_a is parallel with all its neighbors— but the component of the spin angular momentum along a selected axis may change. In the quantum treatment referred to earlier, it is shown that the harmonic oscillator energy levels of the ferromagnet are separated by $\Delta E = \gamma\hbar H_{\text{eff}}$, which allows transitions to occur when the frequency of incident radiation satisfies the condition $\hbar\omega_{\text{res}} = \Delta E = \gamma\hbar H_{\text{eff}}$. This last statement states again the resonance condition Eq. (2.3). Figure 1 illustrates the energy levels.

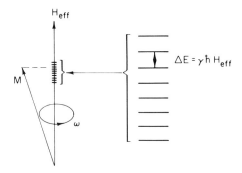

FIG. 1. A schematic representation of the precession described by the magnetization M about its equilibrium direction, H_{eff}. The energy levels along H_{eff} are the proper energies given by the quantum mechanics and are separated by one quantum of angular momentum. Because M itself originates from approximately 10^{22} quanta of angular momentum, the spacings on the scale shown along H_{eff} should actually be much closer together. In fact, on the scale of this picture, the energy levels should be essentially continuous as in the classical picture of the resonance.

It has been pointed out [2, 5] that the spectroscopic splitting factor g and the magnetomechanical factor, g' (g' occurs in magnetization by rotation experiments) should be related, the precise connection between them being

$$g^{-1} + (g')^{-1} = 1 \qquad (3.2)$$

as given by Smit and Wijn [8]. A summary of some of the experimental observations on these factors through 1961 is contained in a recent review article [9].

4. The Effective Field, H_{eff}. The effective field, H_{eff}, that characterizes the "stiffness" of the system has several different physical origins, whose common feature is that they each give rise to a dependence of the system's free energy upon the orientation of the net magnetization. For instance, an applied dc magnetic field, magnetocrystalline anisotropy, and shape-dependent "demagnetization factors" all give rise to an orientation dependence of the free energy. The orientation-dependent free energy of the magnetic system may be described by $E = E(\theta, \phi)$, where the polar angle is θ and the azimuthal angle is ϕ (see Fig. 2). We note first that the magnetization is in equilibrium at a position defined by the conditions that $\partial E/\partial \theta = \partial E/\partial \phi = 0$. By considering small deviations of M from equilibrium and expanding the energy about the zero torque (equilibrium) position, we find that

$$H_{\text{eff}} = (M \sin \theta)^{-1} [\partial^2 E/\partial \theta^2 \, \partial^2 E/\partial \phi^2 - (\partial^2 E/\partial \theta \, \partial \phi)^2]^{1/2}. \qquad (4.1)$$

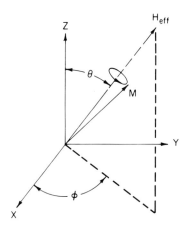

FIG. 2. The geometry used to describe the precessing magnetization M. The free energy of the system comprising M depends upon orientation. The equilibrium direction is H_{eff}, defined as an orientation of zero torque. The resonance phenomena refers to small excursions of M about H_{eff}. The resonant frequency is proportional to H_{eff}.

Equation (4.1) has been described by several authors [10, 11] and has utility in descriptions of the resonance [12, 13] in general, and particularly when the applied dc field is not oriented along a principal direction. The singularity at $\theta = 0$ is usually avoidable.

As an example, we may apply Eq. (4.1) to a simple physical situation merely to show that it reduces to a more recognizable form. Suppose we choose an ellipsoid-shaped ferromagnetic sample that is small in comparison to the wavelength of the oscillatory magnetic field that we employ to perform the resonance experiment. The coordinate system is chosen to coincide with the principal axes of the ellipsoidal sample whose shape we characterize by the demagnetizing factors N_x, N_y, N_z. The applied dc magnetic field is oriented along the x axis, and we shall assume that this applied field, H_x, is large enough to make the x axis the equilibrium orientation of the specimen's magnetization, M. Neglecting any other source of orientation dependent free energy, we may now write that

$$E(\theta, \phi) = \tfrac{1}{2}[N_x(M \sin \theta \cos \phi)^2 + N_y(M \sin \theta \sin \phi)^2 \\ + N_z(M \cos \theta)^2] - H_x M \sin \theta \cos \phi. \tag{4.2}$$

For the system so described, the equilibrium direction is $\theta = \pi/2$, $\phi = 0$, and after substituting Eq. (4.2) into Eq. (4.1) we find that

$$H_{\text{eff}} = \{[H_x + (N_z - N_x)M][H_x + (N_y - N_x)M]\}^{1/2}. \tag{4.3}$$

Expression (4.3) may be recognized as the Kittel condition [15], first derived to explain the original ferromagnetic resonance observations of Griffiths [14].

From the Kittel condition we may obtain the specific dependence of resonance field upon shape for fixed frequency (the usual experimental condition). A shape that is of constant importance in the resonance study of bulk metallic ferromagnets is the thin plate. The reason for this is that metals have a sufficiently high conductivity to limit microwave penetration to small depths,* and consequently all smooth, bulk, ferromagnetic metals are plates from the microwave resonance point of view. For a plate specimen in the x–y plane with the dc field in the x direction we find from Eq. (4.3) that

$$H_{\text{eff}} = [(H_x + 4\pi M)H_x]^{1/2}. \tag{4.4}$$

For a plate in the y–z plane and the field H_x applied the effective field becomes [from Eq. (4.3)]

$$H_{\text{eff}} = (H_x - 4\pi M). \tag{4.5}$$

When it is experimentally convenient to reorient the specimen and the geometries Eqs. of (4.4) and (4.5) can be satisfied, then the observation of the two separate resonances allows the determination of both γ and M on the same sample and at the same frequency. Orientation dependence is a widely used experimental technique for extracting fundamental parameters of magnetic specimens in addition to the simple example just given; it is particularly useful in measurement of magnetocrystalline anisotropy constants.

5. The presence of magnetocrystalline anisotropy gives rise to an orientation dependence of the applied dc magnetic field required for resonance, and this dependence may be used to determine the constants that are descriptive of the magnetocrystalline anisotropy. It is necessary to use single crystals in such determinations because in polycrystals the presence of many randomly oriented crystallites merely contributes to an "inhomogeneous" broadening of the absorption line, each crystallite resonating at a field that is dependent upon its orientation. The general shape of the absorption curve will depend upon the size of the equivalent "anisotropy field," the value of the magnetization of the specimen, and the distribution of crystallite orientations. Treatments [16, 17] of the problem give the detail-

---

* In iron, for example, at room temperature and 10 GHz, the "skin depth" is about 1000 Å.

ed behavior observed in certain cases involving polycrystalline specimens. When single crystals are examined, useful measures of the anisotropy may be obtained by resonance techniques. The physical quantity observed is usually, though not necessarily, identical with the quantity that is observed by standard mechanical torque experiments. One interesting and important difference is that a resonance experiment is sensitive to the curvature of the energy surface (i.e., the second derivative) as evidenced in Eq. (4.1), whereas a mechanical torque experiment measures a first derivative. This difference makes resonance measurements important in determining higher-order anisotropy constants. Although torque- and resonance- determined anisotropy constants are usually in accord, there are certain problems that arise in comparison of the two measurements. It has been pointed out [18] that, at least in one case, the difference at elevated temperatures between torque- and resonance-determined anisotropies can be connected with the fundamentally different averages that enter the two measurements.

The magnetocrystalline anisotropy in a cubic ferromagnet is usually described in terms of the direction cosines of the magnetization with respect to the cube axes; thus, the energy is represented as

$$E_k = K_1(\alpha_1^2\alpha_2^2 + \alpha_2^2\alpha_3^2 + \alpha_3^2\alpha_1^2) + K_2(\alpha_1^2\alpha_2^2\alpha_3^2) + \cdots$$

These are the terms of lowest order that are consistent with cubic symmetry. When the total orientation-dependent energy is treated as prescribed by Eq. (4.1), the following expressions are found for the principal planes indicated and the directions within those planes:

(100):

$$H \parallel \langle 100 \rangle: \left(\frac{\omega}{\gamma}\right)^2 = \left[H_{100} + \frac{2K_1}{M} + 4\pi M\right]\left[H_{100} + \frac{2K_1}{M}\right] \quad (5.1a)$$

$$H \parallel \langle 110 \rangle: \left(\frac{\omega}{\gamma}\right)^2 = \left[H_{110} + \frac{K_1}{M} + \frac{K_2}{2M} + 4\pi M\right]\left[H_{110} - \frac{2K_1}{M}\right] \quad (5.1b)$$

(111):

$$H \parallel \langle 110 \rangle: \left(\frac{\omega}{\gamma}\right)^2 = \left[H'_{110} - \frac{K_1}{M} - \frac{K_2}{6M} + 4\pi M\right]\left[H'_{110} + \frac{K_2}{3M}\right] \quad (5.1c)^*$$

$$H \parallel \langle 112 \rangle: \left(\frac{\omega}{\gamma}\right)^2 = \left[H_{112} + \frac{9.45}{64}\frac{K_2}{M} + 4\pi M\right]\left[H_{112} - \frac{K_2}{3M}\right] \quad (5.1d)$$

* A recent treatment which includes one higher order of anisotropy and corrects an error in Eq. (5.1c) has recently appeared [18a]. Equation (5.1d) is also subject to reexamination but Eqs. (5.1a), (5.1b), and (5.1e–g) are unchanged.

(110):

$$H \parallel \langle 110 \rangle : \left(\frac{\omega}{\gamma}\right)^2 = \left[H''_{110} - \frac{2K_1}{M} + 4\pi M\right]\left[H''_{110} + \frac{K_1}{M} + \frac{K_2}{2M}\right] \quad (5.1\text{e})$$

$$H \parallel \langle 100 \rangle : \left(\frac{\omega}{\gamma}\right)^2 = \left[H_{100} + \frac{2K_1}{M} + 4\pi M\right]\left[H_{100} + \frac{2K_1}{M}\right] \quad (5.1\text{f})$$

$$H \parallel \langle 111 \rangle : \left(\frac{\omega}{\gamma}\right)^2 = \left[H_{111} - \frac{4}{3}\frac{K_1}{M} - \frac{4}{9}\frac{K_2}{M} + 4\pi M\right]$$
$$\times \left[H_{111} - \frac{4}{3}\frac{K_1}{M} - \frac{4}{9}\frac{K_2}{M}\right] \quad (5.1\text{g})$$

The experimental information required to solve these expressions for K_1/M, K_2/M, and g (contained in γ) are the frequency, the magnetization, and the appropriate shift caused by exchange effects. The last-mentioned term, the exchange shift, is added to the externally applied dc field in the expressions (5.1). While this "exchange field" is not a dominant term, it is necessary to know this shift in order to obtain precise deductions of the g factor. The importance of the exchange shift is dependent upon the resonance linewidth and is often negligible for very broad linewidths.

6. It is tacitly assumed that the samples are truly thin compared to their extent. If this is not so, then the term in $4\pi M$ becomes $4\pi(1-n)M$ and in the second bracket an additional $4\pi nM$ must be added, where n is the normalized transverse demagnetizing factor of the rotationally symmetric sample. The reason for this is that while the rf demagnetization behavior is that of a very thin plate, the average internal dc field depends upon the sample shape. The specific examples given in Eqs. (5.1) are along interesting directions of the energy surface and some correspond to situations with H parallel to M. In general M will deviate from parallelism with H by an amount ϵ and will modify the character of the orientation dependence of the field for resonance. It is possible to calculate the features of this departure as Artman [19] has shown and it is necessary to do so when one departs from a symmetry axis in samples where K/M is not very small compared to H. In a disk sample containing the unique axis of a uniaxial anisotropy energy, the resonance condition becomes

$$(\omega/\gamma)^2 = (H\cos\epsilon - 4\pi nM + M^{-1}\partial^2 E/\partial\theta^2)$$
$$\times (H\cos\epsilon + 4\pi(1-n)M + H\sin\epsilon\cot\theta) \quad (6.1)$$

where $\sin\epsilon = (MH)^{-1}(\partial E/\partial\theta)A$; E_A is the anisotropy energy (e.g., for a uniaxial case $E_A = K_1^u\sin^2\theta + K_2^u\sin^4\theta + K_3^u\sin^6\theta + K_3'^u\sin^6\theta\sin^6\phi$);

and θ and ϵ measure the angles of the applied field direction with the unique axis and the magnetization, respectively. The term ϕ is the azimuthal angle of the magnetization with respect to an easy direction in the basal plane. Along the unique axis the condition becomes

$$(\omega/\gamma)^2 = (H - 4\pi nM + 2K_1/M)(H + 4\pi(1-n)M + 2K_1/M). \qquad (6.2)$$

As a passing remark it is worth noting that in expressions for the resonance condition the magnetocrystalline anisotropy terms sometimes enter in precisely the same way in each term. This occurs along true energy extrema and these are the orientations for which the concept of an anisotropy field is appropriate.

7. Damping and Linewidth. Thus far we have described an oscillatory system that has no damping, and we recognize immediately that this cannot be physically correct. A simple way to demonstrate the necessity for damping is to point out that because of the gyroscopic nature of the magnetization, an applied field in the absence of damping cannot accomplish an orientation of the magnetization, but would result in an unending precession of the magnetization about the field direction. Landau and Lifshitz [1] noted this dilemma and introduced a damping term to establish equilibrium. The form of the damping is such as to supply a torque that drives the magnetization toward the direction of the effective field and may be expressed as

$$-(\lambda/M^2)\mathbf{M} \times (\mathbf{M} \times \mathbf{H}_{\text{eff}}). \qquad (7.1)$$

The complete equation of motion including damping will then be Eq. (2.2) with Eq. (7.1) added to its right-hand side. The "relaxation frequency" λ is a phenomenological description of the damping that exists. A useful treatment of a formulation of this kind is given by Yager et al. [20] and shows the specific dependence of linewidth and resonance field shift on the value of the damping parameter. The motion of domain walls in ferromagnetic materials is subject to damping that must be accounted for by mechanisms that are related to the same damping just considered. In connection with problems in the behavior of remagnetization of magnetic bodies, Gilbert [21] has suggested a modification of Eq. (7.1) to be

$$-(\alpha/M)(\mathbf{M} \times \partial \mathbf{M}/\partial t) \qquad (7.2)$$

The latter form will give, for a finite damping parameter, α, a minimum in the time to reverse the magnetization, a feature that is discussed in detail by Kikuchi [22].

Another description of the dissipative process of the resonance is that introduced by Bloembergen [23] who adapted the Bloch equations [24] of nuclear resonance for use in ferromagnetic resonance. This description uses two parameters or "relaxation times," as they are called; one characterizes the time it takes transverse magnetization to equilibrate, while the other refers to the longitudinal component of the magnetization. The objection to this formulation is that the magnetization is not conserved. In the limit of small deviations from equilibrium (as we assume here), the two damping forms described above are equivalent except for the predictions of frequency dependence. Available evidence, cited below, supports the Landau-Lifshitz form.

8. There are at present only three clearly established physical sources of linewidth in ferromagnetic resonance. One source of linewidth arises in a class of insulating ferrimagnetic materials with nonequivalent magnetic ions. An example of this class is nickel iron ferrite [25] (a magnetic oxide with spinel structure), and the physical mechanism for relaxation is the rearrangement of valence electrons between Fe^{2+} and Fe^{3+} ions as a consequence of the reorientation of the magnetization during its motion. By reducing the amount of divalent iron present the loss is shown to decrease.

The second source of linewidth thus far established was pointed out by Clogston *et al.* [26]. These authors indicate that the presence of imperfections in an otherwise regular lattice can "scatter" energy from the usual "uniform precession" to other oscillatory modes of the magnetic system. The other oscillatory modes are excitations in which the individual magnetic moments that comprise the net magnetization no longer maintain parallelism with each other. These "spin waves," as they are called, abscond with the energy of the uniform precession and then dissipate that energy by interaction with lattice vibrations. Experimental evidence for the theoretical treatment mentioned is to be found in some of the ferrimagnetic garnet materials whose crystal structure possesses only equivalent sites for magnetic ions and removes the "disorder" that may be present in materials like the ferrites. More particular evidence is that concerning surface imperfections reported initially by LeCraw *et al.* [27] in the yttrium–iron garnet. There also appears to be implicit evidence in the case of iron "whisker" crystals and in whiskers and platelets of nickel metal [28].

The third established source of linewidth occurs in good conductors and arises from the limited penetration of the high-frequency fields used to excite the resonance absorption. This source of linewidth is observed in metals and is related to the exchange shift mentioned earlier. It has been established that this source of linewidth coupled with a Landau-Lifshitz damping term is both necessary and sufficient to account for observations in highly perfect single crystal samples [28]. It remains at present a problem to understand the physical basis of this damping term.

9. *Example I: A Study of Cobalt Precipitates in Copper.* Aside from intrinsically magnetic properties that ferromagnetic resonance may be used to measure (such as magnetocrystalline anisotropy or "g" factor, magnetization, and exchange constants [29]), the sensitivity of resonance behavior to the shape of very small particles makes an interesting metallurgical study. The metallic particles of interest are so small that the microwave field completely penetrates them; in essence, the particle's dimensions are very small compared to their characteristic "skin depth." This means that at the frequencies we employ the particles must be 1000 Å or less in diameter. We have seen that for complete penetration of an individual ferromagnetic specimen the resonance condition is sensitive to the shape of the particle [see Eq. (4.3)]. This fact has enabled studies to be made of the shapes of precipitated magnetic particles [30]. A schematic illustration of the experimental situation in such a study is shown in Fig. 3. The material that has received much attention is an alloy of copper and cobalt in the low cobalt ($\sim$2%) composition range. By proper heat treatment the cobalt can be made to precipitate, and the resultant collection of ferromagnetic cobalt precipitate particles in their nonferromagnetic copper matrix can be examined by many techniques, one of which is ferromagnetic resonance. The interesting thing about resonance is that its sensitivity to particle shape is sufficient to enable the particle shapes to be examined at various stages in their "coarsening" or growth during heat treatment subsequent to the initial precipitation. The resonance experiments showed that the precipitated cobalt particles in a polycrystalline copper matrix are initially equiaxed, i.e., probably spherical, and during further growth the particle shape becomes platelike, not in the sense of "pancakes" but with mild deviations from sphericity. Figure 4 briefly summarizes the polycrystalline results. When single crystals of copper containing precipitated cobalt particles were examined [31, 32], the polycrystalline results were confirmed and, in addition, the crystal symmetry of the precipitate particles was determined. The single-crystal measurements may be interpreted to determine the magnetocrystalline anisotropy—both as to magnitude and symmetry. The cobalt particles in this study were found to be face-centered cubic and have a coincident lattice relationship with the copper matrix. The fact that the very small precipitate particles may be obtained in a matrix that keeps them separated,

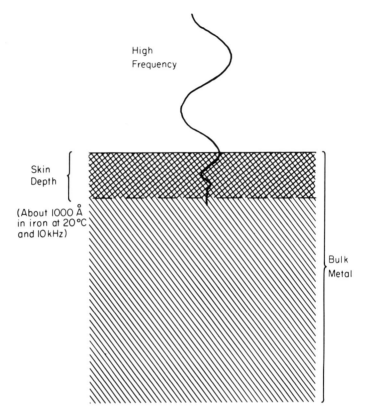

Fig. 3. A pictorial representation of the penetration of a high-frequency electromagnetic wave into a bulk metal sample. The skin depth is defined as the depth at which the incident field intensity is diminished to 37% of its value at the surface. This depth is about 5000 Å in copper at room temperature and at 10 GHz. The representation above includes the presence of a precipitated phase which, if magnetic, may be studied by resonance techniques.

as well as oriented crystallographically, allows certain interesting experiments to be performed. For example, the dependence of the magnetocrystalline anisotropy of the cobalt particles on particle size was examined by resonance techniques, as well as by mechanical torque measurements.

There have also been many "nonresonance" studies of this very convenient system of small particles; one that seems particularly appropriate to discuss here was a study of precipitation hardening [33]. In the course of that experiment, Livingston and Becker found that when a sample of copper containing the precipitated cobalt particles was plastically deformed, the cobalt particles, on the average, deformed to the same extent as the bulk

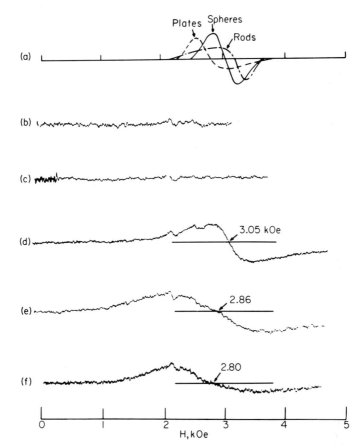

FIG. 4. The derivative of the 9 GHz power absorbed as a function of the applied dc magnetic field for the various conditions listed: (a) expected from a random assembly of plates ($a/c = 1.1$), of spheres, and of rods ($c/a = 1.1$); (b) empty cavity; (c) copper–cobalt specimen as quenched (cobalt in solution); (d) aged 30 min at 600°C (average particle radius, 25 Å); (e) aged 15 hr at 600°C (estimated average particle radius, 65 Å); (f) aged 20 min at 800°C (estimated average particle radius, 300 Å).

deformation experienced by the whole sample. It was not possible from their measuring techniques to obtain more than the average behavior. It is interesting to inquire whether some particles deform a great deal and others very little or do the particles deform homogeneously—all with the average deformation. Ferromagnetic resonance was used* to examine this question

---

* Author's results, previously reported at the Am. Soc. Metals Seminar on Resonance and Relaxation in Metals, Chicago, 1959.

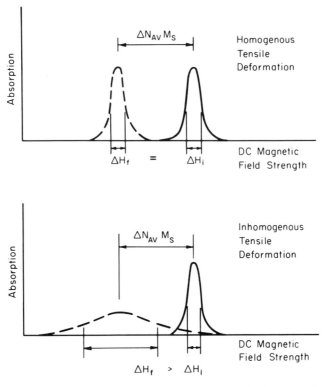

FIG. 5. A pictorial representation of two possible changes in the magnetic resonance absorption line of a collection of initially spherical particles subjected to a tensile deformation applied in the same direction as that of the external dc magnetic field used for subsequent resonance studies. The upper sketch shows the result of a homogeneous deformation; every particle is elongated by the same fraction that the sample as a whole has been deformed. The resonance line is then shifted but not broadened. The lower sketch corresponds to an inhomogeneous deformation in which some particles are deformed greatly and others not at all with the result that the line is shifted and, in addition, broadened. A compressive deformation would shift the lines into higher fields for otherwise similar conditions.

and provided a definite, though not unexpected, answer. The experiment is as follows. A sample containing initially spherical particles is examined in ferromagnetic resonance, i.e., looking at the particles within the skin depth. The position and width of the resonance absorption are noted, the sample is then deformed, and the resonance linewidth and position are reexamined. We expect from Livingston and Becker [33] that the average shift in position of the line will correspond to the average deformation, but that there are two possibilities for the linewidth. On the one hand, if the

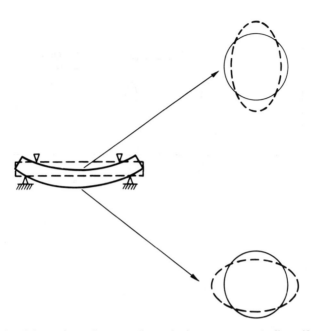

Fig. 6. The deformation scheme used to obtain a *macroscopically* uniform tensile (bottom) and compressive (top) deformation of a 98% copper–2% cobalt crystal containing a precipitated cobalt-rich phase in the form of initially spherical particles of approximately 100 Å diameter. The magnetic resonance experiment performed on top and bottom faces (individually) examines whether the deformation is *microscopically* uniform or not.

deformation is homogeneous, every particle will experience the same change of shape and the total absorption line, which is the sum of contributions from each magnetic particle, will retain its shape (and width) and simply shift its average position to a field determined by Eq. (4.3). On the other hand, if the deformation is inhomogeneous and some particles are deformed greatly while others are not deformed at all, then the resonance line will be smeared out because the shape demagnetizing factors are changed for the deformed particles while the undeformed particles are not shifted at all. Figure 5 shows schematically the two possibilities in the case of a tensile deformation. Figure 6 shows a schematic representation of a sample and the deformation scheme that yields both tensile and compressive deformation. The sample is a single crystal of copper containing 100 Å diameter precipitated cobalt particles. The orientation of the crystal is intended to give "single slip," and is a $\langle 321 \rangle$ crystallographic direction.* The deforma-

---

* I am indebted to J. D. Livingston for the crystal used in these experiments and for informative discussions regarding plastic deformation.

tion scheme employed enables both tensile and compressive deformation to be obtained on the same sample; by proper "masking" we examine first the top face and then the bottom face. Figure 7 shows the results of some experiments. The results indicate that indeed the average deformation is adequate to account for the average shift, but the deformation is not at all homogeneous. This is as we might expect in a model where slip bands accomplish the deformation, i.e., the precipitate particles near a slip plane

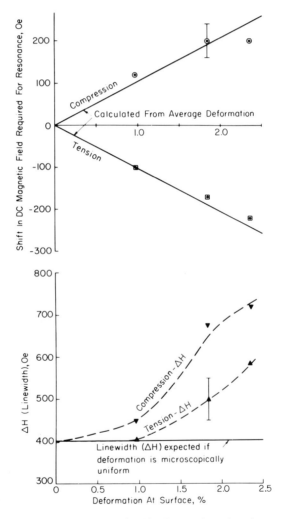

FIG. 7. Results of the experiment described in the test show that the average deformation correctly describes the shift of the resonance line position, but that the deformation is not uniform on a microscopic basis, as evidenced by the increase in resonance linewidth.

are heavily deformed while those particles remote from slip planes are essentially undeformed. Note that the initial deformation appears to be quite homogeneous (particularly the tensile surface) but that the inhomogeneity of deformation becomes more apparent with increasing deformation. This behavior is qualitatively that which might be expected from the theory of slip band formation and shows how magnetic resonance has been used to augment other methods of study.

Antiferromagnetic Resonance

10. To the present time very little new information about metals has been obtained by antiferromagnetic resonance experiments. This situation may change in the near future with the availability of good single crystals of metallic antiferromagnets and especially the rare earth metals, many of which have interesting magnetic structures at low temperatures.

The resonance condition (more properly conditions) for an antiferromagnet depend upon the nature of the antiferromagnet. For the simplest (uniaxial, two-sublattice) case there are two modes that are degenerate in the absence of an external magnetic field but may be separated by the application of a dc field. When such a resonance is observed, one may deduce from it a characteristic that is (approximately) a product of the exchange and anisotropy fields operative in the system. An instructive treatment (including the more general case of uncompensated antiferromagnets, i.e., ferrimagnets) is given by Geschwind and Walker [34]. More complicated systems exist also, e.g., multisublattice and spiral structures of spins, that may be encountered. Some discussion of these are to be found in the current literature [35, 36].

Nuclear Resonance

11. The angular momentum connected with atomic nuclei and the concomitant magnetic moments there associated may be made to exhibit resonance absorption just as do electronic magnetic moments. Since the nuclear magneton is much smaller than the electronic one (1/1826), the resonance frequencies are usually much lower for nuclei than for electrons. This, of course, depends on the effective field acting upon the magnetic moments. In ferromagnetic solids the effective field at the nucleus is often huge compared to the fields normally considered (e.g., local dipole or Lorentz fields), being typically several hundred kilogauss and making resonances which occur in the range of tens to hundreds of megahertzes. The detailed reason for the huge size of this field is beyond the scope of the present chapter but has to do primarily with the finite probability of an s electron being right at the nuclear position (e.g., the Fermi "contact" term; for discussion see Freeman and Watson [37]). Thus, one may examine the nuclear resonance spectrum of ferromagnets and measure the nuclear hyperfine field thereby. This kind of measurement is greatly facilitated by the enhancement of applied rf field strengths that results from a domain wall mechanism [38, 39]. This useful enhancement requires the presence of domain walls; if one wishes to use that feature, it means that no large externally applied dc magnetic fields are used.

12. *Example II: An NMR Study of Cobalt and Cobalt Alloys.* It is also a useful fact that for nuclear spin $S > \frac{1}{2}$, the resonance is sensitive to the symmetry of surrounding, i.e., the nature of the crystalline electric field and its departure from cubicity. This point may be illustrated by the following example of experimental results with cobalt metal. Following the discovery by Gossard and Portis [38] that nuclear resonance of fcc cobalt could be readily observed, Street et al. [40] examined powders of cobalt metal and found several resonance absorptions in addition to the one associated with the fcc phase.

The NMR traces obtained with a sample of cobalt* at 300 and 77°K are shown in Fig. 8a and b, respectively. Fig. 8c is an index to prominent parts of the spectrum. The main resonance corresponds to that reported by Portis and Gossard [38].

The following observations established that the lowest resonance (frequencies 1 and 2) and the highest resonance (frequencies 5 and 6) are associated with the face-centered cubic and hexagonal close-packed phases, respectively. Figure 9a was obtained with a different sample[†] of cobalt powder than that used for Fig. 8. X-ray analysis showed that the ratio of hcp/fcc for the sample of Figure 9 was 0.19, whereas for the powder sample of Fig. 8, the ratio was 1.38. The spectrum of MCB powder, taken with identical apparatus settings, is included for comparison in Fig. 9b. The reduction of the hcp/fcc ratio is accompanied by a decrease in the intensity of the highest-frequency resonance.

To confirm the identification of the NMR resonances, the relative proportion of hcp phase in the MCB cobalt was increased by cold-working the powder [4]. The MCB powder was compressed at 80,000 psi into briquettes

---

* Obtained from Matheson, Coleman, and Bell (−325 mesh).
† Fisher Scientific Company.

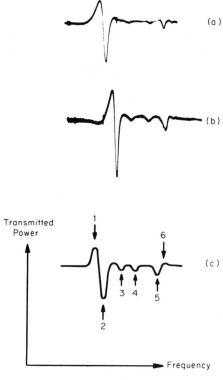

FIG. 8. The nuclear resonance spectrum of cobalt in metallic powders (Matheison, Coleman, and Bell, −325 mesh). The figure displays transmitted power versus frequency. The peak-to-peak variation of the largest signal is 5% of the average transmitted power. (a) The spectrum at room temperature (300°K); (b) the spectrum at liquid nitrogen temperature (77°K); (c) the index to the tabulated frequencies. We note without explanation that the shift in resonance frequencies between 300 and 77°K is linear with frequency.

Position	Frequency (MHz)	
	300°K	77°K
1	211.8_4	215.4_0
2	212.9_7	216.8_3
3	215.4_3	219.8_7
4	218.1_7	223.3_0
5	220.8_0	226.8_7
6	221.9_9	227.9_3

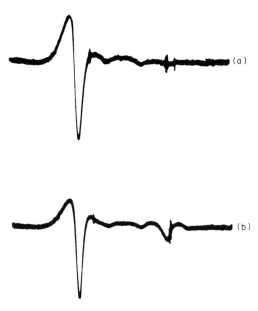

FIG. 9. (a) Room temperature NMR spectrum of a fine grained powder sample of metallic cobalt (Fisher Scientific Company) which has an hcp/fcc ratio of 0.19 as determined by x-ray analysis. (b) Room-temperature NMR spectrum of the MCB sample in which the hcp/fcc ratio is 1.38. The same frequency index as in Fig. 8 caption applies.

which were then powered and annealed for 2 hr at 300°C; the spectrum in Fig. 10b was then obtained. The decreased signal intensity required a higher gain setting than the previous data. A trace of the untreated MCB powder (Fig. 10a) is included for comparison. The intensity of the highest-frequency resonance has increased relative to the lowest frequency resonance. The hcp/fcc ratio for the cold-worked cobalt has increased to 4.9. One may therefore conclude from the data presented that the lower- and higher-frequency resonances are associated with the fcc and hcp phases of cobalt, respectively.

In considering the relative intensities of the resonances of the fcc and hcp phases and allowing for the x-ray-determined volume fractions of these constituents, the hcp resonance is weaker by almost an order of magnitude than that of the fcc phase. The relative weakness of the hcp resonance derives from the interaction of the nuclear quadrupole moment of the ^{59}Co nucleus with the noncubic crystalline field of the hcp phase. In a crystalline electric field of cubic symmetry the ^{59}Co nucleus ($I = \frac{7}{2}$), subject to a strong hyperfine field, H_{eff}, has $(2I + 1)$ distinct energy levels which are equally spaced by $\varDelta = g_N \beta_N H_{\text{eff}}$. However, in a noncubic crystalline field,

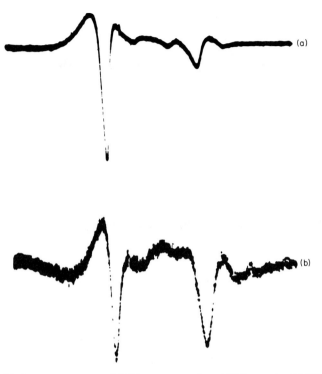

FIG. 10. (a) Room temperature NMR spectrum of an MCB sample. (b) NMR spectrum of the same sample after a deformation and heat treatment that converted sufficient fcc to hcp to give an x-ray-determined hcp/fcc ratio of 4.9, where (a) is the upper trace of the figure and (b) is the lower. The same frequency index as in Fig. 8 caption applies.

there is an electric field gradient at the nucleus that interacts with the nuclear quadrupole moment. The quadrupolar interaction shifts the energy levels and only the transition $-\tfrac{1}{2} \to +\tfrac{1}{2}$ is till given by Δ.* The other transitions are shifted to higher or lower frequencies dependent upon the relative orientation of the crystalline field gradient and the effective magnetic field. The order of magnitude of these shifts may be as much as 20 MHz, which value is observed for the Co nucleus in CoF_2 [43]. In addition to the shift of the energy levels caused by the noncubic crystalline environment, the relative orientations of the local effective hyperfine field are randomized by the domain walls that "drive" the resonance. Within the domain walls the effective magnetic field scans all directions between the positive and negative

* This statement is only correct to first-order perturbations. In the second-order treatment there is an additional shift, but more important. There is also a broadening of the levels. For an analogous effect in electron spin resonance see Bleaney [42].

c axes and this randomization broadens all but the $-1/2 \to +1/2$ resonance line because of the anisotropy of the energy levels. Consideration of the transition probabilities shows that the relative intensity of absorption from just the $-1/2 \to +1/2$ transition is 4/21. Thus, one expects that the hcp cobalt resonance intensity should be 20% of the intensity associated with an equal volume of fcc cobalt. This is just the observed intensity ratio.

From the relative positions in frequency of the resonances, we note that H_{eff} for the hcp phase is 5% higher than that for the fcc phase. The various contributions to H_{eff} have been discussed [38], and it is clear that small changes in the contributing terms could account for the magnitude of the difference observed. A proper calculation of this point seems lacking to date.

It was suggested that the origin of the smaller resonances at frequencies 3 and 4 is associated with stacking faults. Both growth and deformation faults are observed in hexagonal cobalt [41]. Growth faults are sequences of three planes, and deformation faults are sequences of four planes of cubic close packing within the hexagonal phase. If the identification of the smaller resonances with faulted material is correct, the resonance at frequency 3, nearer the fcc absorption, probably is associated with deformation faults, whereas the resonance at frequency 4, nearer the hexagonal absorption line, is associated with growth faults. Investigation of these weaker resonances as well as resonances that appear upon alloying have been made in some detail [39, 44], and show the sensitivity of nuclear magnetic resonance to local properties in metals. It is important to add here that although these investigations of cobalt and ferromagnetic alloys thereof (as well as other ferromagnetic metals) use the enhancement feature of ferromagnetic domain walls to aid in observation of the resonance, that is not necessary. Not only can unenhanced resonances be seen but extensive studies of non ferromagnetic metals have been made and show the importance of nuclear magnetic resonance in connection with alloy studies. A recent example is the study by West [45] of a series of intermetallic compounds in which the intensity of absorption of a $S > 1/2$ nucleus is used to measure the change in local order about that atom, i.e., through the noncubicity of the local environment. The work of Jackson *et al.* [46] shows the local symmetry (hyperfine field distribution) in a series of ferromagnetic alloys. The range in properties that may be studied is illustrated by considering that one may examine properties ranging from the electronic structure of metals (from Knight shift measurements [47] which give electronic density determinations) to the location of magnetic ordering temperatures of alloys (by noticing when the resonance signal disappears with temperature [48]). Studies of alloys [49] and impurities [50] by nuclear resonance give helpful information about the detailed local environment of an atom in a metal. Relaxation time studies yield information about spin coupling mechanisms (see, for instance, Portis and Lindquist [39]). In addition to these studies many nonmetals lend themselves to nuclear resonance experimentation; a particularly valuable set of these are described in Jaccarino [51].

13. Conclusions. Magnetic resonance is no longer a laboratory curiosity to be examined for the understanding of the phenomena of magnetic resonance. It is an effective (and in some cases unique) method to study some material properties and should be viewed, in association with more commonly used methods of studying metals, as a proper tool of the metallurgists trade. This chapter has outlined some of the experiments that can be readily done, described some in great detail, and shown the kinds of information that can be obtained from the analysis of such experiments. As equipment for resonance experiments is as readily available as tensile testing machinery, there is no reason why any serious metallurgical laboratory should avoid resonance techniques when their use is indicated.

REFERENCES

1. L. Landau and E. Lifshitz, *Physik. Z. Sowjetunion* **8**, 153 (1935).
2. C. Kittel, *Phys. Rev.* **76**, 743 (1949).
3. D. Polder, *Phil. Mag.* **40**, 99 (1949).
4. J. M. Richardson, *Phys. Rev.* **75**, 1630 (1949).
5. J. H. Van Vleck, *Phys. Rev.* **78**, 266 (1950).
6. J. M. Luttinger and C. Kittel, *Helv. Phys. Acta* **21**, 480 (1948).
7. R. K. Wangsness, *Phys. Rev.* **91**, 1085 (1953).
8. J. Smit and H. P. J. Wijn, "Ferrites." Wiley, New York, 1959.
9. A. J. P. Meyer and G. Asch, *J. Appl. Phys.* **32**, 330S (1961).
10. H. Suhl, *Phys. Rev.* **97**, 555 (1955).
11. J. Smit and H. G. Beljers, *Philips Res. Rept.* **10**, 113 (1955).
12. P. E. Tannenwald and M. H. Seavey, Jr., *Phys. Rev.* **105**, 2 (1957).
13. J. O. Artman, *Phys. Rev.* **105**, 62 (1957).
14. J. H. E. Griffiths, *Nature* **158**, 670 (1946).
15. C. Kittel, *Phys. Rev.* **71**, 270 (1947).
16. E. Schlömann, *J. Phys. Radium* **20**, 327 (1959).
17. K. J. Standley and K. W. H. Stevens, *Proc. Phys. Soc. (London)* **B69**, 993 (1956).
18. J. D. Livingston and C. P. Bean, *J. Appl. Phys.* **30**, 318S (1959).
19. J. O. Artman, *Proc. IRE* **44**, 1284 (1956).
20. W. A. Yager, J. K. Galt, F. R. Merritt, and E. A. Wood, *Phys. Rev.* **80**, 744 (1950).
21. T. L. Gilbert and J. M. Kelly, *Proc. AIEE Conf. Magnetism Magnetic Materials, Pittsburgh, June 1955,* p. 253. New York (1955).
22. R. Kikuchi, *J. Appl. Phys.* **27**, 1352 (1956).
23. N. Bloembergen, *Phys. Rev.* **78**, 572 (1950).

24. F. Bloch, *Phys. Rev.* **70**, 460 (1946).
25. W. A. Yager, J. K. Galt, and F. R. Merritt, *Phys. Rev.* **99**, 1203 (1955).
26. A. M. Clogston, H. Suhl, L. R. Walker, and P. W. Anderson, *J. Phys. Chem. Solids* **1**, 129 (1956).
27. R. C. LeCraw, E. G. Spencer, and C. S. Porter, *Phys. Rev.* **110**, 1311 (1958).
28. D. S. Rodbell, *Physics* **1**, 279 (1965).
29. T. G. Phillips, *Phys. Letters* **17**, 11 (1965).
30. C. P. Bean, J. D. Livingston, and D. S. Rodbell, *Acta Met.* **5**, 682 (1957).
31. D. S. Rodbell, *J. Appl. Phys.* **29**, 311 (1958).
32. C. P. Bean, J. D. Livingston, and D. S. Rodbell, *J. Phys. Radium* **20**, 298 (1959).
33. J. D. Livingston and J. J. Becker, *Trans. AIME* **212**, 316 (1958).
34. S. Geschwind and L. R. Walker, *J. Appl. Phys.* **30**, 163 (1959).
35. B. Cooper and R. Elliott, *Phys. Rev.* **131**, 1043 (1963).
36. F. Rossol, B. Cooper, and R. V. Jones, *J. Appl. Phys.* [2]**36**, 1209 (1965); J. Liesegang and D. Bagguley, *Phys. Letters* **17**, 96 (1965).
37. A. J. Freeman and R. E. Watson, Hyperfine Interactions in Magnetic Materials, *in* "Magnetism" (G. T. Rado and H. Suhl, eds.), Vol. II A, Chapter 4. Academic Press, New York, 1965.
38. A. C. Gossard and A. M. Portis, *Phys. Rev. Letters* **3**, 164 (1959); *J. Appl. Phys. Suppl.* **31**, 205 (1960).
39. A. M. Portis and R. H. Lindquist, Nuclear Resonance in Ferromagnetic Materials, *in* "Magnetism" (G. T. Rado and H. Suhl, eds.), Vol. II A, Chapter 6. Academic Press, New York, 1965.
40. R. Street, D. S. Rodbell, and W. L. Roth, *Phys. Rev.* **121**, 84 (1961).
41. C. Houska, V. Averback, and M. Cohen, *Acta Met.* **8**, 82 (1960).
42. B. Bleaney, *Phil. Mag.* **42**, 441 (1951).
43. V. Jaccarino, *Phys. Rev. Letters* **2**, 163 (1959).
44. Y. Koi, T. Hihara, and T. Kushida, *J. Phys. Soc. Japan* **16**, 574 (1961); R. C. LaForce, S. F. Ravitz, and G. F. Day, *J. Phys. Soc. Japan Suppl.* **17**, B1, 99 (1961).
45. G. W. West, *Phil. Mag.* **9**, 979 (1964).
46. R. F. Jackson, R. G. Scurlock, D. B. Utton, and T. H. Wilmshurst, *Proc. Intern. Conf. Magnetism, Nottingham, 1964*, p. 384.
47. C. Froideueaux, F. Gautier, and I. Weisman, *Proc. Intern. Conf. Magnetism, Nottingham, 1964*, p. 390.
48. R. G. Barnes and T. P. Graham, *J. Appl. Phys.* **36**, 938 (1965).
49. S. Kobayashi and J. Itoh, *J. Phys. Soc. Japan* **20**, 1741 (1965).
50. M. Kontani, K. Asayama, and J. Itoh, *J. Phys. Soc. Japan* **20**, 1737 (1965).
51. V. Jaccarino, Nuclear Resonance in Antiferromagnets, *in* "Magnetism" (G. T. Rado and H. Suhl, eds.), Vol. II A, Chapter 5. Academic Press, New York, 1965.

Author Index

Numbers in parentheses are reference numbers and indicate that an author's work is referred to, although his name is not cited in the text. Numbers in italics show the page on which the complete reference is listed.

A

Aaron, H. B., 809 (108A), *814*
Abov, J. G., 545 (28), *574*
Ahern, S. A., 529 (4), *573*
Ainslie, N. G., 738 (67), *747*
Akulov, N., 726 (7), *746*
Albert, P. A., 725 (5, 6), 741 (91), 744 (91), *746, 748*
Allen, M. P., 669 (62), *686*
Anderson, P. W., 528 (1), *573*, 824 (26), *838*
Andrew, E. R., 713 (51), *721*
Aoyagi, K., 583 (21), 591 (25A, 25B), 592 (25A, 25B), 593 (25B), 595 (37), *617, 618,* 795 (79), 803 (97, 98), 807 (79, 98), 808 (79), *813*
Arrott, A., 563 (71, 72), 564 (71, 72), *575*
Artman, J. O., 819 (13), 822 (19), *837*
Asanuma, M., 703 (23), *720*
Asayama, K., 534 (16), *573*, 836 (50), *838*
Asch, G., 818 (9), *837*
Aspden, R. G., 741 (89), *748*
Assmus, F., 740 (87), 741 (90), 744 (106), *748*
Auwärter, M., 573 (99), *575*
Averbach, B. L., 555 (56), *574*
Averback, V., 836 (41), *838*

B

Bacon, G. E., 571 (96), *575*
Baer, G., 741 (93), *748*
Bagguley, D., 831 (36), *838*

Bailey, J. E., 696 (14), *720*
Balthesen, E., 797 (85), 798 (85), 809 (85, 109), *813, 814*
Bangert, L., 514 (6), 517 (6), 518 (6), *521*
Baran, W., 732 (24), *746*
Barnes, R. G., 836 (48), *838*
Barnier, Y., 651 (54), 671 (75), *686, 687*
Barrett, C. S., 724 (1), 728 (1, 13), 732 (1), *745, 746*
Basinski, Z. S., 626 (14), 627 (21), 635 (21), *685*
Bass, J., 785 (53), *812*
Batterman, B. W., 601 (49), *618*
Batz, W., 799 (86), *813*
Bean, C. P., 558 (62), *575*, 750 (5), *811*, 821 (18), 825 (30, 32), *837, 838*
Beck, K., 734 (36), *746*
Beck, P. A., 696 (13), *720*
Becker, J. J., 557 (58), *574*, 616 (67), *619*, 750 (4), *811*, 826 (33), 828 (33), *838*
Becker, R., 553 (46), *574*, 579 (8), *617*, 624 (5), 628 (5), 668 (5), 679 (5), *685*, 691 (4), 693 (6), 698 (4), 699 (17, 18), 704 (27, 28), *720, 721*, 774 (37), *812*
Beljers, H. G., 819 (11), *837*
Bennett, W. D., 563 (74), *575*
Berghout, C. W., 750 (5), *811*
Berkowitz, A. E., 750 (5), *811*
Berner, R., 625 (11), *685*
Beshers, D. N., 580 (116), *617*
Bhandary, V. S., 745 (115), *748*
Bijvoet, J., 767 (32), 770 (32), 772 (32), *812*

Bilger, H., 667 (61), 668 (61), 669 (61), 670 (61), 675 (61), *686*
Bindels, J., 767 (32), 770 (32), 772 (32), *812*
Biorci, G., 711 (47), *721*, 757 (26), 794 (74), 797 (83), 798 (83), 800 (74), *811, 813*
Birch, F., 529 (6), *573*
Birchenall, T. E., 799 (86), *813*
Birkenbeil, H. J., 591 (29), 596 (29), 598 (29), *618*, 795 (81), 798 (81), 799 (81), 800 (81), *813*
Birnbaum, H. K., 809 (108A), *814*
Bitter, F., 646 (49), *686*, 736 (41), 739 (76), *747*
Blade, J. C., 697 (15), *720*
Blake, L. R., 726 (10), *746*
Bleaney, B., 835 (42), *838*
Bloch, F., 824 (24), *838*
Bloembergen, N., 714 (53), 717 (53), *721*, 824 (23), *837*
Boas, W., 704 (25), *720*
Boesono, G. J., 798 (85B), *813*
Boll, R., 740 (87), 744 (106), *748*
Boothby, O. L., 745 (116), *748*
Boser, O., 627 (17), 634 (36), 650 (53), 651 (53), 657 (53, 57), 661 (53), 671 (36, 53), 672 (36, 57), 676 (57), 677 (36), 681 (36, 53), 682 (36, 53), *685, 686*
Bosman, A. J., 751 (14), 753 (22), 755 (22), 762 (29), 777 (29), 778 (44), 781 (44), 783 (29, 44, 47), 784 (29, 50, 51), 785 (22, 29), 796 (22), *811, 812*
Both, E., 745 (114), *748*
Bozorth, R. M., 548 (33), 553 (44), 555 (44), 570 (86), 572 (98), *574, 575*, 578 (3, 4, 6, 7), 579 (7), *617*, 646 (50), 654 (50), 679 (50), *686*, 709 (39), *721*, 735 (40), *747*, 750 (6), 792 (69), 794 (73), *811, 813*
Bradley, A. J., 561 (69), 562 (69), *575*, 596 (40), *618*
Braun, E., 520 (23), 521 (23), *522*
Brener, W., 732 (24), *746*
Brenner, R., 514 (8, 9), *521*
Brinkman, J. A., 623 (1, 2), *685*
Brissonneau, P., 751 (11), 767 (33), 771 (33, 34), 772 (34, 35), 773 (34), 776 (33, 40, 41), 778 (33, 40), 795 (80), 796 (33, 80), 797 (80), 798 (80), *811, 812, 813*
Brommer, P. E., 753 (22), 755 (22), 778 (44), 781 (44), 783 (44, 47), 784 (50, 51), 785 (22), 796 (22), 797 (81B), 798 (81B), *811, 812, 813*
Broom, T., 695 (10), *720*
Brown, J. R., 737 (58), *747*
Brown, N., 560 (64), 561 (64), *575*
Brown, W. F. Jr., 642 (40, 41, 42, 43), 661 (42), 667 (40, 41), *686*, 690 (2), 698 (2), *720*
Bruchatov, N., 726 (7), *746*
Buchanan, D. N. E., 534 (17), *573*
Bumm, H., 702 (21), *720*
Bunge, G. I., 726 (11), *746*
Bunge, H. J., 604 (62), 616 (62), *619*
Burbank, R. D., 601 (48), *618*, 805 (105, 106, 107, 108), *814*
Burgers, W. G., 602 (57), *618*, 728 (14), 732 (28), *746*
Burr, A. A., 730 (15), *746*

C

Cable, J. W., 570 (89), 571 (95), 572 (95), *575*
Cahn, R. W., 591 (39), 596 (29), 598 (29), *618*, 795 (81), 798 (81), 799 (81), 800 (81), *813*
Carpenter, V. W., 736 (42, 44), *747*
Carr, W. J., 556 (57), *574*
Catalano, E., 784 (52), *812*
Chamberod, A., 554 (48), 555 (48), *574*
Chambron, W., 549 (37), 554 (37), *574*, 602 (54), *618*, 799 (111), 809 (111, 112), *814*
Chevenard, P., 750 (1), *810*
Chikazumi, S., 581 (17, 20), 583 (17), 585 (23), 591 (20, 30), 593 (36), 594 (36), 595 (30, 36), 596 (30), 599 (45), 602 (23), 604 (61A, 61B, 61C), 606 (61A, 61B, 61C), 607 (61A, 61B, 61C), 608 (61B), 609 (61B), 610 (61B), 611 (61B), 612 (61B), 613 (61B), 614 (61B, 64), 615 (64, 65), *617, 618, 619*, 690 (1), 693 (6), 698 (1), 711 (43), *720, 721*, 787 (59, 60, 61),

790 (67), 795 (61), 799 (88), 800 (92), 803 (67), *812*, *813*
Chin, G. Y., 616 (66B, 66C), *619*
Clare, W. H., 697 (15), *720*
Clarebrough, L. M., 717 (62), *721*
Clogston, A. M., 824 (26), *838*
Cohen, M., 836 (41), *838*
Cohen, M. H., 715 (58), *721*
Cole, G. H., 736 (44, 49), 738 (62), *747*
Collins, M. F., 534 (13, 14, 15), 546 (30), 550 (30), *573*, *574*, 595 (386), *618*
Conradt, H. W., 603 (58), *618*, 710 (41), *721*, 732 (31), *746*
Cooper, B., 831 (36), *838*
Corcoran, R., 737 (57), *747*
Crangle, J., 570 (85, 87), 571 (85, 96), *575*
Crede, J. H., 733 (32), 736 (48), 745 (114), *746*, *747*, *748*
Cronk, E. R., 732 (22), *746*
Cullity, B. D., 745 (115), *748*
Custers, J. F. H., 732 (27), *746*
Czerlinsky, E., 699 (18), *720*

D

Dahl, O., 578 (5), 603 (58), *617*, *618*, 710 (41), *721*, 732 (30, 31), *746*
Dautreppe, D., 549 (36, 37), 554 (36, 37, 47), *574*, 601 (51), 602 (53, 54), *618*, 795 (78), 797 (110), 799 (111), 809 (110, 111, 112, 113), 810 (78), *813*, *814*
Davidson, R. L., 736 (44, 49), 738 (62), *747*
Davis, D. D., 570 (86), *575*
Day G. F., 836 (44), *838*
Detert, K., 737 (61), 739 (78), 740 (87), 741 (90), *747*, *748*
de Vos, K. J., 732 (18), *746*
de Vries, G., 581 (13), 593 (13), 598 (42), *617*, *618*, 751 (14), 753 (23, 24), 754 (24), 755 (24), 756 (24), 757 (23, 24), 764 (23), 795 (23), 798 (85B), *811*, *813*
De Wit, H. J., 665 (58), *686*, 697 (16), *720*
De Wit, R., 629 (30), *686*
Dietrich, H., 668 (64), 684 (82), *686*, *687*, 705 (32, 33), *721*
Dietze, H. D., 764 (30), 795 (30), 798 (30), 809 (109), *812*, *814*

Dijkstra, L. J., 514 (5), 515 (5), 516 (5), 518 (5), 519 (5), 520 (5), *521*, 656 (56), 657 (56), *686*
Dillinger, J. F., 578 (3, 4), *617*
Döring, W., 553 (46), *574*, 579 (8), *617*, 624 (5), 628 (5), 668 (5), 679 (5), *685*, 691 (4), 693 (6), 698 (4), 699 (17, 18), 704 (27, 28), *720*, *721*, 774 (37), *812*
Doyle, M. V., 783 (48), *812*
Dressel, H., 795 (74A), *813*
Dunn, C. G., 708 (38), *721*, 725 (3), 726 (12), 727 (12), 736 (45), 737 (52, 53, 55, 56), 738 (53), 739 (79, 80, 81, 82, 83, 84), *745*, *746*, *747*
du Pré, F. K., 752 (19), *811*

E

Ebeling, D. G., 730 (15), *746*
Elliott, R., 831 (35), *838*
Ellis, W. C., 544 (23), 545 (23), *574*
English, A. T., 616 (66C), *619*
Ernst, G. J., 798 (85B), *813*
Eshelby, J. D., 639 (38), *686*
Essmann, U., 625 (15), 626 (12, 15), 627 (15), 635 (15), *685*
Ewing, J. A., 693 (5), 701 (5), 710 (5), *720*
Eykelenboom, L. C. H., 783 (47), 784 (51), *812*

F

Fahlenbrach, H., 732 (24), *746*, 798 (84), 804 (103), *813*, *814*
Fallot, M., 571 (94), 572 (94), *575*
Farcas, T., 530 (11), *573*
Fast, J. D., 738 (64), *747*, 781 (45), *812*
Faulkner, E. A., 715 (57), 716 (57, 61), 717 (57, 63), *721*
Fedotov, L. N., 530 (7), *573*
Feldtkerller, R., 778 (43), *812*
Ferguson, E. T., 591 (27), 592 (34), *617*, *618*, 788 (62), 791 (62), 795 (62), 802 (62), 804 (62), *812*
Ferro, A., 711 (47), *721*, 757 (26), 794 (74), 797 (83), 798 (81C, 83), 800 (74), 802 (96), 809 (96), *811*, *813*

Fiedler, H. C., 737 (59), 738 (68, 69, 70, 73), 739 (74, 75), 741 (92, 102), 744 (92), 747, 748
Fishbach, D. B., 800 (91), 801 (91), 813
Fisher, J. C., 605 (63), 619, 783 (49), 812
Flanders, P. J., 750 (5), 811
Flinn, P. A., 534 (17), 564 (17), 573
Forrer, R., 529 (6), 545 (26), 546 (31), 573, 574
Forsyth, J. B., 546 (30), 550 (30), 574
Foster, K., 741 (95), 744 (105), 748
Frank, F. C., 639 (38), 686
Freeman, A. J., 832 (37), 838
Frey, A. A., 736 (41) 747
Frischmann, P. G., 741 (92), 744 (92), 748
Froideueaux, F., 836 (47), 838
Fujimori, H., 571 (93), 575, 594 (38A), 595 (38A), 618, 757 (100), 797 (100), 798 (100), 799 (100), 803 (100, 101), 814
Fullerton, L. D., 601 (49), 618

G

Galt, J. K., 758 (28), 759 (28), 775 (28), 811, 823 (20), 824 (25), 837, 838
Ganz, D., 740 (87), 741 (93), 744 (104, 106), 748
Garvin, S. J., 732 (22), 746
Gautier, F., 836 (47), 838
Geisler, A. H., 745 (114), 748
Gengnagel, H., 569 (80), 575, 591 (28), 592 (35), 595 (28), 617, 618, 795 (74A), 799 (90), 813
Gersdorf, R., 581 (13), 593 (13), 617, 753 (24, 25), 754 (24), 755 (24), 756 (24), 757 (24), 792 (70), 811, 813
Gerstner, D., 805 (104), 806 (104), 814
Geschwind, S., 831 (34), 838
Gilbert, T. L., 823 (21), 837
Goldman, J. E., 538 (18), 541 (22), 548 (18, 22), 573
Goldman, N. J., 750 (6), 811
Gorski, W. S., 751 (17), 811
Goss, N. P., 734 (39), 747
Gossard, A. C., 832 (38), 836 (38, 39), 838
Gould, J. E., 732 (21), 746

Grabbe, E. M., 540 (20), 550 (20), 553 (20), 573
Graf, L., 570 (92), 572 (92), 575
Graham, C. D. Jr., 557 (58, 59, 60), 558 (59, 60), 559 (60), 574, 590 (24), 617, (68), 668 (67, 68), 687, 745 (110), 748, 751 (10), 757 (10), 811
Graham, T. P., 836 (48), 838
Greiner, E. S., 544 (23), 545 (23), 574
Grenoble, H. E., 741 (92, 103), 744 (92), 748
Grewen, J., 724 (2), 728 (2), 745
Griffiths, J. H. E., 820 (14), 837
Gruner, L., 750 (5), 811
Guillaud, C., 555 (53), 574
Gutowsky, H. S., 714 (56), 721

H

Hahn, R., 555 (55), 560 (55), 561 (55), 574, 616 (72), 619, 750 (5), 811
Hall, R. C., 548 (32), 569 (81), 574, 575
Ham, R. K., 717 (63), 721
Hansen, M., 514 (15), 521, 550 (39), 551 (39), 570 (84), 574, 575
Hargreaves, M. E., 717 (62), 721
Harris, E. S., 733 (34), 746
Haworth, F. E., 549 (34), 574
Hazzledine, P. M., 626 (13), 685
Heck, J. E., 736 (47), 747
Heidenreich, R. D., 601 (48), 618, 805 (106), 814
Heller, W., 514 (12), 515 (12), 516 (12), 519 (12), 520 (12), 521
Heller, W. R., 787 (54), 812
Henke, R. H., 736 (48), 747
Hibbard, W. R., Jr., 732 (23), 741 (92, 96), 744 (92), 746, 748
Hihara, T., 836 (44), 838
Hinsley, J. F., 730 (17), 746
Hirone, T., 668 (63), 686
Hirsch, P. B., 627 (19), 635 (19), 685, 696 (14), 720
Hofmann, U., 677 (78), 680 (78), 687
Holcomb, D. F., 751 (7), 811
Holstein, T., 666 (60), 686
Hooft, H. A.'t., 797 (81B), 798 (81B)
Hopkins, B. E., 669 (62), 686

Hoselitz, K., 730 (16), *746*
Houska, C., 836 (41), *838*
Howard, J., 738 (71), *747*
Hsun, Hu., 741 (99), *748*
Hubert, A., 647 (90), *687*
Hutchinson, T. S., 712 (48), *721*

I

Ibe, G., 740 (87), 741 (90), *748*
Iida, S., 803 (102), *814*
Ishikawa, Y., 564 (76), *575*
Ito, A., 564 (76), *575*
Itoh, J., 534 (16), *573*, 836 (49, 50), *838*
Iwata, T., 592 (33), 595 (33), 604 (61B), 606 (61B), 607 (61B), 608 (61B), 609 (61B), 610 (61B), 611 (61B), 612 (61B), 613 (61B), 614 (61B, 64), 615 (64), *618*, *619*, 788 (64, 65, 66), 790 (65, 66), 793 (64, 65), 808 (66), *812*

J

Jaccarino, V., 835 (43), 837 (51), *838*
Jack, K. H., 517 (17), 520 (24), *521*
Jackson, J. M., 736 (46), *747*
Jackson, R. F., 836 (46), *838*
Jacobs, I. S., 557 (59), 558 (59), *574*
Janssen, K., 732 (24), *746*
Jay, A. H., 561 (69), 562 (69), *575*, 596 (40), *618*
Jessen, K., 570 (91), 571 (91), *575*
Jones, R. L., 577 (1), *617*, 796 (82), *813*
Jones, R. M., 563 (70), 564 (70), 569 (70), *575*
Jones, R. V., 831 (36), *838*
Josso, E., 550 (41), *574*

K

Kadykova, G. N., 741 (94), *748*
Kasper, J. S., 559 (63), *575*
Kaya, S., 544 (25), 545 (25), 555 (49, 50), *574*, 579 (9), *617*, 750 (6), *811*
Keh, A. S., 628 (24), 668 (24), *686*
Kelly, J. M., 823 (21), *837*
Kelsall, G. A., 578 (2), *617*
Kernohan, R. H., 601 (50), *618*

Kerr, J., 514 (13), 517 (13), 520 (13), *521*
Kersten, M., 514 (1, 10, 11), 520 (1), 521 (1), *521*, 654 (55), 677 (55), *686*, 703 (22), 704 (30), 705 (31), *720*, *721*
Kikuchi, R., 823 (22), *837*
Kimura, Y., 732 (25), *746*
Kittel, C., 758 (28), 759 (28), 775 (28), *811*, 817 (2, 6), 818 (2), 820 (15), *837*
Klein, M. V., 795 (76), 810 (76), *813*
Klein, M. W., 528 (1), *573*
Kneller, E., 549 (38), 555 (55), 560 (55), 561 (55), *574*, 616 (72), *619*, 668 (64, 65), 671 (74), 684 (82), *686*, *687*, 705 (32, 33), *721*, 750 (5), 774 (38), 775 (38), 778 (38), 805 (104), 806 (104), *811*, *812*, *814*
Kobayashi, S., 534 (16), *573*, 836 (49), *838*
Koch, A. J. J., 732 (18), *746*
Koehler, J. S., 629 (29), *686*
Koehler, W. C., 570 (89), 571 (95), 572 95), *575*
Köster, W., 514 (6), 517 (6), 518 (6), 520 (23), 521 (23), *521*, *522*, 553 (45), *574*, 669 (23), *687*, 750 (1), *810*
Koh, P. K., 737 (54, 55), 738 (54), *747*
Kohler, D., 739 (85), *748*
Koi, Y., 836 (44), *838*
Komar, A., 555 (52), 561 (52), *574*
Kondo, J., 528 (1), *573*
Kondorskii, E. I., 514 (4), *521*, 530 (7), 552 (43), *573*, *574*
Kōno, T., 744 (109), *748*
Kontani, M., 836 (50), *838*
Kouvel, J. S., 530 (8), 552 (8), 557 (58, 59, 60), 558 (59, 60), 559 (60, 63), 564 (77), *573*, *574*, *575*
Kramer, J. J., 741 (95), 744 (105), *748*
Krause, D., 668 (66), 671 (66), 676 (66), *687*
Krebs, K., 554 (48), 555 (48), *574*
Krén, E., 573 (100), *575*
Kröner, E., 642 (46), *686*
Kronmüller, H., 624 (6, 87), 625 (6, 9, 11, 22, 87), 626 (16), 627 (17, 22), 628 (26, 27), 634 (36, 37), 642 (44, 45, 47, 48), 643 (48), 644 (48), 645 (48), 650 (53), 651 (53), 657 (53, 57), 661 (48, 53), 666 (44, 45), 669 (48), 670 (6), 671 (36, 53),

672 (36, 57), 674 (76), 676 (57), 677 (36), 680 (37, 48), 681 (36, 53), 682 (36, 53), 683 (81), 684 (45, 76), 685 (84, 85, 86), *685, 686, 687*, 699 (19), 701 (20), 706 (19A), *720*, 795 (75), 810 (75, 76), *813*
Kuhlmann, D., 695 (9), 696 (12), 703 (24), *720*
Kunakov, Ya. N., 744 (107), *748*
Kushida, T., 836 (44), *838*
Kussmann, A., 544 (21), 546 (21), 555 (49), 570 (88, 91, 92), 571 (91), 572 (88, 92), 573 (99), *573, 574, 575*, 750 (6), *811*

L

La Force, R. C., 836 (44), *838*
Lamb, H. J., 697 (15), *720*
Landau, L., 817 (1), 823 (1), *837*
Laugier, J., 549 (36), 554 (36, 47, 48), 555 (48), *574*, 601 (51), 602 (53), *618*, 797 (110), 809 (110, 113), *814*
Lawson, W. A., 784 (52), *812*
Lawton, H., 650 (52), 651 (52), *686*
Lazarus, D., 785 (53), *812*
Le Claire, A. D., 581 (15), *617*
Lecocq, P., 802 (95), *813*
LeCraw, R. C., 824 (27), *838*
Lee, E. W., 632 (34), *686*
Leech, P., 549 (34), *574*
Leibfried, G., 639 (39), *686*
Lemmens, M. C., 798 (85B), *813*
Liesegang, J., 831 (36), *838*
Lifshitz, E., 817 (1), 823 (1), *837*
Lilley, B. A., 633 (35), *686*
Lindquist, R. H., 832 (39), 836 (39), *837, 838*
Littmann, M. F., 725 (6), 733 (34), 736 (47), 738 (72), 741 (91), 744 (91), *746, 747, 748*
Litvin, D. F., 545 (28), *574*
Livingston, J. D., 616 (67), *619*, 750 (4, 5), *811*, 821 (18), 825 (30), 826 (33), 828 (33), 829 (32), *837, 838*
Livshits, B. G., 744 (107), *748*
Lomer, W. M., 581 (15), *617*
Low, G. G., 534 (13, 14, 15), 570 (90), *573, 575*

Low, J. R., 668 (70), *687*
Luteijn, A. I., 732 (18), *746*
Lutes, O. S., 798 (85A), *813*
Luttinger, J. M., 817 (6), *837*
Lyashenko, B. G., 545 (28), *574*

M

McCaig, M., 730 (16), 732 (20), *746*
McClelland, J. D., 712 (49), *721*
McHargue, C. J., 737 (56), *747*
McKeehan, L. W., 734 (36), *746*
McLennan, J. E., 669 (62), *686*
Mader, S., 625 (8, 9, 10), 626 (16), 635 (20), *685*
Madono, O., 741 (100), *748*
Makino, N., 732 (19), *746*
Malek, Z., 705 (35, 36), *721*
Marcinkowski, M. J., 560 (64), 561 (64, 68), *575*
Marechal, J., 591 (26), *617*
Marian, V., 529 (5), 530 (5), *573*
Marshall, W., 528 (1), 539 (19A), *573*
Martin, J. P., 733 (32), 735 (114), *742, 748*
Martin, M. J. C., 529 (4), *573*
Masing, G., 695 (8, 9), *720*
Maxwell, D., 517 (17), *521*
May, J. E., 738 (65, 66), *747*
Mazzetti, P., 798 (81C), 802 (96), 809 (96), *813*
Mead, H. W., 799 (86), *813*
Mehrer, H., 683 (81), 685 (86), *687*
Meiklejohn, W. H., 558 (62), *575*, 602 (56), *618*
Menter, J. W., 519 (21), *522*
Merritt, F. R., 823 (20), 824 (25), *837, 838*
Meyer, A. J. P., 818 (9), *837*
Michet, A., 802 (95), *813*
Mitui, T., 616 (68, 69, 70, 71), *619*, 745 (111), *748*
Möbius, H. E., 741 (101), *748*
Moller, H., 737 (60), *747*
Montalenti, G., 711 (47), *721*, 757 (26), 794 (74), 797 (83), 798 (81C, 83), 800 (74), 802 (96), 809 (96), *811, 813*
Morgan, E. R., 530 (9), 531 (9), 555 (9), 556 (9), *573*
Morrill, W., 736 (43, 45), 738 (63), *747*

Moser, P., 795 (78, 80), 796 (80), 797 (80), 798 (80), 810 (78), *813*
Mott, N. F., 696 (11), *720*
Müller, H. G., 702 (21), *720*
Müller, J. G., 604 (62), 616 (62), *619*
Mughrabi, H., 625 (88), 627 (88), 639 (88), *687*
Mullins, W. W., 739 (77), *747*

N

Nabarro, F. R. N., 639 (38), *686*
Nachtrieb, N. H., 784 (52), *812*
Nacken, M., 514 (12, 14), 515 (12), 516 (12), 517 (14), 519 (12), 520 (12, 14), *521*
Nakagawa, Y., 615 (65), *619*
Nakamichi, T., 561 (68), *575*
Nakayama, M., 555 (50), *574*
Nathans, R., 564 (75, 78), *575*, 595 (41), *618*, 802 (94), *813*
Neel, L., 514 (2), 520 (2), 521 (2), *521*, 549 (36, 37), 554 (36, 37, 47), *574*, 581 (12, 18), 584 (18), 590 (18), 592 (12), 599 (46), 601 (18, 51), 602 (52, 53, 54), 603 (18, 60), *617*, *618*, 628 (25), 650 (25), 651 (25), 685 (83), *686*, 690 (3), 698 (3), 711 (44), *720*, *721*, 751 (9), 753 (20, 21), 755 (21), 759 (21), 760 (21), 772 (21), 774 (21), 778 (21), 787 (55, 56), 797 (110), 799 (111), 809 (110, 111, 113), *811*, *812*, *814*
Neronova, V. I., 802 (93), *813*
Nesbitt, E. A., 601 (48, 49), *618*, 806 (105, 106, 107, 108), *814*
Neurath, P. W., 668 (67), *687*
Nilan, T. G., 725 (4), *745*
Norberg, R. E., 714 (55), *721*, 751 (7), *811*
Nowick, A. S., 581 (16), *617*, 787 (54), *812*
Nutting, J., 519 (21), *522*

O

Ohtsuka, T., 550 (40), 553 (40), 555 (54), *574*
Okamoto, T., 668 (71), *687*
Okamura, T., 668 (63), *686*
Ono, K., 564 (76), *575*
Oomura, T., 581 (20), 591 (20), *617*, 787 (61), 795 (61), *812*

P

Paoletti, A., 561 (65), *575*
Passari, L., 561 (65), *575*
Patter, H. H., 668 (72), *687*
Paulevé, J., 549 (36, 37), 554 (36, 37, 47, 48), 555 (48), *574*, (51, 53, 54), 601 (51), 602 (53, 54), *618*, 797 (110), 799 (111), 809 (110, 111, 112, 113), *814*
Pauthenet, R., 554 (47), *574*, 602 (53), *618*, 651 (54), 671 (75), *686*, *687*, 809 (113), *814*
Pavlovskaya, U. S., 716 (60), *721*
Pawlek, F., 578 (5), *617*, 732 (26), 741 (101), *746*, *748*
Paxton, W. S., 725 (4), *745*
Peach, M., 629 (29), *686*
Pender, H., 577 (1), *617*, 796 (82), *813*
Peretto, P., 795 (78), 810 (78), *813*
Pfaffenberger, J., 732 (30), *746*
Pfeffer, K.-H., 635 (89), 641 (89), *687*
Pfeifer, F., 740 (87), 744 (106), *748*
Phillips, T. G., 825 (29), *838*
Pickart, S. J., 564 (78), *575*
Piercy, G. R., 530 (9), 531 (9), 555 (9), 556 (9), *573*
Pigott, M. T., 564 (75), *575*, 595 (41), *618*, 802 (94), *813*
Pitch, W., 519 (22), *522*
Polder, D., 752 (18), 753 (18), *811*, 817 (3), *837*
Poliak, R. M., 561 (67), *575*
Polley, H., 699 (18A), *720*
Porter, C. S., 824 (27), *838*
Portis, A. M., 832 (38, 39), 836 (38, 39), *837*, *838*
Pound, R. V., 713 (52), *721*
Powers, R. W., 783 (48), *812*
Primakoff, H., 666 (60), *686*
Pry, R. H., 737 (50), 738 (69), 741 (92), 744 (92), *747*, *748*
Pulaski, E. L., 736 (48), *747*
Pussei, I. N., 677 (79), 680 (79), *687*
Puzey, I. M., 545 (28), 569 (82), *574*, *575*

Q

Quereshi, A. H., 514 (7), 515 (7), 516 (7), 517 (7), 518 (7), 520 (7), 521 (7), *521*

R

Radavich, J., 519 (20), *522*
Radeloff, C., 631 (33), *686*
Raffelsieper, J., 695 (8, 9), *720*, 750 (1), *810*
Rahmann, J., 514 (15), 517 (14), 520 (14), *521*
Rapp, M., 627 (17), *685*
Rathenau, G. W., 580 (10), 581 (13), 593 (13), 603 (59), *617*, *618*, 732 (27), *746*, 751 (13), 753 (22, 24, 25), 754 (24), 755 (22, 24), 756 (24), 757 (24), 767 (32), 770 (32), 772 (32), 778 (44), 781 (44), 783 (44, 47), 784 (50, 51), 785 (22), 792 (70), 796 (22), *811*, *812*, *813*
Raub, E., 520 (25), *522*
Ravitz, S. F., 715 (59), *721*, 836 (44), *838*
Reekie, J., 712 (48), *721*
Reif, F., 715 (58), *721*
Reits, D., 697 (16), *720*
Ricci, F. P., 561 (65), *575*
Richardson, J. M., 817 (4), 832 (4), *837*
Richter, G., 711 (45), *721*, 751 (15), 774 (36), *811*, *812*
Rieder, G., 629 (31), 631 (31), 633 (31), *686*
Rieger, H., 624 (6), 625 (6), 628 (26), 634 (37), 670 (6), 680 (37, 77), 683 (80), 685 (84), *685*, *686*, *687*, 701 (20), *720*
Rimet, G., 651 (54), 671 (75), *686*, *687*
Rodbell, D. S., 824 (28), 825 (28, 30, 31, 32), 832 (40), *838*
Romashov, V. M., 741 (97), *748*
Rossol, F., 831 (36), *838*
Rostoker, W., 734 (35), *746*
Roth, W. L., 832 (40), *838*
Rowland, T. J., 717 (64), 718 (64), *721*
Ruby, S. L., 534 (17), 564 (17), *573*
Ruder, W. E., 734 (37, 38), 737 (51), *746*, *747*

S

Sadron, C., 530 (10), *573*
Saitō, H., 571 (93), *575*
Salkovitz, E. I., 601 (50), *618*
Sambongi, T., 616 (68), *619*, 745 (111), *748*
Sasakawa, T., 594 (38a), 595 (38A), *618*, 757 (100), 797 (100), 798 (100), 799 (100), 803 (100, 101), *814*
Sato, H., 539 (19), 544 (25), 545 (25), 563 (71, 72), 564 (71, 72), *573*, *574*, *575*
Sato, M., 616 (68, 70), *619*
Sato, T., 615 (66A), 616 (66A), *619*
Sawyer, B., 745 (113), *748*
Scharnow, B., 544 (21), 546 (21), *573*
Schiller, P., 795 (75, 77), 810 (75, 77), *813*
Schindler, A. I., 601 (50), *618*
Schinkel, C. J., 783 (47), 784 (51), *812*
Schlömann, E., 820 (16), *837*
Schmelzer, G., 668 (65), *687*
Scholefield, H. H., 733 (33), *746*
Schreiber, F., 751 (12), 778 (42), *811*, *812*
Schulze, A., 544 (21), 546 (21), *573*
Scurlock, R. G., 836 (46), *838*
Seavey, M. H., Jr., 819 (12), *837*
Sedov, V. L., 552 (43), *574*
Seeger, A., 623 (3, 4), 624 (6), 625 (6, 7, 9), 626 (16), 627 (7, 17), 628 (26, 27), 634 (36, 37), 642 (44, 45), 650 (53), 651 (53), 657 (53, 57), 661 (53), 666 (44, 45), 670 (6), 671 (36, 53), 672 (36, 57), 676 (57), 677 (36), 680 (37), 681 (36, 53), 682 (36, 53), 683 (81), 684 (45), 685 (84, 85, 86), *685*, *686*, *687*, 699 (19), 701 (20), 704 (26), 706 (19A), 718 (64, 65), *720*, *721*, 795 (75, 77), 810 (75, 77), *813*
Seigel, S., 544 (24), 555 (24), *574*
Seraphim, D. P., 581 (16), *617*
Seybolt, A. U., Jr., 738 (67), *747*
Shimizu, S., 572 (97), *575*
Shinohara, T., 569 (79), *575*
Shockley, W., 514 (3), *521*, 646 (50), 654 (50), 679 (50), *686*
Shull, C. G., 531 (12), 532 (12), 544 (24), 549 (12), 550 (12), 555 (12, 24), 564 (75), *573*, *574*, *575*, 595 (41), *618*, 802 (94), *813*
Silverstein, S. D., 528 (1), *573*
Six, W., 602 (57), *618*, 732 (28), *746*
Sixtus, K. J., 591 (32), 595 (32), 603 (58), *618*, 710 (41), *721*, 726 (8), 732 (31), 734 (8), *746*
Slichter, C. P., 693 (7), 714 (55), *720*, *721*, 751 (7), *811*
Slonczewski, J. C., 585 (22), 589 (22), *617*, 751 (8), 786 (8), 789 (8), *811*
Smit, J., 602 (55), *618*, 818 (8), 819 (11), *831*

Smoluchowski, R., 538 (18), 544 (22), 548 (18, 22), *573*, 745 (112, 113), *748*, 750 (6), *811*
Snoek, J. L., 580 (119), 602 (57), 603 (59), *617*, *618*, 693 (6), 711 (42), *720*, *721*, 732 (28, 29), *746*, 751 (16), 752 (16, 19), 766 (31), 776 (31), 777 (16), 796 (31), *811*, *812*
Sommerkorn, G., 798 (84), *813*
Sosnin, V. V., 741 (94), *748*
Špaček, L., 668 (69), *687*
Spachner, S., 734 (35), *746*
Spencer, E. G., 824 (27), *838*
Stablein, H., 737 (60), *747*
Standley, K. J., 820 (17), *837*
Stanley, J. K., 707 (37), *721*
Stark, Yu., S., 716 (60), *721*
Starreveld, R. W., 697 (16), *720*
Stearns, M. B., 534 (16), *573*
Steeds, J. W., 626 (13), 627 (19), 635 (19), *685*
Steinert, J., 801 (92A), *813*
Steinort, E., 732 (22), *746*
Stevens, K. W. H., 820 (17), *837*
Stewart, K. H., 650 (52), 651 (52), *686*
Stoelinga, J. H. M., 753 (25), 792 (70), *811*, *813*
Stoner, E. C., 528 (2, 3), *573*
Street, R., 832 (40), *838*
Streever, R. L., 534 (16), *573*
Stroble, C. H., 736 (48), *747*
Sucksmith, W., 529 (4), 563 (73), 564 (73), *573*, *575*
Sugihara, M., 799 (87), *813*
Suhl, H., 819 (10), 824 (26), *837*, *838*
Suzuki, J., 595 (39), 604 (61A, 61B), 606 (61A, 61B), 607 (61A, 61B), 608 (61B), 609 (61B), 610 (61B), 611 (61B), 612 (61B), 613 (61B), 614 (61B, 64), 615 (64), *618*, *619*, 799 (89), *813*
Sykes, C., 549 (34), *574*
Szabó, P., 573 (100), *575*

T

Taguchi, S., 741 (98), *748*
Takahashi, M., 594 (38a), 595 (38A), *618*, 744 (109), *748*, 757 (100), 797 (100), 798 (100), 799 (100), 803 (99, 100, 101), *813*, *814*
Tamagawa, N., 616 (65, 69, 71), *619*
Taniguchi, S., 569 (83), *575*, 581 (19), 583 (21), 584 (19), 590 (19), 591 (25A, 25B), 592 (25A, 25B), 593 (25B), 599 (47), 600 (47), 603 (19), *617*, *618*, 787 (57, 58), 792 (57, 58), 795 (79), 803 (98), 807 (79, 98), 808 (79), *812*, *813*
Tannenwald, P. E., 819 (12), *837*
Taoka, T., 550 (40), 553 (40), 555 (54), 561 (66), *574*, *575*
Tarasov, L. P., 709 (40), *721*, 726 (9), *746*
Tatsumoto, E., 668 (71), *687*
Taylor, A., 563 (70), 564 (70), 569 (70), *575*
Thieringer, H.-M., 625 (59), 626 (59), 627 (18, 59), 635 (20, 59), *685*, *686*
Thomas, E. E., 745 (116), *748*
Thomas, H., 741 (93), *748*
Thompson, N., 555 (51), *574*
Tiderman, H., 732 (22), *746*
Torrey, H. C., 714 (54), *721*, 751 (7), *811*
Toth, L. E., 715 (59), *721*
Träuble, H., 626 (16), 630 (32), 634 (36, 37), 646 (32, 51), 647 (51), 650 (32, 51, 53), 651 (53), 654 (51), 655 (51), 656 (51), 657 (51, 53, 57), 660 (51), 661 (53), 668 (61), 669 (61), 670 (61), 671 (36, 51, 53), 672 (36, 57), 674 (51), 675 (51, 61), 676 (57, 61), 677 (36), 680 (37), 681 (36, 51, 53), 682 (36, 51, 53), *685*, *686*, 704 (26), *720*
Trapp, R. H., 725 (6), 741 (91), 744 (91), *746*, *748*
Tsou, A. L., 519 (21), *522*
Turchinskaya, M. I., 557 (61), *574*
Turkalo, A. M., 668 (70), *687*
Turnbull, D., 738 (65, 66), *747*, 750 (3), *811*
Turner, R. W., 745 (112), *748*

U

Uchikoshi, H., 599 (45), *618*
Ulmer, R. P., 798 (85A), *813*
Uriano, G. A., 534 (16), *573*
Utton, D. B., 836 (46), *838*

V

Van Bueren, H. G., 628 (23), *685*
Van Daal, A. J., 778 (44), 781 (44), 783 (44), *812*
Van der Steeg, M. G., 732 (18), *746*
Van Geest, D. W., 581 (13), 593 (13), *617*, 753 (24), 754 (24), 755 (24), 756 (24), 757 (24), *811*
Van Langen, M. J., 798 (85B), *813*
Van Vleck, J. H., 788 (63), *812*, 817 (5), 818 (5), *837*
Varlakov, V. P., 741 (97), *748*
Vergne, R., 598 (43, 44), *618*, 795 (81A), *813*
Verrijp, H. B., 781 (45), *812*
Vicena, F., 629 (28), 656 (28), *686*, 704 (29), *721*
Vlasova, Ye, V., 802 (93), *813*
Volkenshtein, N., 558 (52), 561 (52), *574*
Volkenshtein, N. V., 557 (61), *574*
Von Neida, A. R., 616 (66C), *619*
Von Rittberg, G., 570 (88), 572 (88), 573 (99), *575*

W

Waché, X., 750 (1), *810*
Wagner, F. J., 795 (77), 810 (77), *813*
Wagner, H., 591 (28), 592 (35), 595 (28), *617*, *618*, 795 (74A), 799 (90), *813*
Wakelin, R. J., 549 (35), 550 (35), 552 (35), 553 (35), *574*
Wakiyama, T., 591 (30, 31), 595 (30, 31), 596 (30, 31), 597 (31), *618*, 799 (88), *813*
Walker, E. V., 738 (71), *747*
Walker, J. G., 555 (44), 563 (44), *574*
Walker, L. R., 824 (26), *838*
Walter, J. L., 725 (3), 726 (12), 727 (12), 739 (79, 80, 81, 82, 83, 84), 741 (92, 96), 744 (92, 108), *745*, *746* *747*, *748*
Walter, W., 520 (25), *522*
Wangsness, R. K., 817 (7), *837*
Ward, C. E., 733 (34), *746*
Ward, R., 736 (45), *747*
Wassermann, G., 724 (2), 728 (2), 732 (2), *745*
Watai, K., 572 (97), *575*
Watson, R. E., 832 (37), *838*
Weil, L., 750 (5), 784 (52), *811*, *812*
Weisman, I., 836 (47), *838*

Weiss, P., 529 (6), 546 (31), *573*, *574*
Wenny, D. H., 745 (116), *748*
Went, J. J., 550 (42), *574*
Wernick, J. H., 570 (86), *575*
Wert, C., 514 (5, 13, 16), 515 (5), 516 (5), 517 (13, 16), 518 (5, 18), 519 (5, 19, 20), 520 (5, 13), 521 (16), *521*, *522*, 656 (56), 657 (56), *686*, 794 (71), *813*
Wert, G. A., 781 (46), *812*
Wertheim, G. K., 534 (17), *573*
West, G. W., 717 (62), *721*, 836 (45), *838*
Wheeler, D. A., 595 (386), *618*
Wiener, G. W., 725 (6), 737 (57), 739 (86), 740 (88), 741 (91, 95), 744 (91), *746*, *747*, *748*
Wijn, P. J., 602 (55), *618*, 818 (8), *837*
Wilkinson, M. K., 531 (12), 532 (12), 549 (12), 550 (12), 555 (12), *573*
Williams, A. J., 601 (49), *618*, 806 (108), *814*
Williams, H. J., 514 (3), *521*, 646 (50), 654 (50), 679 (50), *686*
Wilmshurst, T. H., 836 (46), *838*
Wilson, R. H., 530 (8), 552 (8), *573*
Wittke, H., 711 (46), *721*
Wohlfarth, E. P., 750 (2), *810*
Wollan, E. O., 570 (89), 571 (95), 572 (95), *575*
Wood, E. A., 823 (20), *837*
Wotruba, K., 705 (34), *721*
Wuttig, M., 809 (108A), *814*

Y

Yager, W. A., 823 (20), 824 (25), *837*, *838*
Yamamoto, M., 561 (68), 569 (83), *575*, 581 (19), 583 (21), 584 (19), 590 (19), 591 (25A, 25B), 592 (25A, 25B, 35), 593 (25B), 603 (19), *617*, *618*, 787 (57), 792 (57), 795 (79), 803 (98), 807 (79, 98), 808 (79), *812*, *813*
Yates, E. L., 549 (35), 550 (35), 552 (35), 553 (35), *574*
Yokoyama, T., 545 (27), 546 (29), *574*

Z

Zener, C., 519 (19), *522*, 581 (14), *617*, 758 (27), 791 (68), 792 (68), 794 (71, 72), 800 (68), *811*, *812*, *813*

Subject Index, Volumes 1 and 2

All elements, alloys, and compounds are listed alphabetically using their chemical symbols.

A

Acoustic spin-wave modes, 40
Actinide elements, 261
Aftereffect
  magnetic, 693, 710–712, 749–810
  of magnetostriction, 757
  mechanical, 752, 758–759
  in permeability, 584
  recovery, 710–712
  recrystallization, 710–712
  relation between magnetic and elastic, 794
Ag, 252, 257
Ag–Al, 281
Ag alloys, 283
  with B elements, 265, 281
  with T metals, 266
Ag–As, 265
Ag–Au, 264
Ag–Bi, 282
Ag–Cd, 265, 281
Ag–Co, 283
Ag–Cr, 283
Ag–Cu, 265
Ag–F_2, 312, 316
Ag–Fe, 283
Ag–Ga, 265
Ag–Hg, 281
Ag–In, 265, 281
Aging, *see also* Diffusion, Directional order
  in ferromagnetic materials, 91
  influences on magnetic properties, 513–521
Aging isotherms, 515
Ag–Mn, 270, 271, 283
  exchange anisotropy of, 271
Ag–Mn–Al, 319
Ag_2MnIn, 319
Ag–Pd, 272, 274
Ag–Pd–Rh, 275
Ag–Sb, 265, 282
Ag–Tl, 265
Ag–Zn, 265, 281
Akulov-Zener theory of magnetocrystalline anisotropy energy, 65
Al
  change of susceptibility by small additions, table, 266
  magnetic behavior, 251
  susceptibility
    change at melting, 257
    temperature dependence, 253
Al alloys with B metals, 266
$AlAu_2Mn$, 319
AlCo, 283, 287, 300
$AlCo_3$, 319
Al–Co–Fe–Ni, 475
Al–Co–Mn, 319
Al–Cr, 280, 285–286
Al–Cu, 265, 281, 282
$AlCu_3$, 317
Al–Cu–Mn, 39
$AlCu_2Mn$, 317

Al–Fe, 286–287, 309–310, 578, 591, 598
  diffusion, 799
  directional order, 799
  induced anisotropy, table, 799
  Mössbauer-effect measurements, 564
  neutron diffraction, 564
  order effects, 561–569
AlFe$_3$, 595, 596, 615, 616
  order effects, 561–569
Al$_3$Fe$_{13}$, 563
AlFe–AlNi, 287
Al–Fe–Ni, 319, 354, 473, 475, 482–489
Al$_2$FeNi, 287, 302
Alkaline earths, 251, 253
Alloys, nonferromagnetic, transition temperatures, 250–293
Al–Mn, 269, 285–286, 317–318
Al–Mn–Cr, 281
AlMnNi, 319
Al–Ni, 299
AlNi, 283
Alnico, 473–512
  coercivity, 474
  crystal-oriented, 475
  effect of additions, 474
  future development, 509–511
  heat treatment, 498–509
  in magnetic field, 473–511
  magnetocrystalline anisotropy, 479
  mechanism of precipitation, 479–490
  microstructure, 477–479
  phase relations, 475–477
  precipitate shape, 490–498
  preferred orientation of precipitates, 490–498
  shape anisotropy, 479
  thermomagnetic treatment, 490–491
  torque data, 479
  X-ray side bands, 477
Alnico 5, 473–474, 490
  demagnetization curve, 474
  electron microscopy of, 478
  energy product, 474
  heat treatment in magnetic field, 478
  phase relations, 476, 477
  precipitates, 478
Alnico 8, 474
  phase relations, 476, 477

AlNi–Fe, 482–489
Alternating current magnetic measurements, 189–242
Aluminum, *see* Al
A metals, 251
Ampère's law, 190
Angular momentum, 11, 13
Anhysteretic magnetic measurements, 353
  on coarse powders, 358–360
Anhysteretic magnetization, 353, 447
Anhysteretic properties, 353–363, 443, 447
Anhysteretic remanence, 353, 447
Anhysteretic susceptibility, 354, 355, 447
  criterion for finite and infinite, 455
Anisotropy, magnetic
  anneal, *see* Magnetic field anneal, Stress anneal
  crystal, *see* Anisotropy, magnetocrystalline
  deformation induced, 602–616, 709–710
  directional order induced, *see* Induced anisotropy
  field anneal, *see* Magnetic field anneal
  induced, *see also* Induced anisotropy
    deformation, 602–619, 709–710
    magnetic field anneal, 577–601, 745, 754–757, 788–791
    stress anneal, 597–599, 758–759
  interaction, 375–377, 418, 445, 455
  magnetocrystalline, 63–68
    biaxial, 140
    constants, table, 66–68
    cubic, 63, 138, 140
    energy, 63–68
    hexagonal, 64, 138
    order effects, 548, 553, 568
    tetragonal, 64, 139
    theory, 65
    triaxial, 140
    uniaxial, 140
  measurement, 115, 137–142
  roll, 602–616, 709–710
  shape, 62, 139–140, 380, 411–420, 436–438, 455–459, 462–463
  slip-induced, 602–616
  stress, 68–73, 139
  stress anneal, 597–599, 758–759
  surface, 76

Anisotropy, magnetic—*cont.*
  uniaxial, 140, *see also* Induced anisotropy, Magnetocrystalline anisotropy, Shape anisotropy, Stress anisotropy
  unidirectional, *see* Exchange anisotropy
Anisotropy field, 373, 385
Anisotropy of magnetostriction, 76–81
Anisotropy of susceptibility
  anhysteretic, 451
  antiferromagnetic materials, 23–24
  Bi, 253
  elements, table, 255
Annealing
  magnetic, *see* Magnetic field annealing
  textures from, 728
  under stress, 597–599
Antiferromagnetic phases, with NiAs structure, table, 292
Antiferromagnetic resonance, 831
Antiferromagnetism, 4, 10
  molecular field theory, 22, 23
Antimony, *see* Sb
Apparent permeability, 195, 231
Approach to magnetic saturation, 128, 690–691
  dependence on deformation, 683–685
  effect of lattice defects, 666
  influence of recovery and recrystallization, 698–701
Arsenic, *see* As
As–Cu, 265
AsMn, 315
AsNi phases, antiferromagnetism of, 291–292
Atomic moments, 11–18, 126–127
  influence of local environment, 531
Atomic order
  Al–Fe, 561–569
  band theory, 528–531
  Co–Fe, 544
  collective electron model, 528–544
  effect on Curie temperature, 540–541, 550, 567
    on magnetic properties, 523–573
    on magnetocrystalline anisotropy, 548, 553, 569
    on magnetostriction, 548, 553, 561
  Fe–Ni, 549–555

  irradiation-induced, 554, 601
  model of ordered alloy, 537
  ordered phases, 308, 309
  theoretical considerations, 528–544
Atomic susceptibility, definition, 124
Au, 257
Au alloys
  with B elements, 281
  magnetic behavior, 283
  susceptibility, 271
  with T metals, 266
Au–Bi, 282
Au–Co, 269, 271, 283, 300, 616
Au–Cr, 271, 283, 288, 308
AuCu, 264, 308
$AuCu_3$, 264
Au–Fe, 271, 283, 308
Au–Mn, 39, 268, 271, 283, 288, 308, 318
Au_2MnIn, 319
Au_2MnZn, 319
Au–Ni, 269
Au–Pd, 272
Au–Sb, 282
Au–Ti, 269
Au–V, 269, 271
Au–Zn, 281

B

Ba, 253–254
Ballistic galvanometer, 168, 210–211
Band theory of solids, 25–33
  atomic order, 528–531
Barium, *see* Ba
B compounds with T metals, 316
B elements, 251
  alloys of, 280, 305
  phases with T metals, 283
Bethe-Peierls method, treatment of atomic order, 21
Bi
  with small additions of Pb, Sn, and Te, 266
  susceptibility
    anisotropy, 253, 255
    change at melting 257
Bi–Cu, 282
BiMn, 315
Bi–Sb, 264
Bismuth, *see* Bi

Bloch-Bloembergen equations, 97, 100–101
Bloch equations, 824
Bloch wall, 81–85
  arrays, stabilization of, 770–773
  displacement, force function, 652–654
  energy, 84
  enhancement of rf field strengths, 832
  motion, 85–88
  pinning by local directional order, 600
  stabilization by directional order, 759–770
  stabilization field, 773–776
  thickness, 84
  types I, II, definition, 630
Bohr magneton, 14
Borides of transition elements, 316
Boron, see B
Bose function, 41
Bosons, 40
Bragg-Williams method, treatment of atomic order, 20
Brillouin function, 16, 21, 22, 130
Brillouin zone, 28–29
Brownian motion, 393
Brown's equations, 371, 372
Brown's paradox, 375, 381, 392
Buckling, 89, 373–374, 412
Burgers vector, 626

C

C, interstitial, 593
Ca, 253–254
Cadmium, see Cd
Calcium, see Ca
Carbide, 317
  χ-type, 344
Carbon, see C
CCo_2Mn, 317
Cd, susceptibility, 255, 256
  change at melting, 257
Cd alloys
  magnetic behavior, 283
  susceptibility of γ phases, 281
Cd–Cu, 265, 281
Cd–Ni, 281, 283
Cd–Pd, 281, 283
Cd–Pt, 281, 283

Ce, 261–262
Cementite, 317, 344
Cerium, see Ce
Cesium, see Cs
Ce subgroup metals, 34–36
C–Fe, 514, 515, 517, 518, 580, 777
  phase analysis of, 343–347
CFe_3, 515, 519, 520
  magnetic behavior, 317
C–Fe–Si, 777
Chain of particles, 455
Chain of spheres, 375–377
  model, 418, 455
Chromium, see Cr
Circular cylinder, 412
Co, 36
  change of moment and Curie temperature by small additions, table, 30
  domain structure, 647
  magnetic data, 295–299
  magnetic moment, 37
  NMR studies on, 832–837
Co alloys, 288, 299, 306–307
  Curie temperature, 304–305
  magnetic data table, 312
  NMR studies, 832–837
  with palladium and platinum group, 37–38
  with rare earths, 320–321
Cobalt, see Co
Co–Cr, 283, 299, 530, 532
Co–Cu, 269, 299, 616
  ferromagnetic resonance, 825–831
  magnetic behavior of, 300
Coercive force, 46, 371, 410, 412, 653–662
  Bloch wall displacement, 655–658
  deformation, 663, 673–677
  dependence on inclusion size, 518
    of packing fraction, 445
    on particle size, 378, 386–393
  dislocation groups, 663
  effect of interactions, 444–445
    of nonferromagnetic precipitate, 513
  influence of dislocation arrangement, 661–662
  lattice defects, 655, 658, 668–677
    of deformed crystals
      Co, 671
      Fe, temperature dependence, 668–670

Coercive force—*cont.*
  Ni, 671
  of mixtures of fine particles, 438
  particle diameter and, 390
  particle volume and, 388
  recovery, 704–707
  recrystallization, 704–707
  temperature dependence, 658–659
  of undeformed crystals, 670
Coercive force ratio, 430
Coercive force remanence, 429–430
Coercivity, *see* Coercive force
CoF_2, 835
Co–Fe, 299, 307–308, 591
  diffusion, 802
  directional order, 802
  dissolved elements in, 302–303
  effects of order, 544–549
  magnetocrystalline anisotropy, 549
  neutron diffraction, 546
  phase analysis, 336
CoFe–Mn, 302
Co–FeNi, 578, 806
CoFe–Ti, 302
CoFe–V, 302
Co–Gd, 320
Coils, table, 200
Columnar structure, 730
Co–Mn, 299, 302, 317, 530
Compensation point, 24
Complex permeability, 194, 195
  measurement, 230–234
Composition fluctuations, 480–481
Co–Ni, 302, 309, 529, 591, 595, 598, 615
  diffusion, 807–809
  directional order, 807–809
  magnetic behavior, 302
$CoNi_3$, 309
Constitution analysis
  by anhysteretic properties, 358–363
  by hysteresis properties, 347–353
  of Fe–Pt system, 349–352
Constitution of multiphase alloys, 331–363
Contact term, 832
Cooperative magnetic behavior, 18–25
  compensation point, 24
  exchange integral, 19

  exchange interaction, 19
  Hamiltonian, 19
  magnetostatic behavior, 18
  magnetostatic interactions, 455
  molecular field Hamiltonian, 20
  molecular field theory, 19–20
Co–Pd, 306, 309
$CoPd_3$, 309
Copper, *see* Cu
Co–Pt, 306, 309, 544, 572
Co_5–rare earth phases, 320
CoS, antiferromagnetism, 292
Co–Si, 299
Cotton-Mouton effect, 144, 145
Co–V, 299
Co–Zn, 283, 284
Cr
  antiferromagnetism, 36, 39, 287, 303
    spin-density wave, 39
  orbital magnetism, 255
  orbital paramagnetism, 33
  susceptibility
    temperature coefficient, 258
    temperature dependence, 259, 260
Cr alloys
  ferromagnetism, 319–320
  magnetic behavior, 278, 280, 287–288, 315
  magnetic data, table, 311
Cr–Co, 287
Cr–Cu, 283
Creep
  magnetic, 385
  magnetostrictive, 794
Cr–Fe, 278, 283, 299
$CrIr_3$, 308
Critical particle dimensions, 377–393
  dependence on packing fraction, 446–447
  for single-domain behavior, 377–381
Cr–Mn, 280, 283, 290–291
Cr–Ni, 299, 301, 530
Cr–Os, 280
Cross slip, 628
Cross-tie wall, 85
Cr–Pd, 287
Cr–Pd series, 288–289
CrPt, 309
$Cr_{1.2}Pt_{2.8}$, 309

Cr–Pt series, 308, 309
Cr–Re, 280
Cr–Rh, 280
Cr–Ru, 280
CrSb, 291–292
CrSbMn alloys, 39
CrSe, 291, 292
CrTe, 315
Cr–V, 278, 280
Crystal defects, *see* Defects
Cs, 257,
Cs–K, 264
Cu, susceptibility
  change at melting, 257
  temperature dependence, 253
Cu alloys
  with B elements, 265, 281, 283
  with T metals, 266
Cube-on-edge texture, 735, 739
CuF_2, 312
Cu–Fe, 514, 515, 517, 518, 616
Cu–Ga, 265
Cu–Ge, 265
Cu_2GeMn, 319
Cu–In, 265, 281
Cu_2InMn, 317
Cu_2Mg–$MgZn_2$, 283
Cu–Mn, 270, 271
  exchange anisotropy, 271
  magnetic behavior, 317–318
Cu_3Mn, 307, 317
Cu–Mn–Sn, 318, 319
Cu–Ni, 272, 283, 298, 299, 529
Cu–Pd, 272
Curie constant, 7, 16
  molar, 126
Curie law, 8, 16, 17, 397
Curie paramagnetism, 15
Curie temperature, 5, 125, 294, 303–304
  effect of order, 540, 550, 567
  measurement, 132
  paramagnetic, 7, 9–10, 294
Curie-Weiss law, 7, 125–126
Curie-Weiss paramagnetism, 21, 23
Curling, 89, 374–375, 380, 391, 412–414
Cu–Sb, 265, 282
Cu–Si, 281
Cu–Sn, 265, 281, 318–319

Cu–V, 269
Cu–Zn, 265, 269, 281, 282
CW–Co mixtures, anhysteretic measurements, 360
Cyclotron frequency, 30
Cylinder, demagnetization coefficient, 129

D

Damping parameter, 823
Defects
  effects on magnetic properties, table, 623
    on magnetization processes, 621–685
  energy, 694
  superposition of, 658
Deformation bands, 627
Deformation textures, 728
Deformation-induced anisotropy, 602–616, 709–710
de Haas-van Alphen effect, 31–32
Demagnetization, thermal, 46
Demagnetization factor
  geometric, 52, 129, 448, 451, 462
    table, 193
  internal, 353–354, 356–358, 449, 462
Demagnetization energy, 61–63
Demagnetization field, 52, 62, 129, 448
Diamagnetism, 4, 8–9
  bound electrons, 11
  conduction electrons, 11, 30
  metals, 30–32
  susceptibility, 12
Differential susceptibility, 691–692
Diffusion, 749–810
  activation energy, 778–781
  data
    Co–Fe, 802
    Co–Ni, 807–809
    Fe–Ni, 803–807
  magnetic detection, 751–785
  measurement, 778–785
  relaxation times, 781
  self-, 751
  short-range, 750
  time constants, 796

Diffusion anisotropy, 583, see also Induced anisotropy
Diffusion phenomena, 749–751
Digital voltmeter, 167
Diode rectifier, 217
Dipole, force on, 150
Dipole interaction, 62–63, 443–445
Direct current magnetic measurements, 123–185
Directional order, 580–602, 603, 751–810
  activation energies, 774
  activation volumes, 784
  Al–Fe, 799
  Co–Fe, 802
  Co–Ni, 807–809
  deformation-induced, see Directional order, slip-induced
  diffusion and, 751
  effect of irradiation, 601
    of oxygen, 601
  experimental results, 590–597
  Fe–Ni, 803–807
  Fe–Si, 796
  interference with long-range order, 792
  interstitial mechanism and interstitial energy, 751–752
  magnetic anisotropy coefficients, 586–587, 754–756, 789–790
  ordering configuration, 785–787
  slip-induced, 602–617
  stabilization of domain walls, 599, 759–770
  substitutional alloys, 785–810
Directional order energy
  temperature dependence, 791–792
  theory, 584–590, 785–810
  time dependence, 793–794
Directional solidification, 730
Disaccommodation, 584, 751, 764, 773, 798, 802
Dislocation dipoles, 635
Dislocation groups, 635, 662, 663
  recovery of coercive force and initial susceptibility, 664
Dislocation group strength, 662
Dislocation network, 627
Dislocation pairs, 604
Dislocation pileup, 627

Dislocations, 624–628
  annihilation of, 696
  climbing of, 696
  screw, 626
Dislocation tripole, 635
Disordered alloy, model of, 534
Domain(s)
  definition, 46
  in ferromagnetism, 81–85
  origin, 371, 392
Domain structure, 5
  changes during magnetization, 646–650
  deformed crystals, 649
  effect of plastic deformation, 646–650
  large crystals, 371
  multiaxial crystals, 647–650
  undeformed crystals, 647–649
  uniaxial crystals, 647
Domain wall, see Bloch wall, Néel wall, Cross-tie wall
Dy, 34, 35, 261–263
DyHo, 321
$DyMn_2$, 293
Dynamic magnetic effects, 194
Dysprosium, see Dy

E

Easy axes (directions), 65–66, 137, see also Anisotropy, magnetic
Eddy current anomaly, 239
Eddy currents, 90, 194, 238
Edge dislocation, 626
Effective field, 445
Electromagnet, 151
Electromechanical fluxmeter, 168
Electron(s)
  effective mass, 29, 33
  itinerant, 39–40
  nearly free, 26–29
  tight binding approximation, 29
  tightly bound, 29–30
Electronic structure
  metals, 25–33
  nonmetals, 10–25
Electron interactions, resonance investigations, 113
Electronic structure, resonance investigations, 113

Electron–nuclear interaction, 114
Electron paramagnetic resonance, 102
Elements
 ferromagnetic, and their solid solutions, 295–310
 nonferromagnetic, survey of magnetic properties, 251–264
Ellipsoid
 ideal, 371, 380, 381
 nonuniformly magnetized, 448–449, 462
 uniformly magnetized, 52–62
  demagnetization factor, 52, 448
  demagnetizing fields, 52
  Helmholtz free energy, 52
  magnetostriction relationships, 57
  magnetothermal effects, 57–58
  thermodynamic potentials, 55–57
  work done in magnetization, 52–55
Elongated particles, 389
Energy product, table, 156
Epstein square, 235
Er, 34, 262
Erbium, see Er
ESD particles, 375, 391
Eu, 18, 261–262
EuO
 Curie temperature, 133, 134
 Faraday rotation, 145
Europium, see Eu
EuS, 131
Exchange anisotropy, 75, 137, 368, 543–544
Exchange energy, 59–61
 classical expression, 59
 exchange interaction, 59
 exchange stiffness, 61
 spin Hamiltonian, 59
Exchange field, 822
 spin-wave resonance, 116
Exchange lengths, 643
 table, 643
Exchange polarization effects, 543–544
Exchange shift, ferromagnetic resonance, 822

F

Fabry's formula, 201

Fanning
 nonsymmetrical, 376
 symmetrical, 375, 376, 418
Faraday effect, 144, 145
Faraday rotation, 144
 table of data, 145
Faraday's law, 149
Fe
 change of moment by additions, table, 306
 coercive force, 670
 with dissolved B metals, 305
 with dissolved T metals, 305
 Faraday rotation, 145
 ferromagnetism, 36
 magnetic data, 295–297
 magnetic moment of, 37
 magnetic specific heat of, 9
Fe alloys
 magnetic behavior, 275, 278, 295, 305–307
 data table, 311–312
 with nontransition elements, magnetic moments, 37
 with palladium and platinum group, 37–38
ϵ-Fe carbide, 344
Fe carbides, 317
Feedback amplifier, 204
Fe–Ga, 281
Fe–Ge, 295
Fe group elements
 atomic moments, 11
 paramagnetism, 17
Fe–Mo, 275
Fe–Mo–Nb, 275, 278
Fe–Mo–Ni, 806
Fe–Mo–Re, 275
Fe–N, 514, 517
Fe_4N, 517
$Fe_{16}N_2$, 517
Fe–Nb, 278
Fe–Ni, 307, 309, 530, 532, 533, 540, 578, 590, 591, 595, 598, 601, 603, 611
 atomic ordering, 549–555
 diffusion, 803–807
 directional order, 803–807
 disordered, 539
 textures, 732–734

FeNi$_3$, 307, 309, 540, 549, 550, 581, 595, 613
Fe–Pd, 306, 308, 309, 544
  order effects, 569–573
FePd$_3$, 309, 544
Fe$_3$Pd, 570
Fe–Pd–Rh, 275
Fe–Pt, 288, 308, 309
  order effects, 569–573
  partially ordered, 572
  phase analyses, 349–352
FePt$_3$, 544, 570
Fe$_3$Pt, 570
  neutron diffraction, 573
  phase analysis, 351
Fe–Re–Ru, 275
Fe–Rh, 310
FeRh, 39, 308
Fe–Rh–Ru, 275
Fermi contact term (NMR), 832
Fermi energy, 26–27
Fermi statistics, nearly free electrons, 26
Fermi surface, 27
Ferrimagnetism, 4, 5
  Curie temperature, 25
  molecular field theory, 19–20
Ferromagnetic alloys, intrinsic properties, 294–322
Ferromagnetic behavior, principles, 4, 5, 45–92
Ferromagnetic metals, intrinsic properties, 294–322
Ferromagnetic phases, intermediate, 310–322
  binary, with T metals, magnetic data tables, 311–312
  with NiAs structures, table, 315
Ferromagnetic resonance (FMR), 102–106, 816–831
  Co–Cu, 825–831
  damping, 823–824
  effective field, 818–820
  experimental techniques, 110
  linewidth, 823–824
  sources of linewidth, 824–825
  relaxation times, 824
  spin pinning, 104
  spin-wave resonance, 104
Ferromagnetic solid solutions, 295–310

Ferromagnetism, basic aspects, 4, 5, 45–92
  molecular field theory, 19–20
  transition temperature in, 24
  weak, 10
Fe–Ru, 275
FeS, 292
Fe–Si, 295, 578, 591, 595, 752
  commercial production, 736–737
  diffusion, 796
    measurements, 799
  directional order, 796
  induced anisotropy, 796
  phase analysis, 336–343
  textures, 734–744
    metallurgy, 737–739
Fe–Sn, 295
Fe–V, 278, 295, 297
Fe–Zn, magnetic behavior, 284
Fe–Zr, 278
Field, *see* Magnetic field
Fine particles, 366–463
  agglomerates, 443
  anisotropy, 409–411
  application to problems of physical metallurgy, 368–370, *see also* pp. 366–463
  arbitrary angle between field and axis, 414–417
  chemically inhomogeneous, 401–405, 408–409
  critical particle dimensions, 377–393
  cubic anisotropy, 411–412
  Curie temperature, 405–406
  determination of particle volumes, 406–408
  distribution functions of properties, 437–438
  freezing curves, *see* Freezing curves
  general ellipsoid, 411
  Gerlach principle of, 438–440
  hysteresis losses, 422–427
    hysteresis integrals, 425–427
  intrinsic magnetic properties of, 367–368, 405–406
  magnetization processes, 370–377
  magnetostatic interactions, 398–401, 442–463
  melting curves, *see* Melting curves

mixtures with different shapes and sizes, 436–437
nucleation field, 372, 411
  arbitrary angle between field and axis, 414–417
  buckling mode, 374, 412
  chain of spheres, 376, 418
  field curling, 374, 412
  internal, 418
  rotation, in quasi-unison in, 413
    in unison, 374
  sphere, prolate spheroid, and plate of, 417–418
nuclei of reverse magnetization, 371, 392
parallel rotation, 375
particle size, dependence of magnetic properties on, 377–393
remanence, 383–385, 427–431
  alternating field demagnetization, 428
  analysis, 432
  anhysteretic, 428
  coercivity, 429–430
  dc field demagnetization, 427
  particle size dependence, 386–393
  thermally stable, 385
remanence curves, 427–431
  areas between, 447
  field shifts between, 447
  with magnetostatic interactions, 458
  mixtures of particles, 440–441
  related quantities, 449
  relations between, 429, 458
remanent torque, 420–421, 441
rotational hysteresis, see Rotational hysteresis
superparamagnetism, see Superparamagnetism
suspensions, 443
temperature dependence of M_S, 405
without thermal agitation, 409–463
thermal fluctuation effects, 382–386, 393–405, see also Superparamagnetism
thermomagnetic treatment, theory of, 490–491
torque curves, 418–422
Flux ball, 161
Flux gate, 164

Flux integration, 209–217
  calibration techniques, 217
  electrical, 211–217
  feedback, table of, 214
  mechanical, 210–211
  passive networks, table of, 212
Flux pickup systems, 206–209
Flux spool, 161
F_3NiRb, Faraday rotation, 145
Foerster ferrograph, 222
Foerster probe, 219
Free electron gas, 26
Freezing curves, 431–436
  ac, 433–434
  effect of thermal agitation on, 434–436
  pulse, 449
Freezing process, anhysteretic, 431–436, 452–453
Frequency factor, 382

G

Ga, susceptibility
  anisotropy, 255
  change at melting, 257
  temperature dependence, 253
Gadolinium, see Gd
Gallium, see Ga
Galvanomagnetic instrumentation, 148–149
Ga–Mn, 281, 310, 312
Gage factor, 142
GaV_3, 287
Gd, 34, 261–262, 297, 303
Gd–Er, 297
$GdMn_2$, 293
Ge, susceptibility
  change at melting, 257
  temperature dependence, 253
Ge–Ni, 295
Germanium, see Ge
g factor, 14, 127, 817
  orbital motion, 13
  spin, 13
γ ratio, see Gyromagnetic ratio
Giant moment, in dilute Fe–Pd, 275
Gilbert's equation, 383, 823
Gold, see Au
Goss texture, 734–735
Grain boundary, 694

Grain growth
  discontinuous, 730
  exaggerated, 730
  inhibition, 730
  surface energy, 741–744
Gram susceptibility, definition, 124
Graphite, anisotropy of susceptibility, 255
Grassot fluxmeter, 210
Gyromagnetic ratio, 11–13, 94, 816–818
Gyromagnetic resonance, 94–100
  angular momentum operator, 96
  magnetic resonance losses, 97–99

H

Hafnium, see Hf
Hall voltage, 148
  extraordinary, 148
  measurement of, 148
  ordinary, 148
Hard axes (directions), 68
Helical ordering of moments, 10
Helmholtz coils, 157–158, 199
Helmholtz free energy, 52
Heusler alloys
  ferromagnetism, 39
  magnetic properties, 317
  orbital magnetism, 255
  susceptibility, 258
Hg, susceptibility
  anisotropy, 255
  change at melting, 257
  temperature dependence, 253
Hg alloys, 283
  with alkali metals, 280
  susceptibility of γ phases, 281
HgMn, 288
Hg–Ni, 283
Hg–Pd, 281, 283
Hg–Pt, 283
Ho, 262–263
Holmium, see Ho
H_3U, 312
Hume-Rothery β phase, 317
Hund's rules, 14, 40
Hydrogen, see H
Hyperfine energy, 114
Hyperfine field, 836
  from resonance measurements, 116

Hysteresis integrals, 425–427
  table of, 426
Hysteresis loop, see also Magnetization curve
  ac measurement, 221–230
  Bloch wall displacement, 654–655
  characteristic features, 692
  dc measurement, 135–137
  fine particles, 368–369, 409–418
  general considerations, 45–51
  major, 135
  minor, 136–137
  particle size effects, 368–369
  Perminvar, 599–601
  single-domain particles, 409–418
Hysteresis losses, see also Losses
  fine particles, 422–427
    ac, 422
    rotational, 422
  general remarks, 692
  measurement with ac field, 230–242
  recovery, 707–709
  recrystallization, 707–709
  rotational, measurement from torque curves, 422–427
  Warburg's law, 58

I

Impedance, 231–232
  measurement, 232–234
Impedance bridge, 232
In, susceptibility
  anisotropy, 255
  change at melting, 257
  temperature dependence, 253
Incremental susceptibility, 691–692
Indium, see In
Induced anisotropy
  Al–Fe, table, 799
  definition, 75
  deformation, 602–616, 709–710
  dynamics, 749–810
  effect on domain walls, 599
  energy of, 584–597, 754–757
  experimental, 590–597, 795–796
  Fe–Si, 796
  formal theory, 584–590, 785–791

magnetic field anneal, 577–601, 745, 754–757, 788–791
stress anneal, 597–599, 758–759
textures by magnetic field anneal, 745
Induction
  definition, 124, 190
  measurement of, 149, 167–168, 205–217, 221–230
Induction curve tracers, 221–230
  accuracy of, 223–228
  bandwidth limitation errors in, 226–227
  description of, 221–223
  large-amplitude errors in, 227–228
  low-frequency errors in, 223
  sensitivity limits in, 228–230
Induction magnetometer, 167–168
Induction methods, ac, 205–217
Infinitely long cylinder, 372–375, 380–381, 389, 412–418, 424–425, 428–429, 455–459
Inhibition of grain growth, 730
Initial permeability, *see also* Initial susceptibility
  apparent, 193
  complex, 194–196
  definition, 135, 193–196, 691
  disaccommodation, 773–776
  dynamic effects, 194–196
  effect of nonferromagnetic precipitate, 513–521
  grain growth, 701–704
  measurement, 135–137, 193–196
  recovery, 701–704
  recrystallization, 701–704
  time decrease, 773–776
Initial susceptibility, *see also* Initial permeability
  anhysteretic, 447–451, 456–458, 461–463
  Bloch wall displacements, 653, 658–662
  simultaneous with rotation, 660–662
  definition, 691
  dependence on temperature and deformation, 677–682
  table, 659
  grain growth, 701–704
  influence of dislocation arrangement, 661–662
  lattice defects, 658–662

recovery, 701–704
recrystallization, 701–704
$InNi_3$, 312–313
In–Sc, 36
$In_{24}Sc_{76}$, 311
Integrating digital voltmeter, 167
Integrating flux meter, 162–163
Integrating network, 211–212
Integrator, 210
Interaction anisotropy, magnetostatic, 137, 375–377, 418, 445, 455
Interaction domains, 460
Interaction field, 452
  dynamic, 452–455
Interactions
  exchange, 18–25
    antiferromagnetic, 22–25
    ferromagnetic, 18–22, 540–543
    indirect (super), 34
    rare earth metals, 33–36
    spin waves, 40–42
    transition metals, 36–40
  magnetoelastic
    Bloch wall and dislocation, 629–634
      and dislocation configuration, 635–639
      and dislocation group, 639–641
      dislocations and magnetization, 622–625, 628–645
      lattice imperfections and magnetization, 622–625, 628–645
      and nearly uniform magnetization, 641–645
  magnetostatic
    anisotropy, *see* Interaction anisotropy
    cooperative effects, 18, 452, 455, 460
    critical particle size, 446–447
    fine particle assemblies, 442–463
    general survey, 353–363
    between inclusions and magnetization, 513–521
    superparamagnetism, 398–401
Interface thickness, 408
Intermediate phases
  ferromagnetic, 310–322
  general magnetic behavior, 280–283
Internal friction, 514, 794, 800
Interstitial atoms, 580–590, 751–754

Interstitial directional ordering, mechanism and interaction energy, 580–590, 751–754
Ir alloys with rare earths, 321
Iridium, see Ir
Iron, see Fe
Iron magnets, 151
Isoperm, 603, 732

J

Jordan aftereffect, 238

K

K, 257
Kerr effect, 144–146
Kerr rotation, 145
Kersten model, 521
Kinetics of processes involving diffusion, 407, 749–810
Kinetics of transformations, 750
Kittel condition (FMR), 820
Knight shift, 113, 114, 836
Kondo theory, 270
Kundt's constant, 144

L

La, 258
La alloys, 292–293
"λ"-type anomaly, 9
Landau-Lifshitz equations, 97, 101, 102, 823
Landé factor, 127
Landé's formula, 14
Langevin function, 16
Langevin paramagnetism, 16, 395
Langevin's equation for diamagnetic susceptibility, 12–13
Lanthanides, 261
Lanthanum, see La
Larmor frequency, 11–13, 94–97, 384, 816–818
Larmor precession, 11–13, 94–97, 384, 816–818
Larmor's theorem, 11–12
Lattice defects, see Defects
Laves phases, 293

Lead, see Pb
Lecher system, 242
Lenz's law, 11
Li, 257
Lithium, see Li
Lloyd-Fisher square, 192, 235
Localized moments, 275, 308
    of transition metals, 269
Lockin amplifier, 166
Long-range atomic order, see also Atomic order
    development of, 750, 802
    effects on magnetic properties, 523–573
    interference with directional order, 792–793
Loop tracer, see Induction curve tracers
Losses, magnetic, measurement of, 235–240
    calorimetric method, 238
    eddy currents, 238
    hysteresis, see Hysteresis loss
    Jordan aftereffect, 238
    loss separation, 239
    planimetric method, 238
    viscosity, 776–778
    wattmeter method, 236–237
Low-angle boundary, 694, 696

M

Magnesium, see Mg
Magnetic anisotropy, see Anisotropy, magnetic
Magnetic behavior, survey, 4–10
Magnetic circuit, 190–196
    closed flux structures, 191–192
    dynamic magnetic effects, 194–196
    open flux structures, 193–194, 353–354, 448–449, 451
    yoke structure, 192
Magnetic field anneal, 75, 577–601, 745, 754–757, 788–809
Magnetic field gradients, uniform, 154
Magnetic field measurement
    ac fields, 205–220
        ferromagnetic probes, 218
        Hall effect, 218
    dc fields, 160–166
        electron beam, 163–164
        Hall effect, 163

induction method, 160–163
magnetooptical methods, 165–166
magnetoresistance, 163
mechanical methods, 165
nuclear magnetic resonance, 165
permeability variation, 164–165
Magnetic fields
  ac, production of, 196–205
    air core coils for, 198–202
    closed flux structures for, 196–197
    power generation, 202–205
    tuned coils for, 201
  dc, production of, 151–160
    field regulation, 160
    iron magnets and pole piece designs, 151–155
    power supplies, 160
Magnetic field uniformity, 152–154, 198–201
Magnetic flux, definition, 124
Magnetic measurements
  ac, 189–242
  dc, 123–185
  field measurements, *see* Magnetic field measurements
Magnetic moments, *see also* Atomic moments
  data for metals and alloys, 249–322
  definitions, 124
  experimental determination, 127–131
  gyromagnetic properties, 94
  theory, 4–42
Magnetic reluctance, 191
Magnetic resonance, 93–117, 815–837
  anisotropy field, 100
  electron paramagnetic resonance, (EPR), 102
  exchange field, 100
  ferromagnetic resonance (FMR), 102–106
  gyromagnetic resonance, 94–100
  hyperfine field, 100
  linewidth, 97
  Lorentzian resonance line, 98
  losses, 97
  measurements of, 107–112
  in metals, 815–837
  Mössbauer effect in, 106–107, 111–112
  nuclear, *see* Nuclear magnetic resonance
  nuclear quadrupole resonance, 101–102
  sample preparation, 111
Magnetic viscosity, 751, 776–778
Magnetization changes, *see also* Buckling, Curling, Rotation, Wall motion
  in fine particles, 370–377
  by rotation, 88
  simultaneous wall motion and rotation, 657
  by wall motion, 85–88
  work done during, 52–55
Magnetization curve, 45–51, 650–667
  anhysteretic, *see* Anhysteretic magnetization
  demagnetized state and, 46
  dependence on particle size, 368
  of fine particles, 409–411
  hysteresis in, *see* Hysteresis loops
  permeability ratios, 46, 47
  of polycrystalline materials, 46
  Rayleigh region, 47–48, 87
  of single crystals, 50
  of "soft" magnetic materials, 46
  susceptibility ratio, 46–47
  technical saturation, 50
  thin films and fine particles, 50–51
Magnetization measurement, *see* Magnetic measurement
Magnetization reversal
  in fine particles, 425–427
  inhomogeneous, in fine particles, 412
Magnetocrystalline anisotropy, 63–68, 821
  biaxial, 140
  coefficients, table, 66–68
  cubic, 63, 138, 140
  energy, 63–68
  hexagonal, 64, 138
  influence on ferromagnetic resonance, 820–822
  measurement, 115, 137–142
  order effects, 548, 553, 568
  stress dependence, 73–74
  temperature dependence, 65
  tetragonal, 64, 139
  theory, 65
  triaxial, 140
  uniaxial, 140
Magnetoelastic energy, 68–74
  elastic compliances, 71

Magnetoelastic energy—*cont.*
  elastic stiffness coefficients, 71
  elastic stiffness constants, 69
  equilibrium magnetization direction
    under fixed stress, 74
  free energy, strain dependence, 68–69
  magnetostriction constants 71–72,
    see also Magnetostriction
  stress energy, 72–73
Magnetomechanical ratio, 13, 816
Magnetometers
  analytical balance (force), 174–178
  calibration, 184
  horizontal displacement, 179–180
  induction, 167
  integrating digital voltmeter, 168–170
  rotating sample, 174
  torque, 180–183
    calibration, 182
  torsion balance, 178
  vibrating sample, 170–173
Magnetomotive force, 190
Magnetooptical effects, 144–146
Magnetoresistance, 146–147, 163
  measurement of, 148
Magnetostatic energy, 61–63, 463
  demagnetization energy, 61–63, 463
  interaction energy, 62–63, 444
  self-energy, 81
Magnetostatic interactions in fine particle
    assemblies, 357, 442–463
  basic considerations, 442–447
  cooperative phenomena, 452
  experimental approach, 447–451
  experimental results, 450–451
  mean field approximation, 455
  multidomain particles, 357, 462–463
  theoretical approach, 452–463
  two-particle model, 443–445
Magnetostriction, 76–81
  aftereffect of, 757–758
  anisotropy, 76–79
  of cubic crystals, 76
    constants, table, 78
  ΔE effect in, 79–80
  effect of order on, 548, 553, 561
  forced-type, 80–81
  form effect in, 80–81
  of hexagonal crystals, 77–78
    constants, table, 79

  measurement, 142–144
  polycrystalline material, 78–79
  tetragonal crystals, 77–78
  Young's modulus and, 79
Magnetostrictive coupling, 622
Magnetothermal effects, 57–58
  magnetic specific heat, 9, 21, 42
  magnetocaloric effect, 58
  Warburg's law, 58
Magnetothermoelectric effect, 147
Magnons, 40
Manganese, *see* Mn
Melting curves, 431–436
  ac, 431–432
  effect of thermal agitation on, 434–436
  thermal, 431–432
Mercury, *see* Hg
Metals
  definitions of A, B, and T types
  diamagnetism of, 30–32
  electronic structure of, 25–33
    band theory, 25
    energy bands, 25
    localized states, 26
    nearly free electrons, 26–29
  ferromagnetic, 294–322
  magnetic resonance measurements in,
    815–837
  nonferromagnetic, 250–293
  paramagnetism in, 32
Mg, 253–254
Micromagnetism, 89, 371, 392, 642, *see
  also* Buckling, Curling, Rotation in
  unison, Fine particles, Nucleation
  field
Miscibility gap, 480
Mn, susceptibility
  change at melting, 257
  temperature dependence, 259–260
α-Mn
  antiferromagnetism, 36
  temperature dependence of suscepti-
    bility, 259
β-Mn, 259
Mn alloys, 270–271, 293, 313–315
  binary with noble metals, 39
  ferromagnetism in, 319–320
  with rare earths, 320
  table, 311
Mn–Ni, 299, 303

MnNi$_3$, single-domain behavior in partially ordered, 561
Mn–Pd, 281, 289–290
MnPt, 290
MnPt$_3$, 307–309
Mn$_5$–rare earth phases, 320
MnRh, 290
MnSb, 315
Mn–Sn, 285
Mn$_2$Tb, 293
MnTe, 291–292
Mo
  orbital magnetism, 255
  susceptibility of, temperature coefficient, 258
Molar susceptibility, definition, 124
Mössbauer effect, 106–107, 115
  experimental techniques for, 111
  hyperfine field, 107
Molecular field theory, 19–25
  antiferromagnetism and ferrimagnetism, 22–25
  ferromagnetism, 20–21
  Hamiltonian, 20
  influence of ordering on Curie temperature, 541–543
Molybdenum, see Mo
Moments, see Magnetic moments
Multidomain particles, 392, 462–463
  anhysteretic measurements, 358–360
  coercive force of mixtures with single-domain particles, 440
  interactions among, 357
Multiphase alloys
  anhysteretic properties, 353–363
  constitution analysis, 331–363
  interaction phenomena, 353–363
  magnetic behavior of precipitates, see Fine particles, Precipitates, and specific alloy systems
  phase analysis, 333–347
Mutual energy, 62–63, 444
Mutual inductance bridge, 234

N

N, interstitial. 539
Na, 257
Nb, 255
Nb–Ni, 295
Nb–Ru, 278
Nb–Ta, 264, 278
Nd, 36, 261, 262
NdPr, magnetic behavior of, 321
Néel hyperbola, 126
Néel temperature, 8, 23, 126
  measurement of, 132
Néel wall, 85
Neodymium, see Nd
Neutron diffraction, 10, 531–534, 550
  Al–Fe, 564
  Co–Cr, 532–533
  Co–Fe, 544
  magnetic moments of ordered phases, table, 309
  Ni alloys, 534
  Pd and Pt alloys, 544, 571
Neutron irradiation of Fe–Ni alloys, 554, 601
Ni, 36, 294–298
  approach to saturation, 683–684
  change of moment and Curie temperature by small additions, table, 300
  coercive force, 670–671
  initial susceptibility, 677–680
  orbital magnetism, 255
Ni alloys, 295, 299
  Curie temperatures, 300 (table), 304
  magnetic moments, 300 (table), 309
  with nontransition elements, 37
  with Pd and Pt, 37–38
Nickel, see Ni
NiMn, 302, 303, 307, 308, 530
  influence of atomic order, 555–561
  roll anisotropy, 615
Ni$_3$Mn, 531, 616
  disordered, 556
  exchange anisotropy, 543, 559
  order effects, 555–561
  unidirectional anisotropy, 558
Ni–Mo, 281, 301
Niobium, see Nb
Ni–Pd, 302, 306
NiS, antiferromagnetism of, 292
Ni–Sb, 299
Ni–Sn, 283, 299
Ni–Ta, 295
Nitrogen, see N
Ni–V, 295, 299
Ni–W, 301

Ni–Zn, 281, 283, 299
Noncubic paramagnets, torque expression, 140
Nonmetals
  cooperative magnetic behavior, 18–25
  electronic structure of, 10–25
  magnetic properties of, 10–25
Nuclear electric quadrupole moment, 693
Nuclear magnetic resonance (NMR), 100–102, 165, 534, 693–694, 713–718, 831–837
  deformation effects, 836
  experimental techniques, 108–110
  influence of structural defects, 714
  line broadening in, 114
  magnetic dipole effects, 714
  quadruple effects, 714–715
  spin–lattice relaxation, 101, 113
  spin–spin relaxation, 101
Nuclear magneton, 831
Nuclear quadruple resonance, 101–102, 114
Nucleation field, *see* Fine particles

O

Orbital motion
  contribution to susceptibility of T metals, 255
  diamagnetism of conduction electrons, 30–32
  diamagnetism due to, 11–13
  paramagnetism of conduction electrons, 33
Orbital quenching, 18, 127
Order
  atomic, *see* Atomic order
  directional, *see* Directional order
  spin, *see* Spin order
Order–disorder phenomena, *see* Atomic order
Ordered phases, magnetic moments, table, 309
Orientation, preferred, 475, 724, *see also* Textures, Alnico
Oriented superstructure, 602
Os alloys with rare earths, 321
Osmium, *see* Os
Owen bridge, 233–234

P

Packing fraction, 443
Pair ordering, 584–597, 785–794, *see also relevant listings under* Directional order, Induced anisotropy
  experiment, 590–597, 799–809
  theory, 584–590, 785–794
Palladium, *see* Pd
Para effect, 128
Paramagnetic ions, electronic states of, table, 15
Paramagnetic resonance, 102
  experimental methods, 110
Paramagnetism, 6–7, 32–33, 127
  of conduction electrons, 32–33
  Curie, 15
  Curie–Weiss, 21
  of iron group atoms, 17
  nuclear, 94, 95
  quantum theory, 16
  of rare earths, 17
  Van Vleck, 18, 33
Pauli principle, 13, 14, 18
Pb, 253, 257
Pd, 260
Pd alloys, 272, 293, 309, *see also listings under specific alloys*
  with B metals, 272, 274
  with T metals, 274–275
Pd–Pt, 283
Pd–Re, 274
Pd–Rh, 274
Pd–Ru, 274
Pd_3Th, 293
Pd_3Ti, 293
Pd_3U, 293
Pd–Zn, 281, 283
Permanent magnets, 370
  Alnico, *see* Alnico
  as field source, 155–156
  properties, table, 156
  textures, 730
Permeability, 124, 194
  apparent, 195, 231
  complex, 195
  initial, *see* Initial permeability
  measurement, 230–232
Perminvar effect (loop), 579, 583, 599–601, 760

Perovskite-type structure, 317
Phase analysis, 333–347
 of Fe–C system, 343–347
 of Fe–Pt system, 349–352
 of Fe–Si system, 336–343
 formal theory of, 333–335
 limitations of the method, 335–336
Phase change, in magnetic field, 744–745
Phase-sensitive detector, 166
Plastic deformation, 625–628
Platinum, see Pt
Plutonium, see Pu
Pole figure, 724
Pole piece designs, 151
Polygonization, 696
Potassium, see K
Power generation, ac, 202–205
Pr, 261, 262
Praseodymium, see Pr
Precipitates, see also listings under specific alloy systems, e.g. Co–Cu
 measurement of concentration profile, 408–411
 nonferromagnetic, 513–521
Precipitation, 750
Precipitation alloys, see also listings under specific systems, e.g. Alnico, Co–Cu
 Bitter patterns, 460
 magnetostatic interactions, 459–460
Precipitation hardening, Co in Cu, 826
Prism, infinite, 413
Pt, 255
Pt group elements, atomic moments, 11
Pt–Zn, 281
Pu, 261
Pulsed high fields, 159

Q

Quadrupole resonance, 114

R

Rare earth alloys with T metals, 320–322
Rare earth garnets, 25
Rare earth metals, 261–264
 paramagnetism of, 17
 spin configurations, 33–36
Rayleigh constant, 135
Rayleigh loop, 238
Rayleigh rule, 135

Rb, 257
Recording materials, 370
Recovery, 628, 695–696
 effects in ferromagnetic metals and alloys, 697–712
 in nonferromagnetic metals and alloys, 712–718
 influence on magnetic properties, 689–720
 NMR studies of, 715–718
Recrystallization, 696–697, 728–730, see also listings under Texture
 effects in ferromagnetic metals and alloys, 697–712
 in nonferromagnetic metals and alloys, 712–718
 influence on magnetic properties, 689–720
 in magnetic field, 745
 secondary, 730, 732
Rectifier, contact, 216
Relaxation phenomena in diffusion, 749–785, see also listings under Aftereffect, Directional order, Disaccommodation, Induced anisotropy
 in alternating magnetic fields, 778
 relaxation function, 774
 relaxation times, 781–783
  measurement of small changes in, 784–785
Remanence, 46, 136
 anhysteretic, see Anhysteretic magnetization
 remanence curves, see Fine particles
Remanence coercive force, 429–430
Resonance
 antiferromagnetic, see Antiferromagnetic resonance
 condition, 816
 ferromagnetic, see Ferromagnetic resonance
 magnetic, see Magnetic resonance
Rigid band model, 530
Roll magnetic anisotropy, 602–617, 709–710
 coefficients, table of, 616
 experimental results, 608–616
 long-range order, 605, 606
 of precipitation alloys, 616
 short-range order, coarse slip type, 607

Rotating coil gaussmeters, 161
Rotational hysteresis, 240, 420
  integrals, 425–427
  loss, 422–425
Rotation of magnetization, see also Buckling, Curling
  in bulk alloys, 88
  in fine particles, 89
    parallel, 376
    quasi-unison, 413, 414
    unison, 372–374, 380
Ru alloys with rare earths, 321
Rubidium, see Rb
Russell-Saunders coupling, 13
Ruthenium, see Ru

S

Samarium, see Sm
Saturation magnetization, 5, 50, 690, see also Spontaneous magnetization
  approach to, see Approach to magnetic saturation
  effect of order, 531–543
  influence of recovery and recrystallization, 697
$Sb_\perp$, susceptibility
  change at melting, 257
  temperature dependence, 253
$Sb_{||}$, temperature dependence of susceptibility, 253
Sb–Sn alloys, 267
Sc, 258
Scandium, see Sc
Screw spin configurations, 36
Secondary recrystallization, see Recrystallization
Shape anisotropy, 62, 139–140, 193
  in fine particles, 380, 411–420, 436–438, 455–459, 462–463
Shear strain, 625
Shear stress, 625
Shims
  current type, 154
  ring type, 152–153
Side bands
  of Alnico, 477
  of Cu–Ni–Fe, 477
Silicon, see Si
Silver, see Ag

Single domains, see also Fine particles
  critical particle dimensions, 377–393
  magnetization curves, 409–442
  magnetostatic interactions, 442–463
SiV_3, magnetic behavior, 287
Skin effect, 90
Slater-Pauling curve, 37, 299, 303, 305, 316
Slip band formation, magnetic resonance studies of, 831
Slip-induced anisotropy, 602–616
Slip systems, 626–628
Sm, 18, 261, 262
Sn, 257
α-Sn, 253
β-Sn, 253, 255
Sn alloys, 257, 280
Sodium, see Na
Solenoids, 157–159, 198–202
  pulsed field, 159
  superconducting, 158–159
  table, 200
Solidification
  directional, 730
  textures from, 728
Solid solution series, susceptibility of, 264–280
Solubility curve, magnetic determination, 515–516
Specific heat
  of conduction electron gas, 32
  electronic, 27
  magnetic, 9, 21
Spectroscopic splitting factor, 817
Sphere, demagnetization coefficient, 129
Spheroid, demagnetization coefficient, 129
Spin paramagnetism, 32, 33
Spin quantum number, 13
Spin relaxation, 91, 100–101, 113–115, 823–825
Spin-wave resonance, 104–106, 116
Spin waves, 4, 39–42, 824
  in band model, 39
  bosons, 40
  energy, 42
  magnons, 40, 41
  in MnF_2, 41
  modes, 40
  spectrum, 41
Spinodal curve, 481, 482

Spinodal decomposition, 481–482
  in AlNi–Fe, 482–489
Spontaneous magnetization, 5
  experimental determination of, 127–132
  of ferromagnetic metals and alloys, 294–322
  sublattice, 22
Sr, 253, 254
Stacking fault patterns, 716
Steinmetz's law, 239
Strain gage, 142
Stress anisotropy, 68–73, 139
Stress anneal anisotropy, 597–599
Stress annealing, 597–599
Strontium, see Sr
Subgrains, 696
Sulfur, see S
Superantiferromagnetism, 394
Superconducting magnets, 158–159
Superdislocation, 644
Superlattices of AuCu and $AuCu_3$, 264
Superparamagnetism, 368, 385, 393–409
  of antiferromagnetic particles, 394
  applications of, 405–409
  basic experiments, 397–398
  of chemically inhomogeneous particles, 401–405
  determination of particle volumes, 406–408
  effect of anisotropy, 396
    of magnetostatic interactions, 398–401
  frequency dependence, 385
  identical particles, 394–396
  interactions, 394–396
  nonuniform particle volumes, 397
Superposition principle, 397
Superstructure phases, 272, 307–308
Surface anisotropy, 76, 116, 367
Susceptibility, see also specific metals and alloys for data
  ac, measurement of, 234
  dc, measurement of, 125
  definitions of various types, 6, 7, 124
  of elements
    anisotropy of, 255
    change of melting in, 257
    table, 252
    temperature coefficient for T metals, table, 258
  temperature dependence, table, 253
  T metals, 256, 258
  high-field, influence of recovery and recrystallization, 698–701
  incremental, 691–692
  initial, see Initial susceptibility
  parallel, 23–24
  perpendicular, 23
  reversible, 137
    dependence on deformation, 682–683
  of solid solution series, 264–280
SV, antiferromagnetism of, 292

T

Ta, 255
Tantalum, see Ta
Tb, 261, 262
Te, 257
Tellurium, see Te
Terbium, see Tb
Texture, 723–745
  from annealing, 728–730
  control by surface energy, 739–741
  from deformation, 727–728
  determination of, 725–727
  from directional solidification, 728
  of Fe–Ni, 732–734
  of Fe–Si, 734–744
    [110] <100>, 734–739
    [100] <100>, 739–744
  inhibition of, 732
  magnetic control of, 744–745
  in metals, 724–730
  of permanent magnet materials, 730–732
  specification of, 724–725
Thallium, see Tl
Thermodynamics, of uniformly magnetized ellipsoids, 52–59
Thin film, demagnetization coefficient of, 129
Thulium, see Tm
Ti, 255, 258
Tin, see Sn
Titanium, see Ti
Tl, 253, 255
β-Tl, 257

Tm, 34, 262
Toroid, demagnetization coefficient, 129
Torque
　on dipoles, 150
　expressions for various symmetries, 140
　measurement, 141, 180–183
Torque curves
　assemblies of particles, 420–421
　infinite cylinder, 418–420
　measurement, 141, 180–183
　of mixtures of fine particles, 441–442
　superparamagnetic particle assemblies, 421–422
　texture analysis by, 725–726
　uniaxial particle, 418–420
Transition elements, 36–40, 251, 295–298
　alloys and phases with
　　alloys with B metals, 266–272
　　　NiAs structure, 290–292
　　with lanthanides, 292–294
　　binary phases, table, 311–312
　　borides, 316
　　phases with B metals, 212–213, 283–287
　　solid solutions, 278–280, 295–298
　electronic specific heat of, 38
　localized moments, 37
　nonintegral atomic moments, 38
　paramagnetic susceptibility of, 255–258
　quenching of orbital moment, 38
　spin occupation of bands, 38
　spin paramagnetism in, 33
Transition temperatures, *see* Curie temperature, Néel temperature
　data on, 249–322
　measurement of, 125–126, 132–135
Transmission line measurements, 240–242
Transuranic elements, atomic moments, 11
Tungsten, *see* W

U

U, 258
α-U, 261
U group elements, 11
Uniaxial anisotropy, 75, 409–411, *see also* Induced anisotropy, Magnetocrystalline anisotropy, Shape anisotropy, Stress anisotropy
　torque expressions, 140

Uniaxial antiferromagnets, torque expressions, 140
Unidirectional anisotropy, 75, *see also* Exchange anisotropy
Uranium, *see* U

V

V, 255, 258
Vacancies, disappearance of, 695–696
Valence electrons, 11
Vanadium, *see* V
Vibrating coil, for field gradient measurement, 163
Viscosity, magnetic, 766–770, 776–778, *see also* Aftereffect
　losses, 776–778
Volume susceptibility, definition, 124

W

W
　orbital magnetism of, 255
　susceptibility of, temperature coefficient, 258
Wall, *see* Bloch wall, Cross-tie wall, Néel wall
Warburg's law, 58
Wattmeter, 237
Weiss theory of ferromagnetism, 294, *see also* Molecular field theory
Wheatstone bridge, 142, 143
Whiskers, metal, 389, 390, 392
Work hardening curve, 625
　region I, 626–627
　region II, 627
　region III, 628

X

X-ray diffraction, in texture studies, 725

Y

Y, 258
Yb, 262
Y–Fe–garnet, Faraday rotation, 145
Young's modulus, magnetostriction and, 79

Ytterbium, see Yb
Yttrium, see Y

Z

Zeeman energy, 9, 19
Zinc, see Zn
Zirconium, see Zr
Zn, 255–257
Zn alloys, 283
    susceptibility of γ phases, 281
Zn–Zr alloy
    ferromagnetism in, 36
    magnetic data on, 311, 316
Zr
    orbital magnetism of, 255
    susceptibility of, temperature coefficient, 258